Quantitative Stratigraphic Correlation

INTERNATIONAL GEOLOGICAL CORRELATION PROGRAMME

PROJECT 148

Quantitative Stratigraphic Correlation

Edited by

J. M. CUBITT
Poroperm Laboratories Limited, Chester, UK.

and

R. A. REYMENT
Uppsala Universiteit, Uppsala, Sweden

A Wiley–Interscience Publication

JOHN WILEY & SONS
Chichester · New York · Brisbane · Toronto · Singapore

Copyright © 1982 by John Wiley & Sons Ltd.
Reprinted May 1984.

Library of Congress Cataloging in Publication Data:

International Geological Correlation Programme.
  Project 148.
  Quantitative stratigraphic correlation.

  'A Wiley–Interscience publication.'
  Includes index.
  1. Stratigraphic correlation—Mathematics.
I. Cubitt, John M.   II. Title.
QE721.157 1982   551.7   81-21926
ISBN 0 471 10171 0   AACR2

British Library Cataloguing in Publication Data:

Quantitative stratigraphic correlation.—
  (International Geological Correlation Programme)
  1. Geology, Stratigraphic
  I. Cubitt, J. M.   II. Reyment, R. A.   III. Series
  551.7   QE651

  ISBN 0 471 10171 0

Typeset by Preface Ltd, Salisbury, Wilts.
Printed by Page Bros. (Norwich) Ltd.

List of Contents

III. LITHOSTRATIGRAPHY

List of Contributors

F. P. AGTERBERG   *Geological Survey of Canada, 601 Booth Street, Ottawa, Canada K1A 0E8*

H. BREMNER-TEIL   *Ecole Nationale Supérieure des Mines de Paris, CGGM—Sédimentologie, 35 rue St Honoré, 77305 Fontainebleau, France*

J. C. BROWER   *Heroy Geological Laboratory, Syracuse University, Syracuse, New York 13210, U.S.A.*

W. A. BURROUGHS   *Heroy Geological Laboratory, Syracuse University, Syracuse, New York 13210, U.S.A.*

R. CARIMATI   *AGIP SPA, Servizio SGEL, 20097 S Donato Milanese, Italy*

I. COJAN   *Ecole Nationale Supérieure des Mines de Paris, CGGM—Sédimentologie, 35 rue St Honoré, 77305, Fontainebleau, France*

J. CUBITT   *Poroperm Laboratories Limited, Chester Street, Saltney, Chester CH4 8RD, U.K.*

E. DAVAUD   *Institut des Sciences de la Terre, 13 rue des Maraichers, 1203 Geneva, Switzerland*

I. DIENES   *Hungarian Central Statistical Office, Kozponti Statisztikai Hivatal, Keleti Karoly U.5–7, 1525 Budapest 11, Hungary*

L. E. EDWARDS   *United States Geological Survey, Mail Stop 971, National Center, Reston, Virginia 22092, U.S.A.*

B. K. GHOSE   *Department of Geology and Geophysics, Indian Institue of Technology, Kharagpur 721302, India*

A. D. GORDON   *Department of Statistics, University of St Andrews, St Andrews, Fife, Scotland*

F. M. GRADSTEIN   *Atlantic Science Center, Bedford Institute of Oceanography, PO Box 1006, Dartmouth, Nova Scotia, Canada B2Y 4A2*

C. M. GRIFFITHS   *Department of Ocean Engineering, School of Marine Technology, University of Newcastle upon Tyne, Armstrong Building, Queen Victoria Road, Newcastle upon Tyne NE1 7RU, U.K.*

M. E. HOHN   *West Virginia Geological and Economic Survey, Morgantown, West Virginia 26505, U.S.A.*

A. MARINI   *National Group for Mathematical Informatics, Via Saldini 50, 20133 Milano, Italy*

M. T. MILLENDORF   *617 Soniat Street, New Orleans, Louisiana 70115, U.S.A.*

S. A. MILLENDORF   *617 Soniat Street, New Orleans, Louisiana 70115, U.S.A.*

R. G. POTENZA   *Centro Alpi CNR, Via Botticelli 23, 20133 Milano, Italy*

R. REYMENT   *Paleontologiska Institutionen, Uppsala Universiteit, Box 558, S-75122 Uppsala, Sweden*

W. SCHWARZACHER   *Department of Geology, The Queen's University of Belfast, Belfast BT7 1NN, Northern Ireland*

B. R. SHAW   *Amoco Production Company, P.O. Box 3092, Houston, Texas 77001, U.S.A.; present address: Samson Resources Co., 1010 Lamar Building, Suite 1575, Houston, Texas 77002, U.S.A.*

Foreword

It gives me great pleasure to introduce this collection of papers on quantitative stratigraphic correlation which is an important product emanating from the working group of IGCP (International Geological Correlation Programme) Project 148. It is virtually impossible to judge the scientific progress of a project without being able to refer to relevant publications such as this book. Through publications, a project develops its proper image and produces lasting results. IGCP has been operational since 1972. In total, 49 research projects were in progress during 1981 involving thousands of participants. IGCP focusses on the worldwide organization and distribution of knowledge about geological events in space and time. It also sponsors basic research on problems dealing with natural resources and environment.

IGCP Project 148 on Quantitative Stratigraphic Correlation Techniques commenced in 1976 as one of several projects dealing with quantitative methods and data processing in geological correlation. Project 148 deals with numerical methods of time determination—ordering the past and refining the geological calendar.

During the first few years of its existence, the participants in Project 148 have concentrated on the development of computer-based techniques for stratigraphic correlation. During its 1981 Session held in Paris, the IGCP Board concluded that Project 148 recently has been very active in expanding its international coverage, and an extension of the project beyond its initially scheduled termination date in 1981, for two further years, was approved. The project has now entered a second phase of work with the redevelopment and application of its methodologies. The organizers are encouraged to continue to find opportunities to explain and recommend their approach to supplement more traditional methods.

One of the objectives of IGCP is the transfer of knowledge to developing countries. Project 148 has held a successful workshop in Kharagpur, India, and it is commendable that further meetings in developing countries are being organized.

I believe that the results of the studies completed to date and their dissemination by means of workshops and short courses will reduce much of the uncertainties connected with the present automated stratigraphy. This book constitutes a milestone in quantitative stratigraphic correlation. It undoubtedly covers a new and challenging field of research of which the potential deserves to be fully explored.

Accra, Ghana   G. O. KESSE
November, 1981   Chairman, IGCP Board

Preface

The very basis of stratigraphy is correlation. Standard methods of correlation are concerned with seeking and identifying identical points or levels in comparable stratigraphical sequences. The basis of the use of zone fossils for correlational purposes was established by Albert Oppel, who formulated his ideas around the richly-fossiliferous sequences of the British Jurassic. Subsequently, the concept of zone fossils has been generalized to encompass minerals, marker beds etc.

The petroleum industry has been the main area of application of nonpaleontological methods of correlation, alongside classical micropaleontology. Although a wide variety of logs are obtained as a routine measure, and at considerable expense, the techniques of evaluation have not altered greatly since Albert Oppel presented his ideas, more than a century ago. Certainly, attempts have been made to use correlational techniques for extracting more information from fossil data in borehole analysis, but these tend to be fraught with statistical weaknesses and fallacies.

In this book, we believe we have made a first step towards a cohesive presentation of quantitative methods in stratigraphical correlation. The reader soon will perceive that the statistical problems involved are far more complex and difficult to solve than one might expect on first meeting the subject. Without recent advances in the theory of mathematical statistics, it would not be possible to study some of the questions raised in the present connexion. We need only mention the matter of identifying atypical observations for this remark to make itself apparent.

It is to be hoped that the results presented in this book will prove useful to a large number of people working in the field of stratigraphical correlation. Also, that they will stimulate further work and fruitful cooperation between statisticians and geologists.

JOHN CUBITT
Chester, England

RICHARD REYMENT
Uppsala, Sweden

Introduction

IGCP Project 148: Background, Objectives, and Impact
F. P. Agterberg, Geological Survey of Canada, Ottawa, Canada

General Background Information

The International Geological Correlation Programme (IGCP) is cosponsored by the intergovernmental United Nations Educational, Scientific, and Cultural Organization (UNESCO) and the International Union of Geological Sciences (IUGS). This successful collaborative venture has been operating since 1972. IGCP aims to refine and calibrate methods used in correlating geological relationships and events, analyze the processes involved in orogenesis, and understand the formation of mineral deposits.

The scientific problems to be solved in order to achieve the aims and objectives of IGCP affect many related fields of science including mathematics and biology. This is not only due to the great variety of research methods and techniques used but also to the diversity of geological problems that are being tackled. These range from those dealing with geochronology and stratigraphy, to studies of palaeontology, volcanism, metamorphism, tectonics, mineral deposits, and include geostatistics and data processing.

The scientific fields of research which are covered by IGCP fall into four Divisions (1–4) as follows: (1) time and stratigraphy with particular emphasis on practical implications; (2) major geological events in time and space and their implications in environmental processes; (3) distribution of mineral deposits in space and time and relation of the processes of ore formation to other events in earth history: and (4) quantitative methods and data processing in geological correlation. Within these four divisions, there is a wide range of possible subjects for individual projects. The following four priority areas (a–d) have been indicated by the IGCP Board in that they should receive special emphasis: (a) ordering of events in time; (b) evolution of the ancient (Precambrian) crust; (c) geological environment of Man; and (d) energy and minerals.

IGCP Project 148 (Evaluation and Development of Quantitative Stratigraphic Correlation Techniques) belongs to Division (4) and priority area (a). It was proposed originally through the International Association for Mathematical Geology. A project on quantitative biostratigraphic correlation under the directorship of J. C. Brower was accepted in 1976. However, later in the same year, this proposal was combined with equivalent aspects of lithostratigraphic correlation and a combined project covering all aspects of quantitative stratigraphic correlation was initiated under the leadership of J. M. Cubitt. In 1979, F. P. Agterberg took over from J. M. Cubitt as project leader. At the end of 1981, the project had 159 participants and correspondents in 25 countries. As a rule, there is one international meeting per year in addition to a meeting for each of the national working groups in Canada, India, the U.S., and the U.S.S.R. Current research activities and future plans are

1

summarized annually in Geological Correlation published through the IGCP Secretariat in Paris and in the semi-annual Project 148 Newsletter. The project will be completed by the end of 1983.

During the past five years, a large number of computer-based techniques for stratigraphic correlation have been developed. The present book is the most comprehensive collection of papers produced by project participants. Earlier collections of publications are the proceedings of the first international meeting held at Syracuse University, New York, in 1977 (Cubitt, 1978) and papers presented during the second international meeting held in Jerusalem, Israel, in 1978 which were included in a book edited by Gill and Merriam (1979). Other publications emanating from the project including computer programs have been published separately. A complete list of papers can be obtained in successive issues of the publication Geological Correlation and in IGCP Catalogues. General bibliographies of publications from before 1980 were published by Brower (1981; with 136 references) for quantitative biostratigraphic correlation, and Mann (1981; with 156 references) for quantitative lithostratigraphic correlation.

Liaison with other IGCP projects, notably with Project 1 (Accuracy in time) has been maintained. In October 1981, a joint meeting of participants in the four quantitative IGCP projects belonging to Division 4 (Nos. 98, 148, 154, and 163) was held in Ottawa, Canada (for a conference report by the author, see the December 1981 issue of Episodes).

Objectives of Project 148

The main objective of the Project is the development of computer-based mathematical theory and analysis of geological information required for the practical application of automated correlation techniques in stratigraphy. Working along parallel lines, researchers in different countries are conducting this work in the fields of biostratigraphy and lithostratigraphy. Specific problems are solved by establishing regional standards of ordered stratigraphic events and performing correlations on the basis of these standards. Comprehensive descriptions are prepared for different techniques which are applied to the same data sets in order to evaluate their respective advantages and drawbacks. Special attention is given to the performance of computer-based quantitative techniques in comparison with the results obtained by conventional qualitative stratigraphic correlation methods. During the first three years of existence, the emphasis within the Project was on method development. During the last few years, the primary activity in the Project has shifted from method development to application for solving specific stratigraphic problems using large data bases for regions in North America, Europe, and India. Deep Sea Drilling Project data sets are being analyzed also.

Research is conducted on the following major problems: (1) creation and definition of the mathematical theory of stratigraphic relationships: (2) establishment of standards and codes for the biostratigraphic, ecological, lithological, and environmental information attainable from well logs, cores, and surface sections; (3) development of a theory for mathematical correlation; (4) development of practical methods of lithostratigraphic correlation concentrating on methods of spectral analysis (frequency domain), methods of zonation (time domain), methods of stratigraphic interpolation, and multivariate statistical analysis; (5) development of practical methods of biostratigraphic correlation concentrating on quantification of assemblage zones, sequencing methods, set theoretical approaches, morphometric chronoclines, and multivariate methodology; (6) regional testing of the developed methodology with standard data sets, and (7) comparative evaluation of computer algorithms. To date, much work on items (1) to (5) has been completed. However, most work on items (6) and (7) was commenced in 1979 and remains in progress.

Impact of Project 148

The 1980's promise to become significant years in the interdisciplinary search for suitable quantitative stratigraphic correlation techniques. On the one hand, there has been time to digest, redevelop, and apply methodologies published during the first years of Project 148. On the other hand, a number of national and international meetings are being organized of which the spin-off in the way of stimulting research, discussion, and publications of proceedings will provide stratigraphers with easy access to new methodology. These also will serve to make geomathematicians more familiar with the types of questions asked in the stratigraphical field.

Although much progress has been made in the field of quantitative stratigraphy during the past few years, it should be kept in mind that a consensus among geoscientists on the utility of this approach does not yet exist. In the field of biostratigraphy, this was discussed recently by Raup (1981) who noted that during the 1970's, development of the computerization of biostratigraphy has been disappointing with 'much potential but little real progress'. Brower (1981) has argued repeatedly that most nonquantitative biostratigraphers have resisted numerical methods for two reasons. Many of the quantitative methods are rather complex whereas most biostratigraphers prefer simple methods. Secondly, most of the quantitative methods utilize methodologies that basically are foreign to biostratigraphers.

Brower's remarks could be applied equally to developments in lithostratigraphic correlation. Pitfalls in this field have been pointed out by Robinson (1978). These especially occur when too much reliance is placed upon outputs resulting from automatic data processing techniques which are not necessarily suitable for solving the problem at hand.

During the final years of its existence, participants in Project 148 are addressing the problems outlined in the preceding paragraphs. In the redevelopment of the original techniques more attention is being paid to the problems of stratigraphic reasoning. More interaction between stratigraphers and mathematical geologists is required to achieve this aim. Members of Project 148 now are conducting workshops and short courses in order to demonstrate the new concepts and methods to increasingly larger groups of academic and nonacademic stratigraphers accustomed to deal with real practical problems (cf. Agterberg and Gradstein, 1981).

In the transfer of knowledge to developing countries, it should be kept in mind that it is not useful to recommend techniques which cannot be applied because of lack of computing facilities. Project 148 is stimulating research and capabilities in these countries. Participation has been on the increase in India where a number of quantitative stratigraphic correlation projects were commenced in 1980. This was one of the main reasons that Project 148 has been extended beyond its original termination date in 1981, for two further years.

The rapid growth of information in the applied and academic geological sciences has led to an increased demand for quantification of information for machine handling or graphic display. As a result, there is a rapidly-increasing need for numerical models to organize and explain specific geological problems. Studies in the fields of biostratigraphy, lithostratigraphy (especially well logs), and sedimentology make successful use of the quantitative modelling approach. Simple statistical and other numerical techniques can be used in correlation of biostratigraphic events, biozonations, classification, and matching of lithofacies in well logs or sections, lithofacies pattern recognition, and determination of the rate or magnitude of geological processes relative to the numerical time scale.

For many of these studies, little mathematical knowledge is required. However, stratigraphical or sedimentological information is organized in novel ways. Project 148 is now actively seeking opportunities to expose a broad spectrum of stratigraphers to the automated methods of stratigraphic correlation which it has developed.

References

Agterberg, F. P., and Gradstein, F. M., 1981, Workshop on quantitative stratigraphic correlation techniques, Ottawa, February 1980, *Jour. Math. Geol.*, **13**(1), 81–91.

Brower, J. C., 1981, Quantitative biostratigraphy, 1830–1980, in Merriam, D. F. (ed.), *Computer Applications in the Earth Sciences—An Update of the 70s*. Plenum Press, New York, pp. 63–103.

Cubitt, J. M., 1978, Introduction to Proceedings of the 6th Geochautauqua on 'Quantitative Stratigraphic Correlation', *Computers & Geosciences*, **4**(3), 215–216.

Gill, D., and Merriam, D. F., 1979, *Geomathematical and Petrophysical Studies in Sedimentology*. Pergamon Press, Oxford, 266 pp.

Mann, C. J., 1981, Stratigraphic analysis: Decades of revolution (1970–1979) and refinement (1980–1989), in Merriam, D. F. (ed.), *Computer Applications in the Earth Sciences—An Update of the 70s*. Plenum Press, New York, pp. 211–242.

Raup, D. M., 1981, Computer as a research tool in paleontology, in Merriam, D. F. (ed.), *Computer Applications on the Earth Sciences—An Update of the 70s*. Plenum Press, New York, pp. 267–281.

Robinson, J. E., 1978, Pitfalls in automatic lithostratigraphic correlation, *Computers & Geosciences*, **4**(3), 273–275.

I. FORMALIZED STRATIGRAPHY AND CORRELATION

Quantitative Stratigraphic Correlation
Edited by J. M. Cubitt and R. A. Reyment
© 1982 John Wiley & Sons Ltd.

A Short Note on

The Correlation of Geologic Sequences[1]

BRIAN R. SHAW[2]

Amoco Production Co., P.O. Box 3092, Houston, TX 77001, USA

Abstract

The use of stratigraphic units and the ability to demonstrate correspondence of these units are the two basic qualifications in the definition of geologic correlation. Although codes have been formulated to describe stratigraphic units of various kinds, no explicit definition of correlation has been established. The development of modern techniques of correlation necessitate a clear understanding to avoid the association of methodology with interpretation.

Just as stratigraphic units can be formal or informal, the process of establishing geologic correlations can be formal or informal. The principal difference is dependent upon whether the correspondence can be demonstrated physically or traced in outcrop. Any implied correspondence is indirect. Secondly, a procedure which does not compare and contrast stratigraphic units is not a correlation, but becomes a match. Matching is a useful procedure to develop the relationships between sequences of data without respect to stratigraphic interpretation.

'A formation by any other name will map as well'
(Bell, Murray, and Sloss, 1959)

Introduction

The purpose of this note centers on two qualifications of geologic correlation: the use of stratigraphic units and the ability to demonstrate correspondence of these units. Without the incorporation of these concepts in any stratigraphic analysis, interpretations will not yield consistent or reliable results. Although frequent usage has generated informal understandings of the concept of geologic correlation, adaptation of numerical methods necessitates a clear and concise definition.

Correlation

The process of correlation is the determination of geometric relationships between rocks, fossils, or sequences of geologic data for interpretation and inclusion in facies models, paleogeographic reconstructions, or structural models. Therefore, the object of correlation

---

[1] Published by permission of Amoco Production Company.
[2] Present address: Samson Resources Co., 1010 Lamar Building, Suite 1575, Houston, Texas 77002, USA.

is the establishment of equivalency of stratigraphic units. Codes of stratigraphic nomenclature have been established (Hedberg, 1976; American Commission on Stratigraphic Nomenclature, 1972) to define concepts and procedures for creating stratigraphic units; the basic building block of stratigraphy.

A critical point in the development of any stratigraphic analysis is that the object of similarity measures must be stratigraphic units. The nature of the stratigraphic unit, how it is established, and the amount of data determine both the object of analysis and reliability of the interpretations.

Implicit in the choice of stratigraphic unit is the reason for conducting the evaluation. If lithology and structure are primary objectives in an analysis, then, lithostratigraphic units are appropriate. If the time relationships between sections are of interest, a biostratigraphic unit may be employed. However, such clarity of purpose has not always had complete agreement among stratigraphers. In fact, the lingering disenchantment with procedures may be the cause of many problems in attempting to quantify stratigraphic correlation.

The debate over the meaning of geological correlation has roots in the beginning of stratigraphy. Early efforts of geologists were directed at establishing correspondence of strata at different localities largely in relationship to the age of the rocks. The first large-scale attempts at including genesis in stratigraphy are attributable to investigators such as John Smith and Abraham Werner. Unfortunately, they approached the problem from different directions. Smith attempted to correlate by searching for time equivalents through the use of fossil assemblages and their vertical succession. Werner attempted a similar objective through lithic succession. Although the nature of basalt, granite, and other 'crystalline' rocks had begun to supersede that of age connotations, as late as the turn of this century the principle guide to time equivalency remained similarity of sequence and lithologic character. Grabau (1913) listed the fundamental evidence of correlation to be superposition, 'the basis of all stratigraphic work'. Beyond this, stratigraphic continuity, lithology, fossils, unconformities or disconformities, and diastrophism were cited as means of establishing correspondence.

From this beginning, there has been an overall effort to establish the synchroneity of strata in separate regions. Through the development of sophisticated paleontologic methods and radiometric dating, this has become a reasonable task. However, it had become necessary to establish a different objective of correlation. The search for hydrocarbons had generated a need to understand geometry of rock units as well as age, and in some situations, because of the lack of time correspondence. Schenk and Muller (1941) thus revived the Wernerian concept of lithostratigraphic succession without time connotation.

The need to define the concept of lithostratigraphy is a result of viewing time stratigraphy and rock stratigraphy as equivalents. In addition, time correlations do not demonstrate rock correspondence (Shaw, 1964). This raised a semantic question on the meaning of correlation. Is correlation restricted to time equivalence (for example, Rodgers, 1959) or any defined stratigraphic unit (Krumbein and Sloss, 1963)? Although the American Association of Petroleum Geologists and the Society for Economic Paleontologists and Mineralogists held a symposium on this subject (Bell, Murray, and Sloss, 1959), debate has not been settled completely (Beerbower, 1968; Weller, 1960; Shaw, 1964). Both the International Stratigraphic Guide (Hedburg, 1976) and Code of Stratigrahic Nomenclature (American Commission on Stratigraphic Nomenclature, 1972) recognize formal lithic units, only the International Stratigraphic Guide stresses the term 'correlation' as a general procedure without time applications.

In reality, this has been a moot point due to the common usage of the term 'correlation' for lithostratigraphic units in industry and academics. In other words, as long as the object was understood, so was the term, because the procedures, whether lithic or paleontologic were visual. However, with the advent of numerical methods, calculating correspondence is not an arbitrary process and the meaning of the phrase 'it correlates' can have significantly different interpretations. The development of modern techniques of correlation necessitates associating methodology with interpretation of results. Therefore, the concept of correlation must be understood in a framework whereby limitations and advantages of various techniques can be discerned easily. Both the International Stratigraphic Guide and the Code of Stratigraphic Nomenclature have worked to establish formal definitions of stratigraphic units, and, now any correlations involving these units must reflect the nature of the correlation process so that inferences derived from the units are consistent. In other words, the special connotation of correlation representing correspondence of stratigraphic units must be maintained.

The second 'special connotation' of geological correlation concerns the concept of demonstrating correspondence of a stratigraphic unit. Following the philosophy of the American Commission of Stratigraphic Nomenclature (1972), there are formal and informal stratigraphic entities. Krumbein and Sloss (1963) suggest two requirements for formal status, mapability, and attribute consistancy, whereas informal units are defined either by indirect measurements (well logs, sonic velocities) or markers. Thus, the second connotation of correlation can be viewed in a similar way. A formal correlation, one that can be 'demonstrated', must be represented physically, that is, physical tracing of a stratigraphic unit is the only method of showing correspondence unequivocally.

Thus, a two-fold classification of the concept of correlation is possible:

(1) the use of stratigraphic units,
(2) the demonstrability of correspondence.

Unfortunately, the classification manages to exclude the techniques most frequently employed, comparisons of attributes of stratigraphic units such as resistivity, insoluble reside content, or sonic velocity. As it would be ludicrous to imply that well logs cannot be correlated, therefore, correlations should be treated as either formal or informal. An informal correlation is the indirect tracing of units through projections of attributes. In addition, an indirect correlation cannot be demonstrated physically, but has levels of reliability, some methods of indirect correlations being accurate (for example radiometric dating), whereas others (for example, sequence of colors) are not.

As stated previously, the concept of geological correlation involves the use of stratigraphic units and the notion of demonstrability, which leads directly to the question of matching of sequences, as opposed to correlation (Schwarzacher, 1975). A match is defined as correspondence of sections of serial data, without regard to stratigraphic units. Matching is not as uncommon as may be imagined, for example, dipmeter 'correlations' or seismic trace 'correlations' which occur frequently in the analysis of stratigraphic sections.

A simple example will serve to illustrate the difference between matching and correlation. Given two sections with a repeated lithology, the sequences can be aligned so that the correspondence between them is exact. At this stage, the process is matching. However, if the sequence is interpreted stratigraphically, the units may be discreet, although the sequence is identical; the process then is correlation (Fig. 1), therefore, no matter how sophisticated a mathematical routine becomes, unless it includes the concepts of established stratigraphy, computer procedures produce matches and not correlations in a geologic sense.

For this critical reason, it is important to define the term 'correlation' so that differing methods of correlation can be related directly to the degree of stratigraphic inference implied by each process. For example, if a correlation method can be classified as formal,

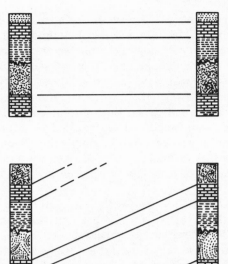

Fig. 1. Two sections with similar rock
and fossil subdivisions. Upper figure
indicates matches, lower figure indicates correlations (from Shaw, 1964)

then direct inferences can be made from the relationship. On the other hand, if a process is matching, the tie-lines do not warrant correlative inference as they do not encompass stratigraphic units.

One final point remains to be made. Apart from physical tracing, there are two basic methods of establishing correlation. Traditionally, indirect correlations have been made arbitrarily, for example visual comparison of well logs. However, with the advent of high-speed computers, systematic measures of correspondence can be made now. As long as stratigraphic units are being compared, consistent criteria may be employed for equivalence, for example, radiometric ages indicate simple numeric correspondence.

Where the criterion is rigid so that possession of a unique set of attributes is both sufficient and necessary for equivalence, then, correlation may be termed monothetic. Where equivalence is determined by statistical measures such as the greatest number of shared characteristics, with no single character being essential or sufficient, the correlation may be termed polythetic, for example, the calculation of similarity of concurrent-range assemblage zones. The relationship of the different approaches to geologic correlation is straightforward (Fig. 2). In addition, this scheme embodies the idea that if the object of an analysis does not employ stratigraphic units, it becomes a matching procedure, and that a correlation can be established only by employing stratigraphic units.

The significance of this concept lies in the design and interpretation of stratigraphic analyses. Many new techniques are flooding journals with sophisticated numerical methods, but how they relate to the theory of stratigraphy is critical. For example, a technique termed seismic-stratigraphy is generating interpretations in areas unreachable by existing sampling methods. However, unless the object of correspondence on the seismic trace, the wavelet, can be related directly to a stratigraphic unit, the resulting interpretations are matches, not correlations. This is not to say seismic-stratigraphy is impossible, but it should not be confused with stratigraphy developed from direct evaluation.

CORRELATION	FORMAL	PHYSICAL TRACING OF STRATIGRAPHIC UNITS		
	INDIRECT	ARBITRARY	SYSTEMATIC	
		VISUAL COMPARISONS	MONOTHETIC	POLYTHETIC
			NUMERIC EQUIVALENCE	STATISTICAL EQUIV- ALENCE
	MATCHING	COMPARISONS OF NON-STRATIGRAPHIC UNITS		

Fig. 2. Systematics of correlation

Conclusion

In conclusion, the need for additional terminology is always debateable. However, it is hoped that this suggestion (Fig. 2) might aid in the evaluation of quantitative correlation methods, and allow a place for their inclusion in formal stratigraphy.

Acknowledgments. These ideas have been developed during investigations conducted under the International Geological Correlation Program Project 148. Drs. John Cubitt, Dan Merriam, and Jim Brower have been influential in shaping many of these concepts. Dr. Gordon Wood and members of the staff at Amoco have contributed much needed criticism.

References

American Commission on Stratigraphic Nomenclature, 1972, *Code of Stratigraphic Nomenclature.* Amer. Assoc. Petroleum Geol., Tulsa, 17 pp.

Beerbower, J. R., 1968, *Search for the Past.* Prentice-Hall, Inc., Englewood Cliffs, N. J., 512 pp.

Bell, W. C., Murray, G. E., and Sloss, L. L., 1959, Symposium on concepts of stratigraphic classification and correlations, *Amer. Jour. Sci.*, **257**, 673.

Grabau, A. W., 1913, *Principles of Stratigraphy.* Dover Publications Inc., New York, 1185 pp.

Hedberg, H. D., 1976, *International Stratigraphic Guide.* John Wiley and Sons, New York, 200 pp.

Krumbein, W. C., and Sloss, L. L., 1963, *Stratigraphy and Sedimentation.* W. H. Freeman & Co., San Francisco, 660 pp.

Rodgers, J., 1959, The meaning of correlation, *Amer. Jour. Sci.*, **257**, 684–691.

Schenk, H. G., and Muller, S. W., 1941, Stratigraphic terminology, *Geol. Soc. America Bull.*, **52** (9). 1419–1426.

Schwarzacher, W., 1975, *Sedimentary Models and Quantitative Stratigraphy.* Elsevier, New York, 382 pp.

Shaw, A. B., 1964, *Time in Stratigraphy.* McGraw-Hill Book Co., New York, 365 pp.

Weller, J. M., 1960, *Stratigraphic Principles and Practice.* Harper and Row, New York, 725 pp.

Quantitative Stratigraphic Correlation
Edited by J. M. Cubitt and R. A. Reyment
© 1982 John Wiley & Sons Ltd.

The Mathematical Formalization of the Geological Relations Identifying the Basic Structure of a Geological Data Bank[1]

ROBERTO CARIMATI,* ALBERTO MARINI,** AND
ROBERTO G. POTENZA†

*AGIP S.p.A., servizio Sgel, 20097 S. Donato Milanese, Italy
**National Group for Mathematical Informatics, via Saldini 50, 20133 Milano, Italy
† Centro Alpi C.N.R., via Botticelli 23, 20133 Milano, Italy

Abstract

Relations among formational units used as main entries in a geological data bank are analyzed and mathematically defined, in order to identify a geological pattern underlying data archived. Explicit languages for the management of data can be based upon this pattern, which also widens the possible uses of the geological data. Examples of relations among formations of the Lombardy (North Italy) and of their automatic processing are provided.

Introduction

Several initiatives for the automatic archivation of geological data have been promoted in Italy in the last three years. An outstanding role in this development has been performed by the Italian Geodynamic Project. Not only were several new, geological, monographic archives started, but also growth of different archives was reoriented towards coordination of standards and procedures, in view of the confluence of archives in a single Geological Data Bank. This aim was pursued by research groups working in the Center for Stratigraphy of the Central Alps and in the Group for Mathematical Informatics, both of the National Council of Research, and in AGIP, the italian petroleum agency.

The Geological Data Bank is growing by stepwise collation of independent monographic archives. However, the archives' structures and treatment procedures are as different as the computing facilities in the sites where the archives are being developed. Therefore, nontrivial problems of compatibility among data and procedures stood in front of the group for coordination (although no formal authority is given to it): some were solved by organizing facilities for the standardization of data, offering portable procedures for implementation on widely-used minicomputers, or even by supporting directly the work of data entry and validation. The study of less-urgent problems is underway.

[1] Paper No. 336 granted by: Finalized Geodynamics Project of the National Council of Research.

The framework of the Data Bank had to be provided in advance also, in addition to the procedures for final validation and collation of data. An internal system for data treatment was outlined to manage modular monographic archives using nonnatural languages; more explicit geological languages are provided for communication between archives and external users. Both classes of languages are based on the use of informatically-defined geological concepts and terms.

Formalized Geological Relationships

The basic reference for data grouping, especially for external use, is represented by Formational units. The Formational unit (Fu, in the following discussion) is a geological entity defined by one, or more, bibliographic or cartographic entries. Each name of Fu therefore is related univocally to a single geological object; the inverse correspondence is not necessarily univocal, because of the concurrence of different interpretations (Carimati and others, in press).

The redundancy of definitions and changes, either in name or geological significance, which Fus can undergo bring troublesome complications to their automatic treatment. In order to overcome these difficulties, the mathematical formalization of geological relationships, proposed for a wide set of stratigraphical occurrences by Dienes and Mann (1978) and by Mann (1978), was simplified and extended to terminological relationships of synonimy and analogy.

Formalization was stated assuming that each Fu is considered as a pair $\langle S,I \rangle$, where S is a set of points on the lithosphere, and I an informative structure formed by a sequence $\langle A_1, \ldots, A_n \rangle$ of attributes. Set S is considered, in principle, to be defined geometrically, as transitional implications at the boundaries appear to be of minor importance and can be ignored in the present context.

In the complex U of the Fu's, the set of binary relationships relevant for the management of archives is identified (Table 1 and Fig. 1; for a self-contained introduction to relationships see, for instance, Preperata and Yeh, 1973).

In this set, each relationship R_i implies its inverse $R_{\bar{i}}$, such that the Fu f is in the relationship R_i with the Fu f', iff f' is in the relation $R_{\bar{i}}$ with f.

Moreover, each relationship R_i can be characterized by one or more of the following properties:

(1) *Symmetry:* the relationship, R_i, coincides with its inverse $R_{\bar{i}}$. This property holds for some relationships of synonimy, for analogy, for nonstratigraphic contacts, as well as for stratigraphic relationships of total and bilaterally partial heteropy. Symmetry does not apply to vertical stratigraphic relationships, referred to as intrinsically polar and nonreversible processes.

(2) *Univocity* (or *functionality*): if the Fu f shows the relationship R_i with the Fu f', it cannot show the same relationship R_i with the Fu f'' if different from f'.

Univocity controls the exclusive character of several relationships. If, for instance, one unit is overlain completely by another, showing, therefore, a relation T1 or T3, it cannot show the same, or alternative relationships T_i, with other Fus; in addition this holds for relationships I1 (inclusion) and J1 (pertinence to one unit of higher rank).

(3) *Multivocity:* if the Fu f shows the relationship R_i with Fu f', at least one another Fu f'' must show the relationship $R_{\bar{i}}$, or a relationship of the same type of $R_{\bar{i}}$, with f.

Multivocity is the opposite of univocity and implies the existence of a plurality of Fus showing relationships of the same type with a given unit. If a Fu is overlain by a

Table 1. Symbols, properties, and short descriptions of geological, terminological, and hierarchical relationships between formational units

f = Formational unit, subject of relationship
f' = Formational unit, object of relationship
Properties: U = univocal, M = multivocal, S = symmetrical, T = transitive

Symbol	Inverse	Properties	Description
T1	L1	U	f is totally overlain by f' and f represents base of f'
T2	L2	M	f is partially overlain by f' and f represents base of f'
T3	L3	U	f is totally overlain by f' and f represents part of base of f'
T4	L4	M	f is partially overlain by f' and f represents part of base of f'
T0	L0	—	f is overlain undefinedly by f'
L1	T1	U	f totally overlays f' and top of f' represents base of f'
L2	T2	U	f partially overlays f' which represents its base
L3	T3	M	f totally overlays f' which represents part of its base
L4	T4	M	f partially overlays f' which represents part of its base
L0	T0	—	f undefined overlays f'
E1	E11	S	f is totally heteropic with f'
E2	E3	M	Whole vertical extension of f is heteropic with part of vertical extension of f'
E3	E2	M	Part of vertical extension of f is heteropic with whole vertical extension of f'
E4	E4	M,S	Part of vertical extension of f is heteropic with part of vertical extension of f'
E0	E0	S	f is undefinedly heteropic with f'
G0	G0	S	f shows nonstratigraphic contact with f'
I1	I3	U	f is totally included in f'
I3	I3	U	f is totally included in f'
I2	I4	M	f is partially included in f'
I3	I1	—	f includes whole f'
I4	I2	—	f includes part of f'
N0	N0	S	f shows undefined spatial relation with f'
S1	SA	T	Part of f is synonimous with f'
SA	S1	T	f is synonimous with part of f'
S2	S2	S	Part of f is synonimous with part of f'
S3	SC	T	Part of area of f is synonimous with f'
SC	S3	T	f is synonimous with part of area of f'
S4	SD	T	Part of series of f is synonimous with whole series of f'
SD	S4	T	Whole series of f is synonimous with part of series of f'
S5	SE	T	Part of serial and areal extension of f is synonimous with f'
SE	S5	T	f is synonimous with part of serial and areal extension of f'
S6	S6	T,S	f is totally synonimous with f'
S7	S7	T,S	f is synonimous with f', but is located in different area or basin
S0	S0	T,S	f is undefinedly synonimous with f'
A0	A0	T,S	Geological analogy can be recognized between f and f'
J0	J1	T,M	f includes f', of lower rank (for example, a group including formations)
J1	J0	T	f is included in f', of upper rank

Fu of minor extension (that is a relationship T2), at least one other unit must lie over it. According to the previous considerations, this property applies frequently to general vertical stratigraphic relationships.

(4) *Transitivity:* if Fu f shows relationship R_i with Fu f', and f' shows the same relationship R_i with f'', relationship R_i holds between f' and f'', also.

This property is peculiar to terminological relationships (synonymy, analogy). If,

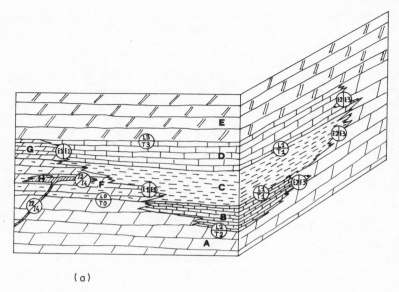

(a)

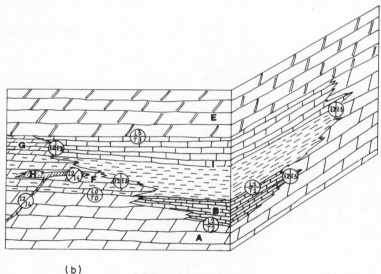

(b)

Fig. 1. Different sets of relationships between Formational units. Different interpretations of same stratigraphic sequence are represented in Figs 1(a) and 1(b) with sets of relationships. To essential sets, redundant relations (not present physically in data base) are added, in parenthesis, to show complete sets of relationships (symbols as in Table 1)

Fig. 2.1(a) Formations C and D are considered two distinct units; set of relationships is: A T2 B, A T0 F, A E3 B, A I4 H, F I4 H, A E3 C, A E3 D, B T1 C, C T1 D, C E1 F, D E1 G, D T3 E, C SA I, D SA I. (B L2 A, FL0 A, B E2 A, H I2 A, H I2 F, C E2 A, D E2 A, C L1 B, D L1 C, F E1 C, G E1 D, E L3 D, I S1 C, I S1 D)

Fig. 2.1(b) Formations C and D are considered one unit I; set of relations becomes: A T2 B, A T0 F, A E3 B, A I4 H, F I4 H, B T1 I, I T3 E, I E2 A, G E2 I, F E2 I, I S1 C, I S1 D. (B L2 A, FL0 A, B E2 A, H I2 A, H I2 F, I L1 B, E L3 I, A E3 I, I E3 G, I E3 F, C SA I, D SA I)

for instance, one unit is completely synonymous of another, and this is, in turn, synonymous of a third one, then the first unit is synonymous with the third one.

In the tables, relationships and their properties are listed with a short explanation as to their geological significance. It must be remarked here, that symbols used to designate relationships have a mere notational significance; greater significance should be attached to the fact that relationships listed here are recognized useful for the management of data archives at this time and that modifications of the set are feasible with minor changes to the structure of the archives.

Mathematical formulation of the relationships between Fus was required for the automatic treatment of data. The main purpose is, in fact, the aggregation of Fus in sets related to geologically-defined domains, as a guide for access to data contained in the monographic archives. The use of the described set of relationships, and possibly of implemented versions of it, enables us to perform these operations, even in the frequent situation of multiple definitions of Fus, bringing to an easier reach of target sets of data into the data bases.

Moreover, the use of relationships, allowing the display and review of the geological context of Fus, implies an interaction between data interpretation and data storage and retrieval, which leads to a generalized view of the activity of data management.

The choice of interchangeable subsets of relationships, applied to particular groups of Fus, can be used to improve comparison between interpretative models. The modular monographic archives now in development are the components of an informative structure, the pattern of which can be seen as a general raw model representing the geology of the Italian region. Each new datum entered in the archives implies a deformation of the general model, which then is updated continuously to new levels of knowledge. The consequent modifications are better evaluated by partial sophisticated models, obtained by simulation techniques from archived data and relationships. It is then possible to verify if the models obtained after updating are consistent with the previous ones: subsequent analysis will state if inconsistencies are to be related either to defective new data entered, or to incorrect assumptions for the model structure.

The possibility of reading by explicit languages the content of the archives, or part of them, and simulating geological models lends itself, finally, to the educational use of the archives. In addition, the growth and extended use of small, cheap, and powerful computers will apply to Earth Sciences also. The realization that formalized stratigraphy can act as a tool for computer-aided teaching of geology will result hopefully in new approaches to the automatic management of geological information.

Example

The described system of geological relations was employed practically in the organization of archives of the Italian Geological Data Bank, and was first applied to the Lombardy region. Four weeks were spent in the recovery of 80 percent of the relationships between approximately 200 Fus; a further three months were required to reach 90 percent. Additional improvements will take longer and will be made in view of the extension of work to the remaining part of the national territory. However, limits of information completeness greater than 95 percent are not considered realistic for reasons of excess costs.

A simple example of output obtained using formalized geological relationships is given in Table 2. The data employed are taken from the Catalog of Formational units of Italy, stored in the memory of an UNIVAC 1100/81 computer, and treated by programs implemented on a PDP 11/34 minicomputer. Table 2 is the translation of the original printout, which is

Table 2.   Example of output (translated in English from the Italian original) obtained treating the set of relationships between Formation Calcare di Angolo (Angolo limestone) and its neighbours

---

Calcare di Angolo

It is undefinedly overlain by REC (Calcare di Recoaro)

It is partly overlain by CDM (Calcare di Dosso dei morti) and by CMR (Calcare di Camorelli), of which it forms the base

It is partly overlain by ESI (Calcare di Esino) and by PRZ (Calcare di Prezzo), of which it forms part of the bottom. It overlies completely GAE (Dolomia del Gaver), which forms part of the base of ANG

It overlies partly BOV (Carniola di Bovegno), CMR (Calcare di Camorelli), ELT (Dolomia di Elto), and SRV (Servino) which form part of the base of ANG

It is heteropic with CRL (Dolomia d'Albiga); part of the vertical extension of CRL coincides with the whole vertical extension of ANG

It is heteropic with CDM (Calcare di Dosso dei morti) and CMR (Calcare di Camorelli); the whole vertical extension of CDM and of CMR coincides with part of the vertical extension of ANG

It is heteropic with ELT (Dolomia di Elto), GAE (Dolomia del Gaver), and SEL (Dolomia del Serla); part of the vertical extension of ELT, GAE, and SEL coincides with part of the vertical extension of ANG

It partially includes ADA (Tonalite dell'Adamello)

---

written in Italian as the data treated are presently of local interest. A choice of different languages is provided however for output of data of general interest; the development of facilities for graphical output is underway.

References

Carimati, R., Gossenberg, P., Marini, A., and Potenza, R. G., 1980, Catalogo delle Unità formazionali italiane, *Boll. Serv. Geol. d'Italia* (in press).

Dienes, I., and Mann, C. J., 1977, Mathematical formalization of stratigraphic terminology, *Jour. Math. Geol.*, **9** (6), 587–603.

Mann, C. J., 1977, Towards a theoretical stratigraphy, *Jour. Math. Geol.*, **9** (6), 649–652.

Preperata, F. P., and Yeh, R. T., 1973, *Introduction to discrete structures for computer science and engineering.* Addison-Wesley, Reading, Mass., USA, 354 pp.

Quantitative Stratigraphic Correlation
Edited by J. M. Cubitt and R. A. Reyment
© 1982 John Wiley & Sons Ltd.

Formalized Eocene Stratigraphy of Dorog Basin, Transdanubia, Hungary, and Related Areas

I. DIENES

Hungarian Central Statistical Office, Budapest, Hungary

Abstract

The faunal lists of densely-sampled, intensely-studied wells from the Eocene of Transdanubian Central Mountains were processed together with coal bed data in order to study the possibilities of formalized stratigraphy. Time scales, local temporal ranges of taxa, and approximating age of samples were computed by formal methods. Comparison of the results with those of previous authors and with stratotype data indicate that these methods give sufficiently reliable results to detect hyati and tectonic elements, to make reasonable sample-by-sample chronocorrelation, and to plot chronoprofiles.

Introduction

In the past years, several efforts have been made by the present author to formulate some stratigraphic notions and problems in a strict language, and to solve them by mathematical methods (Dienes, 1973, 1974, 1977, 1978a, 1978b, and Dienes and Mann, 1977).

In the present study, the tools and theoretical constructions of formalized stratigraphy have been tested against a set of paleontological and lithological data.

Data

The data employed in this study consists of complete faunal lists of samples collected from the Eocene boreholes shown in Table 1. Here, an altitude data for coal beds in the same boreholes. These have been studied together with field descriptions of the core material and traditional age estimates of the beds.

The majority of localities and faunal lists have been published previously by L. Gidai, G. Kopek, E. Dudich, jr., and others and mentioned as representative type-sections. More than two groups of fossils, nannoplankton, pollen-spora, nummulites, ostracodes, foraminifera, or molluscs, have been determined in each borehole. Faunal lists have been reported by M. Báldi-Beke, M. Jámbor-Kness, T. Kecskeméti, A. Kecskeméti-Körmendy M. Kerekes-Tüske, K. Kerner-Sümegy, M. Kedves, G. Kopek, I. Cs. Khogler, I. Korecz-Laky, M. Monostori, L. Rákosi, M. Rumli-Szentai, I. Vitális, and L. Vitális-Zilahy. The field identification of *N. striatus*, *N. perforatus* and some molluscs have been contributed also.

Table 1.   Boreholes and outcrops studied in present paper

Dorog Basin Area	Outside Dorog Basin
Bajna-38, Bajót-19, 24, 26, 31, 32, Csolnok-648, 695, 697, Lábatlan-3, 4, 'part', Mogyorósbánya-75, 82, 83, 89, 93, Nagysáp-1, 67, Nyergesujfalu-19, 27, 28, 29, Sárisáp-31, Tát-4, Tokod-527, Erzsébet-aknai öregmezó, Ebszóny Szabadság lejtakna, Bajót É-i vége, Buzáshegy	Bakonycsernye-12, Balatonhegy, Balinka-252, Csabdi-12, 84, Esztergom-20, 21, 86, Kósd-20, Mány-13, 23, Mesterberek-75, 76, 94, Oroszlány-1838, Solymár-72, Várgesztes-1

The taxa have been revised for the purposes of this study by M. Báldi-Beke, K. Horváth-Kollányi, M. Jámbor-Kness, A. Kecskeméti-Körmendy, and L. Rákosi.

The field description and age determination of the core material has been compiled by E. Albert, E. Dudich jr., L. Gidai, Á. Jámbor, J. Knauer, G. Kopek, Zs. Modrovich-Csajághy, I. Muntyán, Z. Némedi-Varga, M. Sallai, Z. Sipose, I. Szabó, T. Szücs, I. Tóth, I. Vitális, and Gy. Vigh.

All data were stored on a CDC 3300 computer and the computer manipulations were performed by L. B. Kovács and his co-workers. The author is indebted greatly to all the colleagues mentioned for their help in the preparation of the present work.

Formalized Biostratigraphy

Using the collected data, several types of formalized stratigraphic units can be evaluated. One simple objective is the A-type-taxon-range-biozone of a taxon I. This is a geological distribution, consisting of all stratigraphic points containing at least one specimen of taxon I (Dienes and Mann, 1977). Table 2 summarizes data for a number of frequent taxa in the data set. Figure 1 shows the frequency distribution of taxa by occurrence number classes.

The stratigraphy of Dorog Basin has been studied by several authors such as Hantken (1872), Rozlozsnik, and others, (1922), Jámbor-Kness (1973, 1974), A. Kecskeméti-Körmendy (1972), Rákosi (1973). Gidai (1968, 1969, 1971a, 1971b, 1971c, 1972, 1976, 1977) summarized the first hundred years of investigation of the area.

We also need to define an A-type-assemblage-biozone of taxa $1, \ldots, N$ as the set-theoretical union of the A-type-taxon-range-biozones. The A-type-assemblage-biozones can be used to redefine in a formalized way 'traditional' stratigraphic units.

Figure 2 shows the fit of the A-type assemblage biozone of Laevicardium subdiscors, *Bolivana eocaena, Bolivina nobilis, Marginulina granosa, Uvigerina multistriata, Anomalina affinis, Virgulina hungarica, Virgulina schreibersii, Nummulites subplanulatus, Verneuillina tokodensis, Trachycardium gratum, Nummulites burdigalensis, Nummulites subramondi,* and *Globigerina oecaena* to the 'Operculine marl' described by Gidai (1966). The mean deviation between the boundaries of 'Operculine marl' and 'Operculine marl redefined formally by an A-type-assemblage-biozone' does not exceed two samples.

Experiments with other units, described by Gidai proved, that as a rule, these units may be well approximated by formal units. The redefined units are almost disjoint. At the same time those of Hantken overlap. The formal variants of a traditional stratigraphic unit A have been defined by selecting a number of taxa kept characteristic for A by the author of A.

Of course, these formalized biostratigraphic units may not be appropriate in each aspect as a replacement for the traditional units, which can be mapped easily in the field. However,

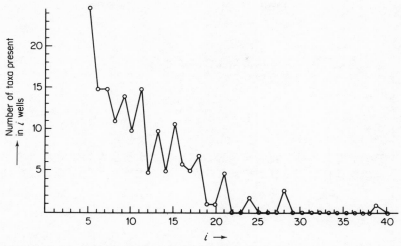

Fig. 1. Frequency distribution of fossil taxa in wells and outcrops of Table 1

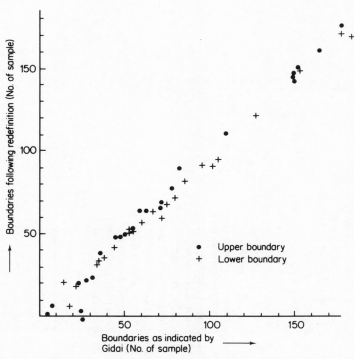

Fig. 2. Boundaries of Operculina marl, as indicated by Gidai and computed from formal definition

Table 2. Characteristic data of taxa, discussed in the present study

No.	Taxon name	Number of holes with taxon for positive wells	Mean thickness (m) for positive wells	Standard deviation (m)	Mean thickness (m) of the A-type zone	Standard deviation (m)
1.	*Leiotriletes microadriennis*	20	155.25	12.34	66.18	4.82
2.	*Punctatisporites lueticus*	18	85.19	13.98	32.62	6.12
3.	*Cicatricosisporites dorogensis*	23	117.63	13.04	57.56	7.85
4.	*Plicatopollis plicatus*	23	88.86	12.58	43.48	6.15
5.	*Plicapollis pseudoexcelsus*	41	94.90	12.43	82.79	9.10
6.	*Tricolporopollenites semiglob.*	18	98.83	13.06	37.85	5.04
7.	*Diporites iszkaszentgyörgyi*	13	53.08	12.96	14.68	3.30
8.	*Pentapollenites pentangulus*	15	60.08	12.67	19.17	4.04
9.	*Milfordia incerta*	17	48.99	9.57	17.72	3.46
10.	*Quinqueloculina juleana*	15	38.98	8.29	12.44	2.64
11.	*Quinqueloculina costata*	12	27.42	7.03	7.00	1.79
12.	*Nonion scaphum*	12	31.35	10.07	8.00	2.31
13.	*Bulimina eocaena*	13	7.84	2.98	2.17	0.82
14.	*Rotalia audouini*	20	29.54	7.31	12.57	3.11
15.	*Lamarckina wilcoxensis*	11	13.83	5.17	3.23	1.21
16.	*Asterigerina rotula*	30	52.48	13.63	33.49	8.70
17.	*Sphaerogypsina globulus*	17	45.43	9.95	16.43	3.60
18.	*Nummulites striatus*	26	74.42	13.66	41.18	7.56
19.	*Nummulites millecaput*	16	19.36	5.00	6.59	1.70
20.	*Nummulites variolarius*	12	33.31	12.97	8.50	3.26
21.	*Turritella vinculata*	11	23.90	7.40	5.59	1.73
22.	*Mesalia elegantula*	14	21.70	7.17	6.48	2.13
23.	*Melanatria auriculata*	11	17.73	7.01	4.15	1.64
24.	*Diastoma roncanum*	16	19.08	5.55	6.49	1.89
25.	*Pyrazus focillatus*	18	29.66	9.06	11.35	3.47
26.	*Calyptrae aperta*	15	59.43	24.88	18.96	7.94
27.	*Cantharus brongniarti*	23	32.20	6.77	15.76	3.31
28.	*Marginella nana*	11	28.27	6.08	6.61	1.42
29.	*Leda striata*	17	34.02	9.92	12.30	3.59
30.	*Brachiodontes corrugatus*	31	53.40	8.56	35.22	5.64
31.	*Arca vértesensis*	19	26.08	6.93	10.54	2.80
32.	*Pteria trigonata*	13	22.14	7.62	6.12	2.10
33.	*Anomia gregaria*	28	44.01	7.65	28.22	4.56
34.	*Anomia tenuistriata*	16	33.25	8.09	11.32	2.75

35. Ostrea supranummulitica	18	17.30	4.71	8.62	1.80
36. Phacoides crassulus	13	10.80	3.29	2.98	0.9
37. Meretrix villanovae	12	19.88	9.72	5.07	2.4
38. Meretrix hungarica	17	30.42	11.08	11.01	4.0
39. Tivelina pseudopetersi	22	54.15	9.39	25.35	4.3
40. Arcopagia mayeri	15	29.48	10.63	9.41	3.7
41. Psammobia pudica	11	30.10	9.74	7.04	2.22
42. Sphaenia hungarica	20	47.09	10.09	20.03	4.2
43. Microfoveolatisporites pseudodent.	19	108.93	14.87	44.03	6.0
44. Laevigatoisporites discordatus	14	49.77	17.32	14.82	5.1
45. Camarozonosporites heskemensis	17	18.82	5.92	6.80	2.50
46. Pleurozonaria concinna	11	24.60	12.55	5.75	2.98
47. Laevigatoisporites haardti	21	106.60	16.67	47.63	7.4
48. Verrucatosporites alienus	18	111.73	17.09	42.79	6.5
49. Momipites punctatus	19	121.00	15.20	48.91	6.50
50. Triatriopollenites minor	19	90.00	17.53	36.38	7.08
51. Cyrillaceaepollenites megaexactus	32	92.88	12.74	63.24	8.67
52. Inaperturopollenites dubius	20	83.74	13.88	35.63	5.9
53. Polypodiidites saalensis	15	46.64	10.24	14.88	3.2
54. Pinus silvestris	15	44.56	10.87	14.22	3.42
55. Porocolpopollenites vestibulum	11	5.91	3.84	1.38	0.89
56. Laevigatoisporites intrapunctatus	17	167.61	18.43	60.62	6.66
57. Concavisporites arugulatus	13	37.66	9.59	10.41	2.65
58. Verrucatosporites favus favus	17	119.06	15.34	43.06	5.54
59. Tetracolporopollenites abditus	13	60.57	18.93	16.75	4.68
60. Tricolporopollenites cingulum pus.	11	166.04	21.54	38.86	5.04
61. Coal	34	135.84	21.82	98.27	15.79
62. Polypodiidites secundus secundus	12	50.54	10.22	12.90	2.63
63. Clavulina parisiensis	13	17.52	6.48	4.84	1.79
64. Quinqueloculina striata	13	22.93	8.78	6.34	2.43
65. Marginulina granosa	27	22.07	4.26	12.68	2.45
66. Lenticulina inornata	11	45.54	30.17	10.68	7.06
67. Nonion commune	20	17.43	4.97	7.41	2.11
68. Uvigerina multistriata	30	40.52	12.67	25.86	8.08
69. Fursenkoina hungarica	25	17.36	3.03	9.23	1.61
70. Rotalia beccarii	12	52.37	35.45	13.37	9.05
71. Rotalia byraminensis	10	30.05	12.68	6.39	2.69
72. Anomalina affinis	15	39.60	23.36	12.63	7.45
73. Cibicides dutemilei	26	39.36	15.21	21.77	8.41
74. Cibicides propinqus	12	23.01	4.60	5.87	1.17
75. Nummulites perforatus	31	14.37	4.17	9.48	2.75

even simple formal definition techniques are flexible enough to define geological bodies, which have important practical applications in geology.

Chronostratigraphy

A linear ordered set of events is understood as a time scale, if the events are ordered according to their time succession.

Time scales studied in this paper

Scale in this study consists of the first and last temporal appearances of most frequent 126 fossil taxa and some coal beds in the geological succession specified. At the first appearance of a taxon I, the highest age of points of the B-type-taxon-range-biozone of taxon I is recognized where the B-type-taxon-range-biozone of a taxon I is a geological body, which consists of the stratigraphic positions containing at least one specimen of taxon I in their original environment. B-type-taxon-range-biozones have been studied in:

(1) the boreholes and outcrops sampled and investigated. At present those are listed in Table 1.
(2) Eocene of the Dorog Basin Area.
(3) Eocene of Transdanubian Central Mountains.

The first (or last) appearance of a taxon I in (1) may not coincide with that in (2) or (3), and so time scales containing these events may differ also, as illustrated in Fig. 3.

The following implicit scales are applied in the text. (1), (2), or (3) are fixed. Then, those events are selected which are not followed by other events. Amongst these, an event with a minimal index will be termed the first event of the scale. Then, this event is deleted and the procedure repeated, resulting in a scale. This is an implicit scale, because one can not say explicitly, which of the appearances stands in item k of the scale.

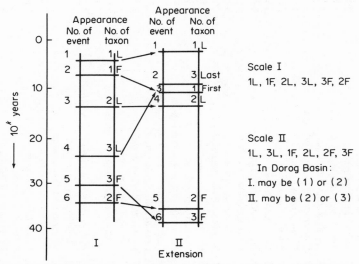

Fig. 3. Hypothetical example: how time scales change when investigation area is extended. The time scales are assumed to be defined by applying the rule for taxa 1–3 described in paragraph 4.1

Relations between standard stages, stratotypes, and the implicit scale applied in this study

If the data set is extended to include data for a number of stratotypes, the temporal range chart obtained by the method described later, should be valid for those stratotypes also. Then, one could define and also determine approximately the temporal range of related standard stages on that scale from those age estimates of stratotypes, which can be obtained by the methods described in the next paragraph. This procedure is illustrated by the Cuisian neostratotype, shown in Fig. 4. Unfortunately, many of the stratotypes have not been investigated by well-stratotypes, sampled, and studied in detail. A faunal list from 'Priabona, dans la partie orientale des Lessini, prov. Vicenza' is not sufficient for exact age estimation or definition. So, present data base does not contain stratotype data with the exception of the Cuisian. When accurately-sampled and adequately-analyzed stratotype data is available, those will be added to data set.

Another approach has been shown on Fig. 5. Figure 6 shows the apparent temporal range of taxa as computed from the data by the method described here.

The age of the rocks 'called in Dorog Basin Priabonian', 'called in Dorog Basin Lutetian' etc. has been estimated and this has been plotted in Fig. 7.

This estimation was performed as follows. A stage 'A' and a borehole have been chosen from those presented in Table 1. Then, the points, nearest to the boundaries of the stage, have been recognized as indicated by traditional methods, where the ages estimated by our computations have been changed. This estimation gives an upper and a lower estimate for the age of the geological body 'called in Dorog Basin stage A'. Repeating the procedure for all boreholes and outcrops, where stage A has been identified traditionally by geologists, and taking the maximum and minimum of ages, an interval-estimate could be obtained. Of

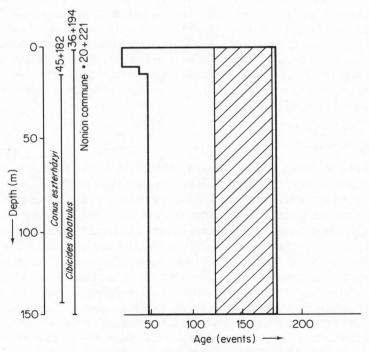

Fig. 4. Estimation of age of Cuisian neostratotype

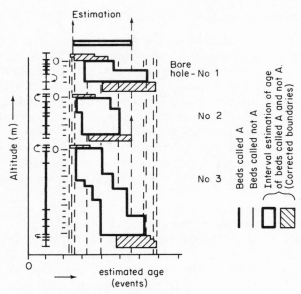

Fig. 5. Hypothetical example for a procedure to estimate age limits of 'beds called A'

course, this is not an exact estimate of the temporal limits of standard stages, and the author
does not wish to suggest the use of these estimates for definition or diagnosis of the
respective stages other than to provide a cross-reference for better orientation.

If the traditional diagnosis of ages is correct, then at least events (1) and (2) must be of
Oligocene age, since the uppermost coal bed is Oligocene in the Dorog Basin. The scale has
no lower limit because sediments older than Sparnacian have not been reported. The age of
the sediments 'called in Dorog Basin Cuisian' is between 118 and 174 and that of Cuisian
neostratotype between 48 and 182. Although the procedure provides estimates larger than
the actual temporal ranges both for Cuisian and 'sediments called in Dorog Basin Cuisian',
the results are not contradictory.

Temporal range charts

The present author (Dienes, 1978a, 1978b) published a method which is suitable for the
construction of temporal range charts of fossil taxa by exact methods.

An upper estimate for the first and a lower estimate for the last appearances may be
obtained regarding the implicit scales defined previously, and temporal ranges estimated
by this method, if a number of conditions are satisfied, contain true temporal ranges. Also, a
lower estimate for the first and an upper estimate for the last appearances can be obtained.
The main idea, forming the basis of the method, is that a deterministically- or
stochastically-correct estimate can be computed for certain elements of the temporal precedence matrix of the respective A-type-taxon-range-biozones. Then, one can count the
number of events which certainly precede or follow an event. The method and theory
together with an example have been published by Dienes (1978a, 1978b).

The possible disturbing factors, the temporal range chart, which has been obtained and its
possible development will be discussed later.

Possible disturbing factors

Definition of taxa. In most fossil groups, the concept of biological species forms the theoretical background for the present methodology of description of new taxa. However, the majority of taxa have no rigorous definition unlike that which has been established for pollen and spora morphotaxa. Furthermore, the technology of stratigraphic investigations may not be based upon repeated study and systematic revision of fossil specimens identified by previous authors. In practice, no stratigrapher can verify the validity of all faunal lists he handles. Therefore, erroneous data may be selected when accepting a working hypothesis in respect of spatial and temporal distribution of fossil taxa and sorting unusual faunal elements. This critical revision may lead to simplification and distortion of fossil spectra, when having extracted not only misinterpreted, but also unexpected fossils.

This is why only formal taxa have been viewed throughout this study. A specimen belongs to formal taxon called ABC, if specialist A determined it in report B as a specimen of taxon C. The formal taxa are not disjoint, even the geographical distribution of the taxa mentioned as C may be predetermined by activity areas of the paleontologists. However, if formal taxa are applied in stratigraphic studies, the curiosities, fuzzy taxa, and their non-consequently used names tend to lose importance, because their spatial and temporal distributions show less regularity. This is illustrated by *Nummulites striatus.* The latter is a frequently-cited taxon in areas (1) to (3). Each of the field geologists knew the name, and several, not only experts, thought that they were able to recognize this taxon.

The assumption is that as a consequence, the taxon became fuzzy and several specimens of taxon N. sp. ex gr. *N. striatus* were identified as *N. striatus.* Consequently, the temporal range estimated by our methods was extended virtually to the end-points of the scale and, hence, the formal taxon 'specimens called *N. striatus* by an arbitrary person in any report' lost its stratigraphic importance.

Of course, this does not exclude practical use of a formal taxon of the type 'specimens called *N. striatus* by specialist A in a report'.

The rejection of all specimens of *N. striatus* may be inadmissible from a paleobiological or evolutionary point of view, but implementation of formal taxa provides a 'clarification mechanism' for stratigraphy.

Throughout this study only formal taxa of the type 'called A by one of the geologists in one of the reports' have been applied. These taxa are well defined. Consequently, the events of our scale, defined on their basis of these taxa, also are well defined.

Normal deposition. One of the conditions necessary for application of the range-construction-method is that of the normal deposition. A geological body is accepted here as of normal deposition, if along any vertical line the age of its points is a monotonous decreasing function of the point elevation.

Normal deposition may be distributed by reverse or normal faults. In order to avoid difficulties with faults, the boreholes studied were divided into sections, which, taking into account the field description of the cores, were fault free, and these sections were handled separately. Undetected faults are probably small and do not distort normal deposition, because they frequently are steep normal faults.

Density condition. The density condition provides a possible extension of the implication made during computation of the temporal precedence matrix. It is supposed that the first appearance of taxon I on the area, where taxon J is not present, does not precede that of taxon I on the area, where both are present. This condition, called the density condition

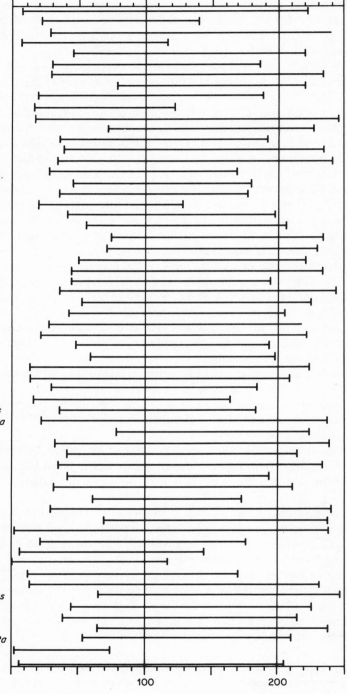

Asterigerina rotula
Asterigerina bimammata
Anomalina affinis
Asterocyclina stellata
Bulimina eocaena
Bulimina elongata
Bolivina nobilis
Bolivina reticulata
Clavulina parisiensis
Clavulina szabói
Cibicides dutemplei
Cibicides propinquus
Cibicides labatulus
Eponides schreibersii
Fursenkoina hungarica
Fursenkoina schreibersii
Guttulina problema
Gyroidina soldanii
Glandulina laevigata
Lamarckina wilcoxensis
Lagena striata
Robulus arcuatostriatus
Robulus inornatus
Robulus vortex
Robulus limbosus
Marginulina behmi
Marginulina granosa
Marginulina fragaria
Miliolina prisca
Nonion scaphum
Nonion communo
Nonion boueanum
Pyrgo bulloides
Rotalia audouini
Rotalia beccarii
Rotalia byraminensis
Rotalia kiliani
Sphaerogypsina globulus
Spiroloculina canaliculata
Spiroloculina limbosa
Quinqueloculina juleana
Quinqueloculina costata
Quinqueloculina striata
Quinqueloculina prisca
Triloculina trigonula
Triloculina gibba
Uviferina multistriata
Verneuillina tokodensis
Nummulites striatus
Nummulites anomalus
Nummulites millecaput
Nummulites incrassatus
Nummulites variolarius
Nummulites perforatus
Nummulites subplanulatus
Operculina granosa
Operculina ammonea
Operculina subgranosa
Actinocyclina tenuicostata
Operculinella vaughani
Discocyclina papyracea

100   200

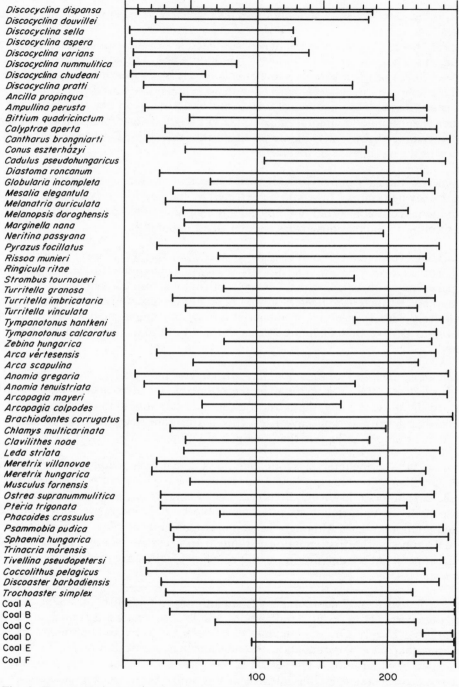

Fig. 6. Temporal range chart of most frequent taxa and lithostratigraphic units in Dorog Basin, Transdanubia, Hungary

(Dienes, 1978) is not fulfilled in each situation but it seems plausible, when area (1) is viewed. This condition is not verifiable directly. The chance, that the density condition is satisfied, is increased by studying only those taxa which have been present in more than 6 boreholes.

However, if area (2) is accepted, it may occur that a few of the estimated temporal ranges are erroneous. This is when taxon I in each borehole is above or below taxa $J_1, \ldots, J_{n_i}$, but there exists in (2) or (3) a vertical succession where it is not the situation, and the empty items in the temporal precedence matrix due to noncomparable taxa do not compensate for the erroneous items. If the number of taxa below or above taxon I in each hole in (1) n_i is large, then the estimated temporal range of taxon I might be incorrect, but that of remainder will vary only slightly. It is highly probable that the majority of temporal ranges are erroneous as a consequence of this factor.

Sample pieces coming from cavings. It is not uncommon, that contamination of cuttings samples by caving occurs. This could be detected by looking for 'very unusual sequences'. Because the paleontologists did not indicate any sign of cavings here, such a procedure has not been followed, but serious errors were avoided hopefully.

Rare sampling. This is a factor that also may influence correctness of results obtained by this method. The effect may be either to shorten or decrease the approximating temporal ranges in comparison with those constructed when sampling is dense enough.

The boreholes which have been discussed here, are representative sections and are sampled relatively densely in terms of industrial norms, and faunal changes. Hence, it seems reasonable to believe that sampling effects did not yield major errors in the range chart computed.

Documentation, sample, and data handling errors. This may lead to unexpected and large errors. Each description in a text may increase the amount of erroneous data by up to 10 percent. In the present study, the most primordial documentation has been applied at the data recording level and data have been checked by repeated collation. Remnant error proportions are hoped to be less than one percent, and not fatal.

Long cores. If the coring interval is long, but the sample investigated inside the interval is of normal size, then a correct range chart can be obtained. The exact position of the sample inside the core interval is not required, but the succession of samples is important. If the sample consists of uniformly-distributed small pieces of a long core, this increases the estimated temporal range. Although the sampling technique employed has not been noted in the field reports, the first situation is thought to occur frequently. This assumption is based upon personal communication with project geologists. Mean core interval does not exceed 4 meters.

Incomplete faunal lists. Paleontologists frequently do not publish complete faunal lists, because of uncertainty in taxon identification, and deletion of abundant taxa, or taxa which are reworked evidently. Our experiments with faunal lists created by different paleontologists indicate that in the worst situation, this may result in a vertical shift of up to 60 meters in the estimated boundaries of A-type-taxon-range-biozones. Approximating temporal ranges may increase or decrease in consequence of this shift, when one compares those to the temporal ranges constructed with complete lists. However, this error is not thought to be systematic in each borehole sequence, and particularly when frequent taxa are studied, one should not attribute great importance to this factor.

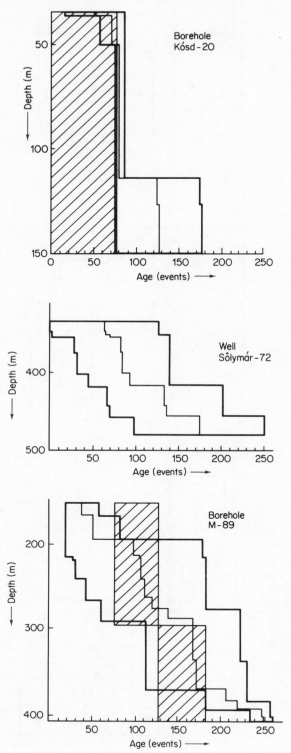

continued overleaf

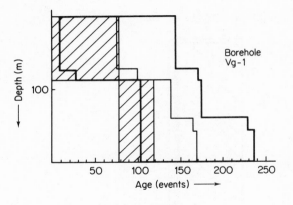

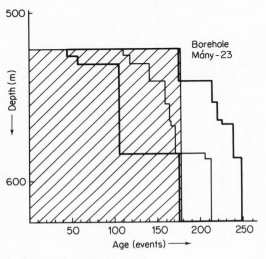

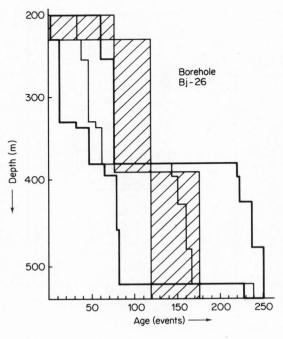

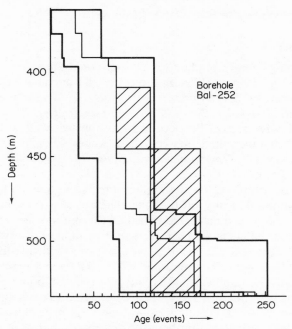

Fig. 7. Age-depth diagrams. Thick line: age-interval estimates. Thin line: mean of the interval. Dashed: age-estimation of previous authors

Heterogeneity investigation. Pollen and spore studies are restricted to fine-grained sediments, and determination of molluscs and foraminifera to dispersible rocks, and this yields artificial barren zones (false gaps) in occurrence of the taxa. Nevertheless, a sample may apply to the study of more than one fossil group, although in practice, samples are not the subject of a joint study of all fossil groups in the majority of the situations.

If the sampling is dense enough for each fossil group, complications are not produced. All possible fossil groups have been studied in a great number of samples from the boreholes, excluding nannoplankton and particularly ostracodes. Hence, ostracodes have been excluded from the computations. Considering the great number of outcrops, wells, and taxa that have been studied and the nonsystematic nature of the gaps, the effect of this factor has been neglected.

Facies dependence. Regional temporal ranges have been defined by apparent first and last appearances which are governed by facies changes. At the same time, accuracy of the computed range chart is not influenced directly by facies changes, only through the density condition, which has been described earlier.

Reworking. As far as apparent occurrences have been studied here, it is thought that reworked sediments and faunal mixing do not give rise to errors in the time scale and range chart computed.

Discussion of the Results Obtained for Dorog Basin Area

Description of the range chart

Figure 6 shows the range chart obtained for Dorog Basin Area.

The maximum length of temporal ranges reaches the maximal value 251 at 'coal not specified'. It is followed by *Brachiodentes corrugatus* 239, *Nummulites striatus, Anomia gregaria* 238, *Cibicides dutemplei* 230, and *Tivelina pseudopetersi* 225.

The best temporal markers are 'the second from upwards coal bed' 22, 'the first from downwards coal bed, if more than one coal bed occurs' 29, *Discocyclina chudeani* 60, *Tympanotonue hantkeni* 66, *Operculinella vaughani* 74, *Discocyclina nummulitica* 77, *Arcopagia colpodes* 105, etc.

It is emphasized that a time-lag on the implicit time-scale may be longer, when expressed in year units, than another disjoint time-lag, which is longer, when expressed in event units of the implicit scale. It is remarkable that the best markers are not fossils. A shortening of range-widths may be achieved by increasing the number of taxa and outcrops.

The range lengths obtained also have been compared with modified RBV values of McCammon (1970). Relative biostratigraphic value (RBV) is a function of facies dependence, horizontal occurrence, and vertical extent. As expected, points are scattered above a straight line of negative slope on a RBV versus estimated taxon-range-width graph. The negative slope follows from RBV definition and the large spread observed is due to two groups of taxa. The first is that of 'thin-bed-like distributed but no isochronous' taxa, and the second is that group which is comparable to only a small number of other taxa, consequently, their ranges seem wide.

The distribution of estimated first and last appearances shows two maxima, as expected in consequence of the statistical character of estimates, but also attributable to rapid faunal changes at stage boundaries.

M. Jámber-Kness, A. Kecskeméti-Körmendi, L. Rákosi, and L. Vitális-Zilahy have published earlier charts for the temporal range of *Nummulites*, molluscs, pollen, and foraminifera in Dorog Basin. Because these charts have been constructed by unspecified methods applying other data sets, standard scales, and different temporal range definitions, one should not expect total coincidence. However, there are no large discrepancies. The present chart is nevertheless more detailed than the earlier presentations.

Possible future changes in temporal range chart obtained in the present study

As has been indicated earlier, if any of the conditions suggested for accuracy of the computations prove to be false, the chart in Fig. 6 may (but not 'must') be wrong. Disturbing factors which may distort the approximating range charts have been discussed previously. The final conclusion is that although these factors might lead to major errors compared to approximating temporal ranges computed from the data which are free from these effects, however, in our estimates, these lead to minor deviations only.

If new boreholes and outcrops are added from areas (2) or (3) and the same taxa determined, then, the set of boreholes and outcrops to be studied for area (1) will be changed and as a consequence, scale (1) and the temporal ranges may change also.

Scales (2) or (3) and the temporal ranges of taxa on (2) or (3) do not change, only the estimated position of the first and last appearances on scale (2) or (3). The time-scale and temporal ranges of (1) approximates those of (2) and (3), when the number of boreholes and outcrops in (2) and (3) tends to infinity. Temporary fluctuation, both increasing and decreasing, of the ranges may occur. If the number of boreholes increases, decrease of range width is more probable.

If a number of boreholes were deleted, scale (1) and temporal ranges of taxa in (1) might change, but the scales (2) and (3) and temporal ranges of taxa therein would not vary.

If the area (3) was extended, for example, to Hungary or south-east Europe, the time-scale of the new area might deviate significantly from that of the Transdanubian Central Mountains, although the estimates made from the same data set (1) might be useful if the density condition held for the new area also. Reliability of temporal range charts computed from (1) is unknown in regard of any subarea of (1).

If new samples are added to the data set from the same boreholes, the temporal range estimates may vary, as has been indicated by scarce sampling.

If new taxa are added, a new time-scale for areas (1), (2), or (3) containing the first and last appearances of the new taxa, may be defined. Temporal ranges of the original taxa may change in both directions on the new scale. The position of the taxa added later may be estimated on the original or new scales.

If a number of taxa are amended or united, the time-scale and temporal ranges on (1), (2), and (3) may change, but the estimated temporal ranges are not additive.

Age Estimation

Once a time-scale, either (1), (2), or (3) has been defined, and the temporal range of taxa has been determined approximately, as discussed in the previous paragraphs, then an estimation for the age of any sample may be computed. An age estimation procedure has been described in Dienes (1978a, 1978b).

The procedure consists of two steps. First, an interval estimation based upon taxa recognized in the sample, is made for the age of each sample. The estimates are computed from the temporal range chart of Fig. 6. The minimum possible age of a sample is the maximum of ages of estimated last appearances and the maximum possible age is the minimum of ages of estimated first appearances of the taxa present in the sample.

Numeration of our scale is directed backwards in time. Suppose a sample B lies below sample A and the age of B is lower than k, then, the age of sample A also should be lower than k. Similarly, the age of any sample below sample B of age higher than k, must be younger than k. Then, if normal deposition is assumed, further correction is straightforward. This step is shown graphically in Fig. 8. The conditions of computations, disturbing factors, and results are discussed below.

Disturbing factors

Normal or reverse faults. When moving down a well, a fault not interrupting normal deposition may lead to a positive horizontal jump in the depth–age diagram, and to a negative one if it interrupts normal deposition. Hence, if a large positive jump appears on a depth–age diagram, this may indicate a normal fault not preventing normal deposition. Nevertheless, large positive jumps also may occur in consequence of frequent faunal changes and also faults, which interrupt normal deposition, do not appear in diagrams in every situation. A fault interrupting normal deposition, for example a reverse fault, may lead to an empty-interval age estimate, if the correction based upon the assumption of the normal deposition is made. An example is discussed later.

Age estimations based on the normal-deposition hypothesis gave realistical results for all samples of studied outcrops and boreholes, excluding Mogyorósbánya-83. This is near to a

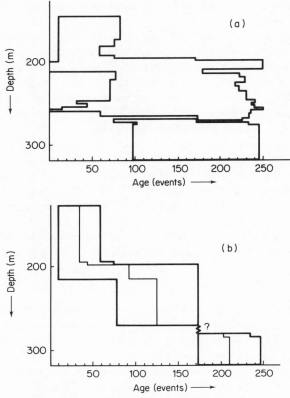

Fig. 8. Age-depth diagram of borehole M-83. (a)
Uncorrected; (b) Corrected for normal deposition

belt, where a large jump appears in the altitudes of all biozone boundaries and indicates a
major fault. The depth–age diagrams have been plotted for M-83 with and without the
normal-deposition hypothesis. The sequence of estimates on Fig. 8 at 269–279 meters
contradicts the hypothesis of normal deposition. At that interval, the corrected lower
estimate is higher than the corrected upper estimate.

Figure 9 illustrates another problem. The estimated age–depth diagram of the samples
taken from the Balatonhegy quarry is plotted with and without the normal deposition
hypothesis. The fact that the differences between Fig. 9(a) and 9(b) are small indicate that
the large dip-angle changes, observed in the rocks, are not associated with a major distortion
of normal deposition.

Long cores. Assume that a core taken from 128.6–142.0 meters has been sampled. If the
sample is a single small volume of rock, then our procedure guarantees a good estimate
irrespective of the core length. On the contrary, if the sample is a mixture of several small
pieces of rock taken from the whole interval, the age of the different particles of the long
core may differ significantly such that either estimated age intervals do not overlap or
the procedure does not give any reasonable result.

Incomplete faunal lists, heterogeneity investigation. If only some of the fossil taxa present in

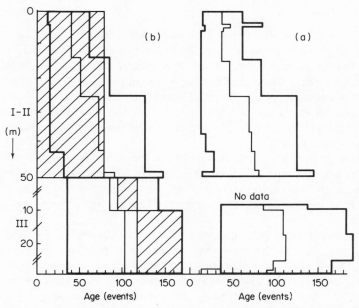

Fig. 9. Age estimation-depth diagram of outcrops Balatonhegy I.
(a) Uncorrected; (b) Corrected for normal deposition hypothesis.
Figure 9(b) has not been corrected at 0 m of outcrop III. / = Fault
indicated in field documentation

a sample have been mentioned in a faunal list, the age estimate might become wider than
that made from complete lists.

Facies changes. Facies changes do not influence directly accuracy of the age estimates made
by the algorithm described. However, abrupt changes of estimated ages at facies changes
may occur due to variations of fauna content and lack of data.

Unconformities, hiatuses, and reworking. Unconformities, hiatuses, and reworking do not
influence the accuracy of the age estimations made by the method described here. Large
jumps, greater than 50 events on the depth–age diagram, may indicate an unconformity or
hiatus. Reworked sediments may not be recognized by apparent time scales.

Errors in the temporal range estimates. In this context, a temporal range estimate is
erroneous, if it does not contain the true temporal range. This may lead to erroneous age
estimates.

Future extension of the time scale validity. If a new well is studied, and its temporal precedence matrix does not contradict that of the range chart, the range chart may be used
mechanically, although event k of the new data set is not the same as event k of the old data
set. If the temporal precedence matrix on the new well contradicts that of the range chart, a
new common temporal precedence matrix and range chart should be computed, and then, if
necessary, a correct estimation made for all wells and outcrops.

Our experience shows that although contradictory matrices logically are possible at any
time, especially when the new borehole is located far from previous ones, contradiction is
not frequent if the range chart has been based on more than 40 boreholes.

Discussion of the results

Description of the depth–age diagrams. Temporal range chart of Fig. 6 and normal deposition depth–age diagrams have been made for boreholes and outcrops of Table 1, and also for some wells in Europe and Africa. Illustrative examples are shown in Fig. 7.

Sample mean ages were obtained by averaging the upper and lower sample age estimates. The distribution is unimodel in the Transdanubian Central Mountains, if small maxima and minima (perhaps 'random') are not taken into account. This is shown in Fig. 10. The minima at 0–30, 180–190, and 240–250 are due to the structure of the range chart estimate and do not bear importance.

The distribution of sample age interval widths in Fig. 10(b) is as expected considering the distribution of temporal range estimates (Fig. 10(a)). The mean interval length is approximately 52 events. No separate samples, only homogenous intervals with a constant estimated age, have been studied. The majority of means are at approximately the means of the scale.

Distribution of jumps on depth–age diagrams can be seen in Fig. 11. The well-expressed peaks in the central part of the scale may indicate real hiatuses. Earlier authors (Gidai, 1966; Kopek, 1965) suggested that the hiatuses were denudation periods within the Eocene. Although the jumps show a regional distribution, their exact well-to-well correlation is not possible because the sample age estimates are too wide.

Coincidence with traditional age estimates. The position of the temporal limits of standard Eocene stages on the implicit scale has been estimated by the method described previously. Applying these estimates, one can compare the boundaries of the standard Eocene chronostratigraphic units, as marked earlier using traditional methods by several geologists and as computed from age estimates of samples and tentative estimates of stage limits. One can compare the traditional and algorithmic estimates sample-by-sample, or section-by-section. The latter can be applied to sections with the same traditional and algorithmic age estimates, I_2 and I_1, for each sample within the section.

The third method was adopted here. The results are summarized in Table 3. I_3 is the mean sample age.

The general coincidence is good, but there are discrepancies. If one accepts the conditions discussed in the earlier paragraphs, then the following situations will be distinguished.

If I_1 and I_2 are disjoint, then I_2 must be erroneous. If I_1 and I_2 are not disjoint, then I_2 may be correct and may be a better estimate than I_1.

Mean estimated age interval width (accuracy) is 29 events for the traditional and 52 for the algorithmic method. Therefore, algorithmic estimates are in mean less accurate by a factor of five-thirds. At the same time the traditional estimates contain more than 5 percent certain (provable and proved) errors plus an unknown amount of such errors that cannot be proved. In addition, 9 percent of the traditional estimates are less accurate than the algorithmic estimates.

Conclusions

A pilot study has been reported here investigating the possibilities of formalized stratigraphy in bio- and chronostratigraphy. The faunal lists of densely-sampled, well-studied sections from the small Dorog Basin, and the Transdanubian Central Mountains have been processed together with coal bed data in order to study the feasibility of formal stratigraphic units, time scales, local temporal ranges of taxa, and approximating age of samples obtained by formal methods. The results have been compared with those of previous authors.

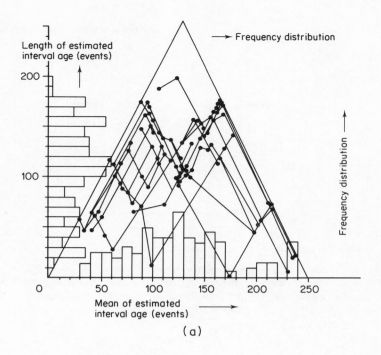

(a)

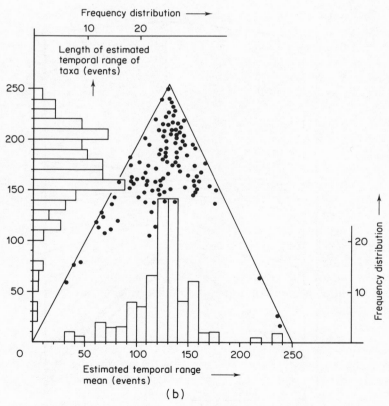

(b)

Fig. 10. Distribution of (a) Estimated temporal ranged of taxa and (b) Estimated interval age of samples in dependence of temporal range mean and mean of estimated sample age

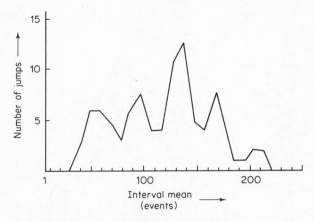

Fig. 11. Frequency distribution of jumps in dependence of mean of age-interval estimate of the sample above the jump. Transdanubian Central Mountains. Smoothed by 20-event window

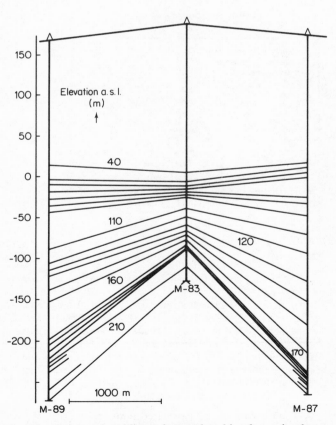

Fig. 12. Estimated and linear interpolated isochrons in plane of wells M-89, M-83, M-87

Table 3. Evaluation of age-interval estimates made for wells and outcrops from Transdanubian Central Mountains

Relations	Number of occurrences	Percentage	Implication
I_1 and I_2 \ are dis-	9	5	I_2 is erroneus
I_2 and I_3 / jointed	54	36	
I_1 contains I_2	67	42	If I_2 is correct, then it is better, but it may be erroneus
I_2 contains I_1	14	–	I_2 is correct but I_1 is better

The results, summarized in Figs 6 and 7, and in Table 3 indicate that by formalized methods, despite complications with data sources and uncertainties concerning the conditions which guarantee the accuracy of the applied algorithms, realistical temporal range charts and age estimates have been obtained. The advantages of formalized procedures, such as exact definition of time scales, detailed definition, and known conditions for accuracy of the manipulations, are to be extended in the near future with smaller length of interval estimates.

Complications with data sources and the effects of disturbing factors are avoidable by standardization of sampling and investigating technique and reorganization of the technological process of stratigraphic activities. At the same time, a larger number of taxa and wells may be studied, and precision of the age estimates may be improved.

References

Báldi-Beke, M., 1971, The Eocene nannoplankton of the Bakony Mountains, Hungary, *Ann. Inst. Géol. Publ. Hung.,* **54** (4), 11–40.

Dienes, I., 1974, Subdivision of geological bodies into ordered parts, in Dienes, I. (ed.), *Matematika és számitástechnika a nyersa-nyagkutatásban I.* Hungarian Geological Society (MFT), pp. 138–147.

Dienes, I., 1977, Formalized stratigraphy: basic notions and advantages, in Merriam, D. F. (ed.), *Recent Advances in Geomathematics.* Pergamon, N.Y., pp. 81–87.

Dienes, I., 1978a, Establishment of optimal complex time scales and their use, *Acta geol.* (in press).

Dienes, I., 1978b, Exact methods for plotting temporal range charts, *Computers & Geosciences*, **4**, 269–272.

Dienes, I., and Mann, C. J., 1977, Formalization of stratigraphic terminology, *Jour. Math. Geology,* **9** (6), 587–603.

Dudich, E., and others, 1968, Quelques problèmes actuels de l'Eocène dans la Montagne Centrale Transdanubie, Hongrie, *Mém. du BRGM,* **58**, 675–682.

Gidai, L., 1968, Geologische Ergebnisse der Bohrung Nyergesujfalu 29, *Jahresberichte d. Ung. Geol. Anstalt für 1966,* 141–148.

Gidai, L., 1969, Les subdivisions stratigraphiques de formations Éocène de la partie NE de la Montagne Centrale de Transdanubie (Hongrie), *Mém. du BRGM,* **69**, 183–192.

Gidai, L., 1971a, Coupe repère de l'Éocène de la region Nord-Est de Transdanubie (Sondage de Tokod—527), *Ann. Inst. Géol. Publ. Hung.,* **54** (4), 101–111.

Gidai, L., 1971b, Les données fournies par la revision géologique de quelques affleurements Éocènes classiques de la region Nord-Est de Transdanubie, *Ann. Inst. Géol. Publ. Hung.,* **54** (4), 82–97.

Gidai, L., 1971c, Les faciès de Eocène dans la région de Mány, de Szomor-Zsámbék et de Bajna, *Rapp. Ann. de l'Inst. Géol. de Hongrie sur l'année 1969,* 111–120.

Gidai, L., 1972, L'Éocène de la region de Dorog, *Ann. Inst. Géol. Publ. Hung.,* **55** (1), 1–40.

Gidai, L., 1976, Stratigraphie des formations Éocènes des environs de Várgesztes (Transdanubie) et leurs possibilités de corrélation, *Rapp. ann. de l'Inst. Géol. de Hongrie sur l'année 1974,* 335–342.

Gidai, L., 1977, Coupe de référence géologique des formations Éocènes des environs de Sümeg et Csabrendek, d'après le sondage No. Cn-850, *Rapp. ann. de l'Inst. Géol. de Hongrie sur l'année 1975*, 229–249.

Hantken, M., 1872, Die geologische Verhältnisse des Graner Braunkohlengebietes, *Jahrbuche der kön. Ung. Geol.*, **1** (1), 5–140.

Jámbor-Kness, M., 1969, Recherces sur les Nummulites dans le bassin de Dorog, *Mém. du BRGM*, **59**, 201–208.

Jámbor-Kness, M., 1973, Eocénkoru Nummulitesek vizsgálata és rétegtani értékelése a Dorogi medence Ny-i részén, *Ann. Inst. Géol. Publ. Hung.*, **55** (3), 378–417.

Jámbor-Kness, M., 1974, Nummulites dâge Éocène inférieur remeniées dans l'éocène moyen de la Transdanubie nord-est, *Ann. Inst. Géol. Publ. Hung.*, **54** (4), 177–185.

Kecškeméti, T., 1971, Appreciation de quelques espéces des Nummulites par rapport a leur valeur stratigraphic avec la prise en consideration des factors paleogeographiques, *Ann. Inst. Géol. Publ. Hung.*, **54** (4), 189–199.

Kecškeméti, T., 1960, A bakonyi eocén szintezése nagy foraminiferák alapján, *Földt. Közl.*, **90** (4) 442–455.

Kecšeméti-Körmendy, A., 1972, Die eozäne Molluskenfauna des Doroger Beckens, *Ann. d. Ung. Geol.*, **55** (2), 141–377.

Kopek, G., and others, 1965, Stratigraphische Probleme des Eozäns im Transdanubischen Mittelgebirge, *Acta geol.*, **9** (3–4), 411–425.

Kopek, G., 1967, Zusammenhänge zwischen der perspektivischen Braunkohlenerkundung und faziologischen und entwicklungsgeschichtlichen Problemen des Eozäns im Transdanubische Mittelgebirge, *Ann. Hist. Nat. Mus. Nat. Hung. Pars. Mineral Paleont.*, **59**, 81–92.

Kopëk, G., 1968, A bakony-vértesi köszénkutatás legujabb eredményei, *Rapp. ann. de l'Inst. Géol. de Hongrie sur l'année 1966*, 105–115.

McCammon, R. B., 1970, On estimating the relative biostratigraphic value of fossils, *Bull. Geol. Inst. Univ. Uppsala.*, **2**, 49–57.

Rákosi, L., 1971, Les associations de la vegétation du Bassin lignitifère de Dorog, *Ann. Inst. Géol. Publ. Hung.*, **54** (4), 263–274.

Rákosi, L., 1973, Palynologie des formations paléogènes du bassin de Dorog, *Ann. Inst. Géol. Publ. Hung.*, **55** (3), 497–697.

Rozlozsnik, P., and others, 1922, Az esztergomi bányaterület bányaföldtani viszonyai, Bp, *M. kir. Ioldt, Int.*, 128 pp.

Strausz, L., 1966, Über die stratigraphische Verteilung der Gastropoden im Eozän Ungarns, *Földt. Közl.*, **93** (3), 349–355.

Szõts, E., 1956, Le Éocène (Paléogène) de la Hongrie, *Geol. Hung, Ser. Geol.*, **9**, 76–106.

Szõts, E., 1961, Remarques sur les niveaux á Foraminiféres du Paléogène en Hongrie, *C.R. Soc. Géol. France.*, **3** (6), 161–162.

Vitális-Zilahy, L., 1968, Foraminiféres planctoniques dans la série de Éocène du bassin de Dorog, *Mém. du BRGM*, **58**, 131–135.

Vitális-Zilahy, L., 1971, Les formations Éocène moyen a Foraminiféres du bassin de Dorog, *Ann. Inst. Géol. Publ. Hung.*, **54** (4), 131–135.

II. BIOSTRATIGRAPHY

Quantitative Stratigraphic Correlation
Edited by J. M. Cubitt and R. A. Reyment
© 1982 John Wiley & Sons Ltd.

Quantitative Biostratigraphy: The Methods Should Suit the Data

LUCY E. EDWARDS

U.S. Geological Survey, 970 National Center,
Reston, VA 22092, U.S.A.

Abstract

The practicing biostratigrapher can choose among numerous quantitative and semiquantitative methods. The various methods have distinct strategies, require different assumptions, and may produce differing results. They need to be compared critically. Important questions to be considered are:

Are all data from stratigraphic sections?
Are all data from MEASURED stratigraphic sections?
How many sections are to be used?
Am I working with two types of events (bases and tops) or one kind of event (bases only or tops only)?
Am I looking for maximum ranges or average (most probable) ranges?
Can I justify the assumptions of linearity/normality/independence required by the various methods?

The biostratigrapher is invited to consider various quantitative methods, to reject, accept, or agree to live with the underlying assumptions, and to experiment with the robustness of the methods. A hypothetical dataset is introduced and applied to one method of quantitative biostratigraphy, no-space graphs. This technique produces accurate results and is robust. I hope that this experiment will prompt biostratigraphers to repeat it using other quantitative methods.

Introduction

At the first meeting of the Biostratigraphic Working Group for the IGCP (International Geological Correlation Program) Project 148 in Syracuse, N.Y., U.S.A. in 1977, three common goals were established (Brower and Millendorf, 1979):

(1) the development of algorithms for quantitative biostratigraphic correlation,
(2) the implementation of computer programs, and
(3) the evaluation of the various methods through case studies utilizing both actual and simulated sets of data.

Substantial progress has been made on goals (1) and (2). Several techniques for quantitative biostratigraphic correlation have been devised and computerized. The third goal, the evaluation of various methods, now requires concerted efforts.

The various quantitative methods in biostratigraphy have distinct strategies, require different assumptions, and may produce differing results. An excellent summary (Bower, 1981) discusses the methods and some of the strategies. In the present paper, I discuss

45

various methods in terms of their applicabilities and underlying assumptions. Some methods are suited more to specific kinds of data. The differing goals of individual practicing biostratigraphers may be served better by some methods than by others.

In addition to the discussion of various methods, I present the application of one method of quantitative biostratigraphy to a hypothetical dataset. I hope that this will stimulate further study.

Methods in Quantitative Biostratigraphy

The various quantitative methods discussed here may be categorized according to underlying biostratigraphic philosophy. Four philosophies are discussed here:

(1) *Tally concept*—Biostratigraphic information is conveyed by the fact that a particular sample contains a particular taxon.

(2) *Event concept*—Biostratigraphic information is conveyed by the fact that a particular taxon has its stratigraphically-highest or stratigraphically-lowest occurrence in a particular sample.

(3) *Morphology concept*—Biostratigraphic information is conveyed by the morphological characteristics of a taxon in a particular sample.

(4) *Ecostratigraphic concept*—Biostratigraphic information is conveyed by the paleoecologic characteristics of stratigraphic sections, as inferred from relative abundances of taxa.

The reader is referred to original publications or various summary papers (Hazel, 1977; Hay and Southam, 1978; Brower and Millendorf, 1979; Millendorf, Brower, and Dyman, 1979; Brower, 1981) for more details of individual methods. Brief discussions of strategies follow.

Tally concept

In methods with the tally concept, the raw data can be presented as entries in a sample by taxon matrix. The entries often consist of notations indicating the presence or absence of taxa, but relative abundances or values weighted according to an estimate of biostratigraphic value may also be used. The data may be analyzed in Q-mode—comparing similarities or differences between pairs of samples to produce a sample by sample matrix, they may be analyzed in R-mode—comparing taxa to produce a taxon by taxon matrix, or they may be analyzed by methods that treat taxa and samples together. Main themes may be extracted from the resulting matrices by a variety of techniques including cluster analysis, principal components, correspondence analysis, and others. Assorted multivariate techniques have been discussed and summarized in recent papers by Hazel (1977), and Millendorf, Brower, and Dyman (1979).

The range-through method. A biostratigrapher using multivariate techniques in Q-mode may wish to count a particular taxon as present in a sample in which the taxon has not been observed if this taxon is recognized both above and below the sample in question in the given stratigraphic section. This procedure was advocated by Cheetham and Deboo (1963) and is discussed and illustrated in Hazel (1977), who termed it the range-through method. From the standpoint of biostratigraphic significance, the method has much to recommend it (Hazel, 1970; Millendorf, Brower, and Dyman, 1979).

The use of the range-through method has important philosophical ramifications, however. When using this method, the biostratigrapher departs from the tally concept. Biostratigraphic information no longer is conveyed by the fact that a sample contains, or does not contain, a particular taxon. Now, only stratigraphically-highest or stratigraphically-lowest occurrences of taxa affect the computed similarities or differences between pairs of samples. Thus, perhaps unaware, the biostratigrapher is operating within the event concept when the range-through method is used. The event concept is discussed further in the following section.

The tally concept, *sensu stricto*, has not been used extensively in biostratigraphy (although it is employed frequently in paleoecology and biogeography). Jekhowsky (1963) and Reyre (1973) used relative frequency and minimal distances to effect correlation. Hazel (1970, 1977) has illustrated the concept in the R-mode examples. Guex (1977) used the concept, but his methods have been modified (Davaud and Guex, 1978) outside the scope of the tally concept.

Event concept

In the event-concept methods, the raw data must come from samples that can be ordered stratigraphically (from stratigraphic sections). For each taxon, two stratigraphic events, the highest- and lowest-observed occurrences, are noted for each section in which the taxon occurs. The relative or absolute positions of each event in each section are noted. Within the event concept, probabilistic and graphic strategies are based explicitly on stratigraphic events; most multivariate, seriation, and relational strategies implicitly are concerned with stratigraphic events.

In PROBABILISTIC strategies, each event is compared with every other event to determine an overall 'most probable' sequence of events. The strategy may be binomial (Hay, 1972; Southam, Hay, and Worsley, 1975; Blank, 1979), modified binomial (Gradstein and Agterberg, in press), or trinomial (Edwards and Beaver, 1978).

In GRAPHIC strategies, the geometry of construction of various graphs or figures is used to determine the absolute or relative positions of events when stratigraphic events in each section used are compared with the events in other sections (Shaw, 1964; Miller, 1977; Edwards, 1979).

MULTIVARIATE strategies using the range-through method implicitly fall within the event concept. These strategies include Q-mode cluster analysis (Hazel, 1970; 1977), principal coordinates (Hazel, 1977), similarity matrix contouring (Christopher, 1978), lateral tracing (Millendorf, Brower, and Dyman, 1979) and others. Note that, in the range-through method, all changes in similarity or correlation coefficients are caused by one or more local lowest or highest occurrences. This may lead to high correlations at unconformities. It definitely leads to the weighting of a taxon (inversely) by the diversity of the flora or fauna associated with it. In a completely different multivariate strategy, Hohn (1978) used principal components analysis to correlate events from several sections. This method explicitly falls within the event concept.

RELATIONAL strategies can be separated philosophically from multivariate techniques. In relational methods, taxon by taxon matrices are constructed, not on similarities or differences, but rather on the mutual occurrence, or lack thereof, between each pair of taxa (Guex, 1977), or by a notation indicating whether the range of one taxon in any pair precedes, succeeds, or overlaps the range of the other taxon (Scott, 1974; Davaud and Guex, 1978; Rubel, 1979). Because these methods deal with the relationships between the ranges of taxa, they implicitly fall within the event concept. To the practicing biostratigrapher, however, these methods appear to share the philosophy of the tally concept because occurrences and co-occurrences of taxa are emphasized, rather than lowest- or highest-occurrence events.

SERIATION of sample by taxon data matrices (Brower and Burroughs, this volume) resembles relational methods philosophically. The range-through method is used and differences between samples occur only at lowest- and highest-occurrence events; thus the method implicitly uses the event concept. In practice, however, by rearranging the rows and columns of the matrix to minimize range zones, the biostratigrapher is emphasizing occurrences, not events.

Morphologic concept

In methods in which the morphologic concept is used, morphologic changes in populations of a taxon within its range can be analyzed by multivariate methods (see, for example, Reyment, 1979). As I have not experimented with these techniques, I will not discuss them further in the present paper except to note that they show potential in biostratigraphic correlation. Statistically-defined taxa (Scott, 1974) can, by themselves, provide high-resolution biostratigraphy or can be used with other taxa in any of the previously-mentioned methods.

Ecostratigraphic concept

In methods in which the ecostratigraphic concept is used, multivariate techniques are used to derive paleoecologic trends for stratigraphic sections. Correlation is effected by comparing these trends (Cisne and Rabe, 1978; Rabe and Cisne, 1980). Ecostratigraphy may be combined with more traditional biostratigraphy, and resolution on the order of 100,000 years is predicted (Rabe and Cisne, 1980). This is another technique that I have not explored personally.

Questions to be Asked

The biostratigrapher seeking to use quantitative methods should be aware that certain methods provide better results for specific kinds of data, and that differing biostratigraphic goals may be served better by some methods than by others. I have compiled a list of questions that I think should be asked. The reader will undoubtedly think of others.

Are all data from stratigraphic sections?

If the biostratigrapher answers this question 'NO', the choice of methods is severely limited. Multivariate methods falling within the tally and seriation concepts may be used. Methods that implicitly use the event concept (multivariate and relational strategies) may be applied if the number of isolated samples is relatively small. Methods explicitly using the event concept cannot be applied.

One might hope that all data come from stratigraphic sections, but often the biostratigrapher must cope with isolated samples. Once a firm biostratigraphic foundation has been laid, of course, isolated samples are seldom a problem. The previously-mentioned methods, however, allow incorporation of isolated samples in the early synthesis. If the majority of the samples lack good stratigraphic control, or are collected from short stratigraphic sections, the biostratigrapher has two choices: to acquire more, better data (everyone's first choice), or to try to obtain as much information as possible from the data at hand (which may be the best that can be achieved given various practical limitations).

If the biostratigrapher answers this question 'YES', any of the various methods can be used. My personal preference is that if the data are sufficient to use a probablistic or graphic strategy, these should be used instead of a multivariate one. (I prefer explicit to implicit use of the event concept.) I point this out as one place where comparative studies using the same dataset are needed. Frederiksen (1980) used multivariate and graphic methods on the same dataset to complement each other.

Are all data from MEASURED stratigraphic sections?

By measured stratigraphic section, I mean a section in which every sample can be accurately placed in both relative and absolute position. To know that Sample X was collected below Sample Y is not sufficient; one must know exactly how much below, in meters, feet, or some other linear measurement. In core material, but not necessarily in cuttings, accurate positioning of samples is generally easy to achieve. In outcrop, this requires careful measurement.

A measured section, as I am using the term, requires additional precision. Any faulting must be recognized and taken into account, and all unconformities must be considered. Composite sections that have been 'pieced together' over even a small geographic area are highly suspect because a thickness of 10 meters in one outcrop may bear no logical relation to 10 meters in a neighboring outcrop.

If the biostratigrapher answers this question 'NO', two methods in quantitative biostratigraphy are eliminated immediately. Shaw's (1964) graphic correlation and Hohn's (1978) principal-components technique both require data from measured sections. Certain methods of calculating relative biostratigraphic value used in some of the multivariate techniques (see discussion in Brower, Millendorf, and Dyman, 1979) require thickness measurements as well.

If the biostratigrapher answers this question 'YES', any of the various methods can be used.

How many sections are to be used?

Although a biostratigrapher may sometimes choose to base a zonation on a single section, most problems in quantitative biostratigraphy involve multiple sections. Increasing the amount of biostratigraphic material studied can have one of three effects: it may increase the biostratigraphic refinement, may increase the effort required on the part of the biostratigrapher, and (or) may cause total chaos. The storage capacity of even large computers may be exceeded if the biostratigrapher tries to use too much data (although the creative programmer can discover methods to avoid this).

In some kinds of methods, as more sections or samples are added, the results of quantitative biostratigraphy will improve. This is especially true of probabilistic and relational methods. These methods use event by event matrices and taxon by taxon matrices, respectively. With the exception of the rare taxa added as more material is studied, the matrices do not increase in size as the number of samples increases. Instead, the reliability of the value in each cell of the matrices increases as more comparisons are possible. Except for the increase in effort caused by compiling the additional comparisons, increasing the number of samples increases the precision of the results without increasing the matrix size, the amount of necessary computations, or the computer storage requirements. The limiting factor, computationally, becomes the number of events or taxa.

For other methods, most notably graphic methods, increasing the number of sections does increase significantly the number of calculations involved. As most graphic methods

Table 1.   The number of sections vs. the
strategy   used   (X's   indicate   general
                applicability)

Strategy	Number of sections		
	0–4	5–10	>10
Probabilistic		X	X
Graphic	X	X	X
Multivariate	X	X	?
Relational	X	X	X
Seriation	X	X	

do not require large matrices needed in the other techniques, computer storage requirements are not as likely to be the limiting factor. However, graphic techniques, especially as
practiced by Miller (1977) and Edwards (1979), require knowledgeable input on the part of
the biostratigrapher. I feel that this input is one of the strong points in favor of graphic
techniques; others may feel that this is a limiting factor when many sections are considered.

Multivariate techniques are likely to become unmanageable when large numbers of
samples or taxa are involved. In Q-mode, sample by sample matrices can exceed the storage
capacity of even the largest computers. Also, with larger datasets, the assumptions required
by the particular multivariate technique to reduce the dimensionality of the problem are
increasingly likely to be violated. In Q-mode multivariate methods, the limiting factor is not
the number of sections, but the number of samples. However, as the number of sections
increases, so does the number of samples. Hohn's (1978) method is creative in its use of
computer storage; theoretically, both the number of sections and the number of events will
eventually become limiting factors.

I have compiled Table 1 as a rough guide to the applicability of various strategies for
the number of sections used. The reader should be aware that for various multivariate techniques, this guide is approximate because the limiting factor will be the number of samples,
or in R-mode, the number of species. The chart is a first attempt at plotting applicabilities.
It requires further experimentation and refinement.

*Am I working with two types of events (bases and tops) or one kind of event (bases only or
tops only)?*

The question of the kind of events involved, applies only to quantitative methods using the
event concept, explicitly or implicitly.

The fact that bases and tops are inherently different cannot be overemphasized. As I have
stated previously (Edwards and Beaver, 1978; Edwards, 1979), the upward or downward
displacement of observed first and last occurrences of a taxon can be the result of two
different kinds of processes. Relative to the true earliest occurrence, the observed lowest
occurrence (base) can be displaced upward because of facies control, vagaries of preservation or sampling, errors in identification, or delay in geographic dispersal of the organism.
However, downward displacement of observed bases can be caused by downward mixing,
downhole caving, or incorrect identification. Relative to the true latest occurrence, the
observed highest occurrence of a taxon (top) can be displaced downward because of preser-
vational effects, incomplete or inadequate sampling, local extermination, or misidentifica-

tion of the taxon involved. Upward displacement of observed tops can be caused by misidentification, reworking or contamination. Unless we can assume reasonably that probabilities of downhole mixing or reworking are as high as probabilities of delayed dispersal, lack of preservation, facies control or insufficient sampling, data concerning lowest- and highest-observed occurrences should be treated in distinctive manners. I have attempted to show these inherent differences, hypothetically, as I envision them, in Fig. 1. Note both that the probability curves are distinctively asymmetrical, and that the probability of observing a first occurrence below the time of evolution or observing a last occurrence above the time of extinction are extremely low compared with probability of observing a first occurrence above the time of evolution or observing a last occurrence below the time of extinction.

For the biostratigrapher working with two types of events (bases and tops), the choices are either methods that specifically exploit inherent differences between bases and tops (graphic and relational strategies) or those that do not exploit these differences (probabilistic and multivariate strategies and seriation). Edwards and Beaver (1978) refrained from

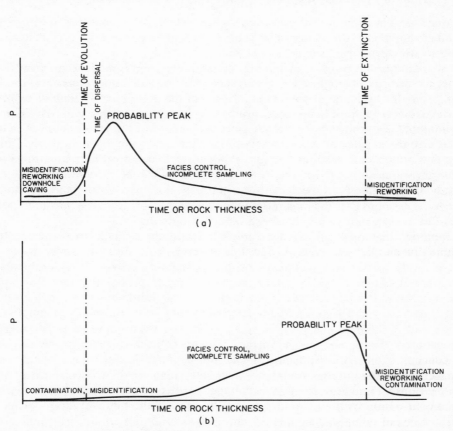

Fig. 1. One attempt at displaying probability (p) of observing lowest- or highest-occurrence event relative to 'true' time of evolution or extinction in outcrop or core material for: (a) first occurrence event; (b) last occurrence event. In each, area under curve must sum to 1. These curves are hypothetical and idealized; details for curves will vary for every individual taxon, and gross shapes of curves will vary with kind of organism (for example, rapidity of dispersal, facies control) and nature of sample material (core, outcrop, cuttings)

applying their probabilistic technique to first and last occurrence events treated jointly because the assumptions appeared to be unwarranted. In certain Q-mode multivariate strategies, bases and tops have different effects on the calculated similarity coefficients, but these differences are not exploited.

The biostratigrapher working with a single kind of event (bases only or tops only) may choose among probabilistic, graphic, or certain multivariate techniques and seriation. All have drawbacks. Probabilistic and Hohn's (1978) multivariate techniques do not take into account the asymmetry shown in Fig. 1. Graphic techniques work better and are easier to use when bases and tops can be used to 'bracket' possible correlations. Empirically, however, all types of methods produce reasonable results when applied to actual data. An interesting course to pursue at this point would be to adapt a probabilistic strategy to use inherent differences between lowest and highest occurrences. Relational strategies, because they study entire ranges, cannot be employed with a single kind of event.

Am I looking for maximum ranges or average (most probable) ranges?

The concepts of maximum or average ranges are distinct philosophies. A biostratigrapher's choice between these two philosophies is likely to depend on the kind of data available and the use to which the results will be put.

In the methods that produce maximum ranges, an attempt is made to discover the lowest first occurrence and the highest last occurrence of each taxon under consideration. In terms of Fig. 1, an effort is made to obtain the location of the dotted lines—time of evolution and time of extinction. Graphic methods produce range charts of this type. By their nature, these methods are nonprobabilistic. A single occurrence of a taxon below its purported first occurrence negates all other observations that led to the old hypothesis. This philosophy has possible detractions. Firstly, it assumes that reworking, contamination, and misidentifications do not exist, or, if they do, that they can be recognized and discounted. In studying microfossils, especially from cuttings, this assumption may be unfounded. Secondly, it assumes that the biostratigrapher wants to know what the maximum range of a fossil is, even if this maximum has been observed only once.

In methods that produce average ranges, an attempt is made to recognize the most probable first and last occurrences of each taxon under consideration. In terms of Fig. 1, an effort is made to discover the peaks on the probability curves. Both probabilistic and multivariate techniques will produce average ranges. Probability models require the assumptions of independence and spatial homogeneity (Southam, Hay, and Worsley, 1975; Edwards and Beaver, 1978). Spatial homogeneity may be a reasonable assumption. If it is not, the data frequently can be divided to make this assumption more reasonable. The assumption of strict independence of the elements of each cell in a probability matrix is not reasonable. As noted by Edwards and Beaver (1978), 'regardless of the calculated probabilities, the last occurrence must occur later than the first occurrence of the same taxon!' Gradstein and Agterberg (this volume) further restrict the probability curves to a normal distribution with all variances equal to one. Of the multivariate methods, Hohn's (1978) principal components method requires assumptions of linearity and normality. Various workers have applied experimentally nonparametric statistics to produce average ranges.

In the relational method of Davaud and Guex (1978), ranges produced are neither average ranges nor precisely maximum ranges. By the rules of the method, a taxon that is recognized anywhere within an *association unitaire* is counted as present throughout the duration of the association; thus, ranges may be extended relative to the true maximum range. These extensions are small typically, however, and the method in effect produces maximum ranges.

The seriation techniques of Brower and Burroughs (this volume) also produce ranges that are neither maximum nor average ranges. The ranges produced by seriation are less than maximum but owing to the nonprobabilistic nature of the method, are longer than average ranges.

The choice of probable or maximum range methods depends on the ultimate goals of the biostratigrapher. One philosophy holds that once the total ranges of all pertinent taxa are known, any unknown sample can be correlated to the smallest overlap produced by the ranges of the taxa present, without any mention or thought of confidence intervals. The degree of resolution of the correlation will depend on the number of taxa present in the unknown and on the 'tightness' of the overlap of their ranges. Biostratigraphers having this philosophy probably will choose a method that produces maximum ranges. A second philosophy holds that once a probable sequence is known, any unknown sequence can be correlated and confidence intervals can be assigned. This second philosophy requires that the unknown be a sequence of samples long enough to contain at least one highest- or lowest-occurrence event.

In the first example, the biostratigrapher seeks to be right, but perhaps not precise (and can be wrong totally, if ranges are not well known). In the second example, the biostratigrapher is willing to admit the possibility of being wrong and will even give a numerical value for that possibility.

Not surprisingly, the first philosophy tends to be more popular with biostratigraphers working with megafossils and (or) outcrop and core material, and the second philosophy may be more popular with biostratigraphers working with microfossils, particularly from cuttings.

Can I justify the assumptions of linearity/normality/independence required by the various methods?

Every method in quantitative biostratigraphy requires the biostratigrapher to make certain simplifying assumptions. I am using the present paper as a forum for an urgent plea to all who publish on quantitative biostratigraphy to please explicitly state the assumptions required. In many papers, though not necessarily for all of the multivariate techniques, a clever reader can deduce most of the underlying assumptions given sufficient time and inspiration. But quantitative biostratigraphy now is moving past the stage where it is practiced only by the inspired and the theoreticians. Many biostratigraphers want a simple set of step-by-step instructions; if we leave out a listing of the assumptions required, we do a disservice to other biostratigraphers and to ourselves.

On the other hand, I have discovered empirically, that although I may not be able to justify assumptions required by a particular quantitative method, it may produce reasonable results. The term for this is 'robustness.' A method is robust if inferences based on it remain valid even if one or more of the basic assumptions required by the theory of the method are violated. (Once the assumptions are violated, however, significance levels and confidence intervals are no longer rigorous.) Further work on the robustness of various quantitative methods is sorely needed. Hohn's (1978) study is a welcome step in this direction.

Case Study With Simulated Dataset

The third goal of the Biostratigraphic Working Group is the evaluation of various methods of quantitative biostratigraphy using actual and simulated sets of data. Comparison studies using actual data have been discussed by, among others, Hazel (1977), Millendorf, Brower, and Dyman (1979), Brower and Burroughs (this volume) and Hudson and Agterberg (1982). When actual data are used, the results produced by the various methods may be compared with each other and with the results derived by conventional methods. When simulated data sets are applied, the results may be compared with each other and with the 'true situation' as specified by the inventor of the dataset. Conclusions drawn from the application of hypothetical data must be evaluated in terms of how closely the artificial situation reflects reality.

With the goal of stimulating comparative studies, I now present a hypothetical dataset. These 'data' in various permutations have been applied to the no-space graph technique (Edwards, 1979). The purpose of this exercise is threefold: (1) To compare the results produced by the no-space graph (NSG) technique with a known 'true situation,' (2) To determine the robustness of the NSG technique under varying conditions, (3) To encourage other biostratigraphers to repeat this exercise using other quantitative techniques.

Description of the 'dataset'

The hypothetical dataset consists of 20 species distributed in space and time as shown in Fig. 2. Of these 20 species, 5 are ideal species (wide geographic distribution, rapid dispersal, nearly instantaneous extinction) (Fig. 2a), 10 are semi-ideal species (geographically restricted, less consistent dispersal and extinction) (Fig. 2(b), (c)) and 5 are less-than-ideal species (Fig. 2(d)).

The time and space represented by the hypothetical dataset have been sampled by 8 'stratigraphic sections' (Fig. 2(e)). The length of the sections and the spacing between sections differ, but they provide fairly complete coverage. The reader should note that in Fig. 2, the vertical axis represents time, rather than stratigraphic thickness. We may assume that time and thickness are related, but not necessarily linearly.

Description of the sampling regimes

In order to compare results produced by the NSG technique with the 'true situation' and to test the robustness of the method, I sampled the data shown in Fig. 2 in three different manners, varying sample interval and probability of recovering a species. In the present experiment, optimal recovery implies that each species was recovered from every sample that intersects its range in space and time. Less-than-optimal recovery is defined as: probability of recovery equals one-half for species F–T and equals one-third for species A–E. For less-than-optimal recovery, a random number table was used and sampling was repeated for added control. The following sample regimes were used to test the NSG technique:

(1) continuous sampling, optimal recovery,
(2) sample interval = 1 unit, less-than-optimal recovery,
(3) repeat of 2,
(4) sample interval = 2 units, less-than-optimal recovery,
(5) repeat of 4.

In the simulated data set, the maximum durations of the species shown range from two to 28 units. The average duration of all species in the sample sections is 4.4 units.

Application of the NSG technique

The procedure involved in the NSG technique is described and illustrated in Edwards (1979). Briefly, this procedure consists of successive rounds of graphic comparisons be-

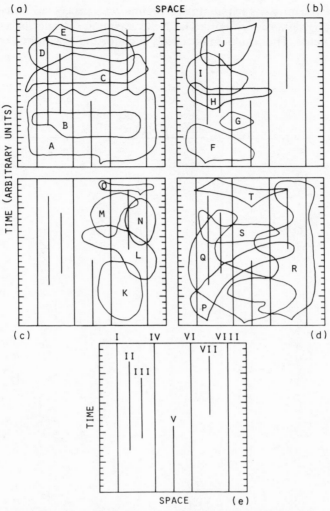

Fig. 2. Ranges of 20 species (A–T) in hypothetical dataset. Vertical axis on all drawings represents arbitrary time units. Horizontal axis on all represents space. Simply for convenience, dataset was created such that all species-range boundaries intersect sampled sections at integral values of time units. (a) Ranges of the best species (A–E) in time and space; (b) Ranges of geographically-restricted species (F–J) in time and space; (c) Ranges of geographically-restricted species (K–O) in time and space; (d) Ranges of the worst species (P–T) in time and space; (e) 8 sections shown in Figs. 2(a)–2(d)

tween a hypothesized sequence of events and observed relative positions of events in each of several sections. These comparisons are repeated until all events are in-place in at least one section and no range revisions are required. Indeterminate relationships may be noted with the final hypothesized sequence. Interpretations of graphic comparisons are based on geometry of the graphs and best judgment of the stratigrapher. In order to test the degree of influence of user judgment, each sampling regime was processed twice, using different 'simulated biostratigraphic judgment.' The NSG technique results in a rank ordering of biostratigraphic events. The results of two trials each for the 5 sampling regimes are shown in Table 2.

Table 2. Rank ordering of biostratigraphic events produced by NSG for two trials of each of the five sampling regimes

'Truth'	1		2		3		4		5	
t	o	o	o	t	t	t				
o	t	t	O	e	j	j				
r	r	r	t	T	e	e	t	t		
j	e	e	r	m	T	T	j	o		
e	j	O	e	d	o	m	m	O	t	j
m	O	j	T	o	r	r	e	m	j	t
O	m	m	m	O	O	d	E	e	T	T
d	d	d	j	r	m	E	M	q	o	d
n	T	T	d	j	d	o	o	j	O	i
T	n	n	E	E	E	O	O	T	m	
E	E	E	c	i	q	q	T	d	e	
s	q	i	n	q	c	s	r	i	r	q
q	i	s	N	c	s	n	n	D	E	o
i	c	q	D	n	i	c	N	E	d	O
c	s	c	M	N	J	N	c	M	i	m
l	l	J	i	D	n	i	s	J	s	r
J	J	l	q	M	D	J	q	n	q	e
D	D	D	s	l	N	D	d	r	J	E
M	M	S	J	h	l	l	i	N	c	c
p	S	M	h	s	M	C	D	c	D	n
h	p	p	l	J	C	M	J	s	n	N
S	h	h	C	C	S	S	l	S	N	p
N	C	C	S	S	h	h	p	l	p	C
a	N	N	a	p	I	I	L	p	C	D
C	a	a	I	a	p	p	C	C	S	M
I	L	I	H	L	g	g	S	C	M	S
k	I	L	f	I	G	G	a	h	l	h
H	H	H	p	H	a	a	h	a	h	a
g	k	k	L	b	H	L	H	I	a	I
b	b	b	b	g	L	H	I	H	I	l
L	g	g	P	G	k	k	P	k	H	L
f	G	G	g	k	b	b	k	K	L	H
G	f	f	G	f	B	B	K	R	b	b
B	B	B	k	B	K	f	g	P	B	R
R	R	K	B	R	R	Q	G	g	R	R
P	Q	R	R	Q	f	K	R	G	f	f
K	P	Q	A	F	Q	R	f	A	Q	P
Q	K	P	Q	A	F	F	Q	f	P	Q
F	A	A	K	P	A	A	A	Q	A	A
A	F	F	F	K	P	P	F	F	F	F

For each species: upper case = first occurrence, lower case = last occurrence, square brackets connect events judged to be equal or indeterminate.

Discussion of results

With continuous sampling and optimal recovery (regime 1), NSG produces good results, consistently. Results of the two trials show only minor differences. Both trials failed to show the maximum range of species L because this was observed in only one section (Section 8) that does not contain either species B or species K, which could have been used to correlate. Species that overlap in time but not space may not overlap in the NSG results (species O/M, T/N, and A/C), may overlap in the NSG results (K/F, Q/L), or may overlap in one interpretation and not in another (species L/J).

With closely-spaced sampling and less-than-optimal recovery (regimes 2 and 3), the results remain generally good. Here, however, results of repeated trials and randomly-varied repeated samples are less consistent. Predictably, the worst species (species P–T) and those species recovered only from one or two sections show the most variation in their resulting ranges.

With more widely-spaced sampling and less-than-optimal recovery (regimes 4 and 5), results produced by the NSG technique are not as good. Here again, the time-transgressive species and those present in few sections are the most variable. In these regimes, judgment of the user plays a more important role. Good results can be achieved by skill (and luck?) in choosing which species are considered to be in-place (regime 5a), but poor results may also be produced (regime 5b).

Differences between sequences of events determined by NSG and the 'true sequence' of events may be caused by one of two factors. The first is incomplete sampling or recovery (the range will be shorter than the 'true' range because the maximum range never was recovered). The second is inaccurate correlation, which may cause ranges to be shortened or extended. Because a hypothetical data set was employed in this example, the two factors may be separated. The sequence produced directly from the actual recovered values for any sampling regime (and assuming perfect correlation) may be compared to the 'true situation' to detect incomplete recovery, and may be compared to the sequences produced by NSG to detect imperfect correlation.

The Spearman coefficient of rank correlation is a convenient method of comparing sequences of events. For any two sequences of events, this coefficient, R, is

$$R = 1 - 6 \sum_{i=1}^{n} D_i^2/(n(n^2 - 1)),$$

where n is the number of events and D_i is the difference between ranks of event i in the two sequences. Table 3 gives the Spearman coefficient of rank correlation between each NSG sequence and the 'true sequence,' between each NSG sequence and the actual recovered values for that sampling regime, between NSG sequences of duplicate trials for each sampling regime, and between the actual recovered values for each sampling regime and the 'true situation.' In the present example, both incomplete recovery and inaccurate correlation contribute to the differences in sequences; inaccurate correlation contributes slightly more.

The preceding example demonstrates that the NSG technique produces generally good results. When the sample interval is less than half the average duration of the species present, the results are good only when the user's judgment is good. The example reveals the method to be relatively robust.

The NSG technique is based on the relative spacing of events. It does not require assumption relating rock thickness to time other than that the relationship must be a direct one. A transformation from time to rock thickness such as proposed by Davaud (1979, Fig. 2) could have been applied to the data set with little change in the results produced. I hope other stratigraphers will try this and other permutations of the data set and compare other methods of quantitative biostratigraphy.

Table 3. Spearman coefficient of rank correlation (R) for two trials of each of the 5 sampling regimes

Number of events	1 $n = 40$		2 $n = 40$		3 $n = 40$		4 $n = 38$		5 $n = 36$	
R comparing resulting sequence to 'TRUTH'	0.9934	0.9930	0.9538	0.9598	0.9665	0.9613	0.9237	0.9266	0.9578	0.8954
R comparing resulting sequence to actual recovered values	0.9935	0.9929	0.9686	0.9779	0.9740	0.9679	0.9422	0.9733	0.9886	0.9316
Between trials, each regime:										
R comparing resulting sequences	0.9959		0.9651		0.9906		0.9369		0.9548	
R comparing actual recovered values to 'TRUTH'	0.9982		0.9884		0.9891		0.9612		0.9721	

Conclusions

Discussion has focused on some of the considerations the biostratigrapher should make before applying one or several quantitative methods. The method or methods selected will depend on the nature of the data and the goals of the biostratigrapher. For any method, the underlying assumptions should be stated explicitly, but violation of assumptions may not lead automatically to rejection of a given method. More experimentation is needed on robustness of the various methods.

The no-space graph technique (Edwards, 1979) is shown to produce accurate results when applied to hypothetical data. The method is robust. When recovery is poor and sampling is incomplete, inaccurate correlation contributes to poor results.

By pointing out areas and directions that need more study, I hope that I have inspired the undertaking of such studies. Much remains to be done in quantitative biostratigraphy. I also hope that the practicing biostratigraphers who may attempt quantitative methods will recognize them for what they are. They are tools that can make biostratigraphy easier and, I think, more fun. They should supplement, not replace, sound paleontology and sound biostratigraphic practices.

Acknowledgments. I express sincere appreciation to the following for sending me preprints and (or) comments and criticisms on earlier versions of this manuscript: F. P. Agterberg, J. C. Brower, Eric Davaud, N. O. Frederiksen, F. M. Gradstein, J. E. Hazel, M. E. Hohn, and C. B. Hudson.

References

Blank, R. G., 1979, Applications of probabilistic biostratigraphy to chronostratigraphy, *Jour. Geology*, **87**, 647–670.

Brower, J. C., 1981, Quantitative biostratigraphy, 1830–1980, in Merriam, D. F. (ed.), *Computer Applications in the Earth Sciences—an update of the 70's*. Plenum Press, New York, pp. 63–103.

Brower, J. C. and Burroughs, W. A., 1982, A simple method for quantitative stratigraphy. (This volume.)

Brower, J. C., and Millendorf, S. A., 1979, Biostratigraphic correlation within IGCP Project 148, *Computers and Geosciences*, **4**, 217–220.

Brower, J. C., Millendorf, S. A., and Dyman, T. S., 1979, Methods for the quantification of assemblage zones based on multivariate analysis of weighted and unweighted data, *Computers & Geosciences*, **4**, 221–227.

Cheetham, A. H., and Deboo, P. B., 1963, A numerical index for biostratigraphic zonation in the mid-Tertiary of the Eastern Gulf, *Trans. Gulf Coast Assoc. Geol. Soc.*, **13**, 139–147.

Christopher, R. A., 1978, Quantitative palynologic correlation of three Campanian and Maestrichtian sections (Upper Cretaceous) from the Atlantic Coastal Plain, *Palynology*, **2**, 1–28.

Cisne, J. L., and Rabe, B. D., 1978, Coenocorrelation: applied gradient analysis of fossil communities and its application in stratigraphy, *Lethaia*, **11**, 341–364.

Davaud, E., 1979, Automatisation des corrélations biochronologiques: un exemple d'application de l'informatique à la résolution d'une problème naturalist complex, *Publication spéciale*, Université de Genève, 30 pp.

Davaud, E., and Guex, J., 1978, Traitement analytique 'manuel' et algorithmique de problemes complexes de corrélations biochronologiques, *Eclogae geol. Helv.*, **71**, 581–610.

Edwards, L. E., 1979, Range charts and no-space graphs, *Computers & Geosciences,* **4**, 247–255.

Edwards, L. E., and Beaver, R. J., 1978, The use of a paired comparison model in ordering stratigraphic events, *Jour. Math. Geology*, **10**, 261–272.

Frederiksen, N. O., 1980, Paleogene sporomorphs from South Carolina and quantitative correlations with the Gulf Coast, *Palynology*, **4**, 125–179.

Gradstein, F. M., and Agterberg, F. P., 1982, Models of Cenozoic foraminiferal stratigraphy—northwest Atlantic margin. (This volume.)

Guex, J., 1977, Une nouvelle méthode d'analyse biochronologique, Note préliminaire, *Bull. Soc. Vaud. Sc. Nat. No. 351*, **73**, and *Bulletin No. 224 des Laboratoires de Géologie, Mineralogie, Géophysique et du Musée Géologique de l'Université de Lausanne*, 309–321.

Hay, W. W., 1972, Probabilistic stratigraphy, *Eclogae geol. Helv.*, **65**, 255–266.

Hay, W. W., and Southam, J. R., 1978, Quantifying biostratigraphic correlation, *Ann. Rev. Earth Planet. Sci.*, **6**, 353–375.

Hazel, J. E., 1970, Binary coefficients and clustering in biostratigraphy, *Geol. Soc. Am. Bull.*, **81**, 3237–3252.

Hazel, J. E., 1977, Use of certain multivariate and other techniques in assemblage zonal biostratigraphy: Examples utilizing Cambrian, Cretaceous, and Tertiary benthic invertebrates, in Kauffman, E. G., and Hazel, J. E. (eds.), *Concepts and Methods of Biostratigraphy*. Dowden, Hutchinson and Ross, Stroudsburg, Pa., pp. 187–212.

Hohn, M. Ed., 1978, Stratigraphic correlation by principal components: effects of missing data, *Jour. Geology*, **86**, 524–532.

Hudson, C. B., and Agterberg, F. P., 1982, Pair comparison models in biostratigraphy, *Jour. Math. Geology*, **14**, 141–159.

Jekhowsky, B. de, 1963, Le méthode des distances minimales, nouveau procédé quantitatif de corrélation stratigraphique, *Rev. Inst. Francais Pétrole*, **18**, 629–653.

Millendorf, S. A., Brower, J. C., and Dyman, T. S., 1979, A comparison of methods for the quantification of assemblage zones, *Computers & Geosciences*, **4**, 229–242.

Miller, F. X., 1977, The graphic correlation method in biostratigraphy, in Kauffman, E. G. and Hazel, J. E. (eds.), *Concepts and Methods of Biostratigraphy*. Dowden, Hutchinson and Ross, Stroudsburg, Pa., pp. 165–186.

Rabe, B. D., and Cisne, J. L., 1980, Chronostratigraphic accuracy of Ordovician ecostratigraphic correlation, *Lethaia*, **13**, 109–118.

Reyment, R. A., 1979, Biostratigraphical logging methods, *Computers & Geosciences*, **4**, 261–268.

Reyre, Y., 1973, Palynologie du Mesozoique saharien: Traitement des donnés par l'informatique et applications à la stratigraphie et à la sédimentologie, *Paris Muséum National d'Histoire naturelle Mémoires, sér. C, Science de la terre*, **27**, 284 pp.

Rubel, M., 1979, Principles of construction and use of biostratigraphical scales for correlation, *Computers & Geosciences*, **4**, 243–246.

Scott, G. H., 1974, Essay review: stratigraphy and seriation, *Newsletters on Stratigraphy*, **3**, 93–100.

Shaw, A. B., 1964, *Time in Stratigraphy*, McGraw-Hill, New York, 365 pp.

Southam, J. R., Hay, W. W., and Worsley, T. R., 1975, Quantitative formulation of reliability in stratigraphic correlation, *Science*, **188**, 357–359.

Quantitative Stratigraphic Correlation
Edited by J. M. Cubitt and R. A. Reyment
© 1982 John Wiley & Sons Ltd.

A Simple Method for Quantitative Biostratigraphy

JAMES C. BROWER AND WILLIAM A. BURROUGHS

*Heroy Geology Laboratory, Syracuse University,
Syracuse, New York 13210, U.S.A.*

Abstract

Most nonquantitative biostratigraphers have resisted stubbornly numerical methods. This is probably due to two causes. First, many quantitative methods are complex whereas most biostratigraphers are not oriented quantitatively. Secondly, quantitative techniques utilize methodologies that are basically foreign to biostratigraphers.

Therefore, we have concentrated our attention on simple methods of quantitative biostratigraphy which share features in common with other biostratigraphic methods. This paper outlines one simple technique, archaeological seriation of a data matrix consisting of the presence-absence of m species taken from n samples in p stratigraphic sections. The basic problem is to arrange data in the form of a range chart with species in columns and samples in rows. This is achieved by concentrating presences along the matrix diagonal to minimize range zones of taxa. This generates a range chart and groups similar samples in adjacent rows of the matrix. Two types of solution can be determined. In an unconstrained solution, data on superposition of samples within the stratigraphic sections are ignored and samples are free to group in any order. Typically, groupings of species and samples, similar to those obtained from multivariate analysis of assemblage zones, are produced. The constrained solution employs information on stratigraphic position of samples within individual stratigraphic sections, such that samples are forced to remain in stratigraphic order within the final seriated matrix. This generates usually a 'better' range chart for biostratigraphical correlation.

Introduction

Most biostratigraphers have resisted, stubbornly and successfully, the use of numerical methods. Basically, the intended consumers have rejected quantitative biostratigraphy as unpalatable despite the fact that quantitative methods have produced excellent results with large and complex sets of data. We believe that this situation is mainly a result of two factors. Firstly, many of the quantitative methods are complex whereas most biostratigraphers are not mathematically or statistically inclined. Secondly, most of the algorithms employ techniques, such as binomial probability, cluster analysis, and multidimensional scaling, that are foreign to biostratigraphic practice. The reader is referred to Brower (1981), Brower and Millendorf (1978), Hay and Southam (1978), Hazel (1977), and Miller (1977) for detailed reviews of numerical methods in biostratigraphy.

Consequently, our efforts have been focused on simple methods of quantitative biostratigraphy over the past few years. Some of these techniques virtually eliminate the numbers. We hope that these simple algorithms will be accepted and employed more widely by practicing biostratigraphers than the elaborate schemes that have been formulated to date. Also, in designing simple techniques, features have been incorporated that are similar to practices frequently employed by biostratigraphers.

The purpose of this paper is to present one technique, that is, archaeological seriation of an original data matrix.

Previous Work

For many years, archaeologists have faced problems which are closely allied to those of biostratigraphy. Numerous samples containing various objects must be arrayed into a one-dimensional sequence which represents a time or 'evolutionary' series. Examples of such case studies are numerous and include such fascinating subjects as artifacts in graves, Iron Age brooches, sentence structure and word frequencies in manuscripts written over the ages, and attributes of hominid skulls. Mathematical solutions to seriation in archaeology date back to Flinders Petrie (1899) and have been reviewed by Cowgill (1972), Doran and Hodson (1975), Gelfand (1971), Hodson, Kendall, and Tautu (1971), Johnson (1972), Kendall (1971), Marquardt (1978), and Wilkinson (1974).

Two types of matrices can be seriated. In the first, a matrix of similarity or distance coefficients is calculated between objects or samples. Then, the desired variety of seriation is performed on this matrix. In effect, the similarities or distances serve as 'middle-men' in the process. For 'ideal' data, seriation of certain types of similarity or distance matrices is equivalent to seriating original data (for example, Kendall, 1971). Unfortunately, data are rarely ideal. Furthermore, use of similarity or distance coefficients raises numerous questions such as: What sort of coefficient is appropriate? What is the metric nature of the data? Seriation is studied generally in this form.

The second type of seriation directly manipulates the original data matrix which typically has n rows and m columns. Although working with a rectangular block of data can be awkward computationally, elimination of the 'middle-men' in the form of similarity or distance coefficients is an advantage. Despite the direct approach, this type of seriation is applied infrequently. Seriation of an original data matrix is the technique of interest here.

Strangely enough, few of the seriation techniques have been noticed by biostratigraphers and other geologists. Similarity between artifacts in graves and fossils in rocks is obvious. Scott (1974) first suggested that archaeological seriation of both types could be usefully applied to biostratigraphic data. He also advocated that only key taxa should be employed to ensure consistent results (a key taxon displays an invariant relationship with other key taxa in all stratigraphic sections studied.) Although the data were discussed in the context of networks, Smith and Fewtrell (1979) adopted a nonquantitative seriation approach to original data matrices of microfossils. Doveton, Gill, and Tipper (1976), Tipper (1977, 1980), and Doveton (unpublished manuscript) seriated ecological data as well as a paleoecological example of Pennsylvanian conodonts.

Algorithm for Seriation

Introduction

The data matrix consists of the presence/absence of m species or other taxa taken from n samples in p stratigraphic sections. Hereafter, the organisms examined will be referred to as species or taxa; the majority of case studies discussed later examine species. It must be noted that a biostratigrapher has one major advantage compared to the typical archaeologist in that the sequence of taxa is known in individual stratigraphic sections; only rarely do archaeologists possess analogous information. Presences can be scored as 1.0 and absences as 0.0 or data could be viewed in black and white.

The species analyzed include key taxa, in the sense of Scott (1974), as well as other species which do not fit the criteria for key taxon. For general biostratigraphic studies, use of only key taxa would eliminate the majority of species in the data set. However, the range-through method of data recording may be employed so that each taxon is listed as present in samples within its local range zone for each stratigraphic section. This aids in eliminating sampling problems and minimizes bias toward factors, such as environmental parameters, that could dictate the absence of a particular species. In these studies, the time duration of the taxon is of interest, not the vagaries of its distribution, and therefore this type of data is used in most biostratigraphic research (Hazel, 1977; Brower, Millendorf, and Dyman, 1978; Brower, 1981).

Frequently, biostratigraphers omit species which occur in only a few of the stratigraphic sections under study and eliminate taxa which are recognized in less than half of the sections because such species provide correlation information for relatively few samples. For the seriation technique, this practice is not strictly necessary. However, elimination of taxa, which occur in a small number of samples and sections as well as samples that contain a few species, does improve the speed and quality of seriation because nondefinitive and ambiguous information is removed from the data.

One basic problem is to arrange data in the form of a range chart with species or taxa in the columns and samples in the rows. However, by concentrating presences along the matrix diagonal in order to minimize the range zones of the taxa, a range chart is provided and similar samples are grouped in adjacent rows of the matrix. This is termed the concentration principle by archaeologists (for example, Cowgill, 1972; Johnson, 1972; Kendall, 1971; Marquardt, 1978). For ideal data, the final seriated matrix should have a solid block of presences along the diagonal and absences should be located in off-diagonal elements. Goodness-of-fit between the observed matrix and a matrix which is seriated perfectly can be measured directly or indirectly by methods discussed later.

Two types of solutions can be determined for the generation of these charts. In an unconstrained solution, data on superposition of samples within any given stratigraphic section are ignored and the samples are free to group in any order. Typically, groupings of taxa and samples are produced that are similar to those obtained from the multivariate analysis of assemblage zones. However, these groupings are frequently not in a systematic stratigraphic order. The alternative constrained solution employs information on stratigraphic position of samples within individual sections such that samples are forced to remain in stratigraphic order within the final seriated matrix. For biostratigraphical correlation the constrained solution usually generates a more appropriate range chart. However, both solutions preserve the integrity of the original data matrix and although data are rearranged, samples still contain the same suite of taxa and species appear in the same stratigraphic sections and samples.

The unconstrained solution

Wilkinson (1974) proposed a simple method for the analysis of an original data matrix of presences and absences which is adopted here for unconstrained seriation. The heuristic scheme orders alternately matrix rows and columns based on the average position of presences. In our computer program (Burroughs and Brower, in press), the positions are numbered from top to bottom for each column and from left to right for each row; however, other conventions could be employed if desired. The computations are performed in a series of iterations, each of which includes the following four steps:

(1) calculate mean position of presences in rows;
(2) order rows according to these means;
(3) calculate mean position of presences in columns;
(4) order columns according to these means.

Additional iterations are attempted until no change is recorded in the positions of rows and columns.

The unconstrained solution ignores all information concerning stratigraphic position of samples and taxa within the sections. Examination of numerous case studies indicates that samples that are grouped together share similar or identical faunal compositions. In addition, species in adjacent columns are characterized by similar patterns of occurrences in the samples. Therefore, it is important to note that neither taxa or samples will necessarily be grouped in stratigraphic order. Groupings of taxa and samples are similar to those derived from multivariate analysis of assemblage zones in which clusters of species and samples are frequently not in stratigraphic order; here, one must examine the original data to determine the stratigraphic significance of the various groups, clusters and/or ordinates (Hazel, 1977; Brower, Millendorf, and Dyman, 1978; Millendorf, Brower, and Dyman, 1978). Also, it is critical to observe that this type of seriation represents a simple averaging process. Therefore, resulting zonation of species and samples will be an average rather than a conservative zonation. Whereas, a conservative zonation is intended to give the stratigraphically-highest estimate of the top of a range zone and the stratigraphically-lowest estimate for the base of that range zone, an average zonation would place each biostratigraphic event (base or top of a range zone) at approximately the mean position. However, both types of zonation are subject to mixed advantages (see Brower and Millendorf, 1978, and Brower, 1981, for further details).

The constrained solution

Data on stratigraphic position of samples within individual sections are introduced into the analysis for a constrained solution to force samples to remain in stratigraphic order. This is not true of unconstrained seriation in which two samples in a single section can be seriated out of stratigraphic order.

The input data matrix, with samples in rows and species or taxa in columns, is read in order of stratigraphic sections numbered from 1 to p; within each stratigraphic section, samples are numbered from 1 at the top of the section to the last consecutive sample number at the bottom. Each sample is identified, therefore, by two numbers, one, for the stratigraphic section and, a second, for the sample number within that stratigraphic section. In addition, each sample is given a number from 1 to n within the total sequence of all samples (see case studies).

The program operates in a series of iterations, each of which consists of the 5 following steps:

(1) Calculate mean position of presences in rows (samples);
(2) Arrange rows into order according to means;
(3) Rows of this matrix are scanned to determine if samples are seriated in stratigraphic order within individual sections. If all samples are sequenced correctly within respective stratigraphic sections, algorithm moves to next step. If samples are out of strati-

graphic order within an individual section, samples are interchanged within seriated matrix to arrange them in proper stratigraphic order for that section;

(4) Mean position of presences is determined for each column (taxon or species);
(5) Columns are ordered according to means.

The test criterion

Both types of seriation employ a simple index to measure the concentration of presences along the diagonal of the seriated matrix. The basic criterion favored by archaeologists is that the optimum arrangement of the presence-absence data matrix occurs when the range of presences of characters (taxa for a biostratigraphic data set) is minimized across the samples. Kendall suggests that the optimum configuration is produced when

$$\Sigma R_j \qquad = \text{a minimum (Kendall, 1963)}$$

or

$$\Sigma N_j \log R_j = \text{a minimum (Kendall, 1971),}$$

where N_j is the total number of occurrences of the jth character or taxon and R_j is the range in rows or samples in which character or taxon j is present. The coefficient adopted here follows the same line of reasoning and our index is 1 minus the ratio of embedded absences in all columns to the range of presences in all columns (an embedded absence or blank is any absence between the highest and lowest presences in a particular column.) Expressed algebraically, the index equals

$$1 - \left(\sum_{j=1}^{m} A_j \bigg/ \sum_{j=1}^{m} R_j \right)$$

where A_j is the number of embedded absences in column j and R_j is the total range of presences in column j. The maximum value is 1.0 for perfect seriation in which all presences would be concentrated along the matrix diagonal with no embedded blanks. Progressively lower values of the index indicate higher frequencies of embedded absences and less perfect seriations. In constrained and unconstrained solutions, the index serves as one indicator that a stable solution has been obtained. When the index reaches a maximum or begins to oscillate around a high value, a stable level of concentration of presences about the diagonal has been obtained. Seriation then terminates with the configuration that produced the highest index value. Values of the test index obtained for the case studies range from 0.506 to 0.866 for both types of seriation.

Number of iterations

Iterations continue until the position of rows and columns stabilizes in a configuration with presences concentrated along the matrix diagonal. The number of iterations depends on the size of the data matrix as well as complexity of the information in the data. Both constrained and unconstrained seriations converge rapidly. Initial iterations reduce the number of embedded absences rapidly, whereas later iterations produce only small changes in the data. The number of elements (number of rows times number of columns) in the matrices examined range from approximately 40 to 4000 and solutions required 4 to 42 iterations.

The unconstrained solution always obtains a point of 'best' ordering and concentration where means of rows and columns are in order; consequently, no rearrangement is called for. The iteration with the highest test criterion represents the stopping point or terminal iteration for the unconstrained solution.

On the other hand, the constrained solution does not converge absolutely, but may oscillate at a certain level of concentration. This problem would not be recognized in 'ideal data' where a perfect constrained seriation could be found. Unfortunately, the typical biostratigraphic data set is far from 'ideal.' As expected, the most troublesome species and samples are poorly-defined stratigraphically and contain a few taxa. In addition, the samples are not similar to any other samples and the species involved are rare and/or reveal erratic patterns of distribution. In this situation, the constrained solution first arranges rows or samples according to their means, and then rearranges samples according to positions within individual sections. Therefore, rearrangement of rows and columns can take place even if a stable or near stable configuration has been achieved. If this effect is compounded by poorly-defined samples or taxa with dubious ranges, an oscillating series of iterations with different test values and configurations of rows and columns generally results; however, most iterations are similar and all oscillate around the best or most stable configuration. For constrained solutions, the logic of our computer program examines the oscillations and selects the stable iteration with the highest test criteria and the most appropriate concentrations along the matrix diagonal.

Comparison of results from unconstrained and constrained seriations

It is informative to compare results of constrained versus unconstrained solutions on the same data set. Study of final test criteria and inspection of orders for row and column means displays the 'cost' of the stratigraphic constraint. The unconstrained solution relentlessly groups samples with similar faunal composition and taxa with similar patterns of occurrence until an optimum ordering of position means for rows and columns is obtained. Frequently, a higher degree of concentration along the diagonal results than seen in constrained seriation.

The unconstrained seriation ignores stratigraphic 'ground truth' inherent in biostratigraphic data, namely relative position of taxa and samples in a stratigraphic section. As previously mentioned, the vagaries of taxon distribution (for example, due to facies control, nonpreservation, or misidentification) may impose an apparent ordering of events in an unconstrained solution which may violate the actual order of occurrence of the samples or taxa in a given section, while maximizing the degree of concentration of presences along the matrix diagonal. The constrained solution preserves stratigraphic 'ground truth' but frequently decreases the concentration. The unconstrained solution is handicapped also by a lack of linkage between groups of samples and taxa which may be found in nonoverlapping faunal zones. As no stratigraphic information is included in unconstrained seriation, groups of samples and taxa are not usually in stratigraphic sequence and overall results are similar to those obtained in cluster analysis. To ascertain the stratigraphic distribution of clusters, one must inspect original data.

It is important also to note that the unconstrained solution is sensitive to initial arrangement of input data matrix because adjacent rows or columns with equal or tied means remain in original order. To a certain extent, this is true of constrained seriation but the effect is less marked because the stratigraphic constraint results in fewer tied means in adjacent rows and columns. Consequently, stratigraphically-constrained seriation produces range charts that are 'better' for correlation.

Case Studies in Seriation

Case studies are presented to indicate how the two types of seriation work and to compare results derived from the various techniques. The first simple examples are hypothetical in order to illustrate the computations. The third study is a research example comparing results of constrained seriation, unconstrained seriation, and cluster analysis. Example 4 consists of a more complicated research example dealing with Cambrian fossils, that emphasizes the comparative performance of seriation and other algorithms. Lastly, numerous other case studies will be listed briefly.

Case Studies 1 and 2

These case studies are concerned with a simple hypothetical example of 5 species distributed among 9 samples in three stratigraphic sections. Species are numbered in the columns from 1 to 5. Rows contain samples numbered as follows. The first digit represents the section number whereas the subsequent digits designate the location number for that sample within each stratigraphic section; numbers increase from stratigraphically highest to lowest in each section. As noted previously, data are pictured in black and white: presences are black and absences white. These conventions for numbering and illustration will be followed for all matrices presented in this paper.

The hypothetical data were derived from a range chart constructed for the 5 taxa. Each species was recorded as present in all samples within its range zone as determined from a small *a priori* correlation chart. To obtain the hypothetical data matrix, some of the presences were eliminated from the data so that a perfectly-constrained seriation would not be possible. The data are tabulated in Fig. 1(a) and have 18 embedded absences or blanks.

Both types of seriation converge rapidly. Only three iterations were required for unconstrained seriation and four iterations for constrained solution. There are 7 embedded absences after the first iteration of the constrained data yielding a test value of 0.731. Study of Fig. 1(b) indicates that an approximate stratigraphic order has been achieved. The final iteration (number 4, Fig. 1(c)) reduces further the number of embedded blanks to 5 and raises the test criterion to 0.792. As mentioned previously, the embedded absences are intentional, the presences involved having been deleted deliberately from the data.

Three embedded absences are detected following the first iteration of unconstrained seriation versus 18 embedded absences in the original data (Fig. 1(d)). The second and third iterations are identical perfect seriations with no embedded absences and test values of 1.0 (Fig. 1(e)). Comparison of Figs. 1(c) and 1(e) illustrates differences between the two solutions. Neither species or samples are in stratigraphic order in the unconstrained solution. However, adjacent rows and columns are characterized by similar patterns of presences and absences among taxa and samples. Note that although the unconstrained solution violates known information about stratigraphic distribution of samples and taxa, it produces nevertheless a perfect test criterion of 1.0 as opposed to the lower value of 0.792 for the constrained solution. Here, the 'cost' of the stratigraphic constraint consists of 5 absences.

Case Study 3

This small research example treats 11 species of Lower Triassic ammonites from the Salt Ranges of Pakistan (data from Guex, 1978, and Kummel, 1966). 31 samples were obtained from four stratigraphic sections which occur over a distance of approximately 180 km. Only species known from two or more of the sections are included. Figure 2(a) illustrates a time-correlation chart which has been compiled from Guex (1978) and Kummel (1966) using classical biostratigraphic methods. The main purposes of this case study are to compare results of the two seriations with the time-correlation chart and with cluster analyses performed on the same data. Species are numbered as in Table 1 and designations for samples are presented on Fig. 2(a) (see previous discussion for numbering conventions).

ORIGINAL DATA

(A)

CONSTRAINED SOLUTION

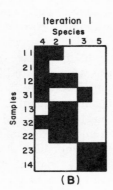

(B)

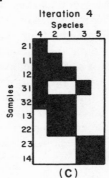

(C)

UNCONSTRAINED SOLUTION

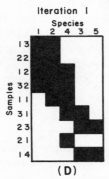

(D)

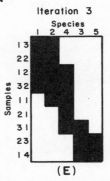

(E)

Fig. 1. Results for hypothetical data set. (A) Original data; (B) First iteration for constrained seriation; (C) Final iteration for constrained seriation; (D) First iteration for unconstrained seriation; (E) Final iteration for unconstrained seriation

The stratigraphically-constrained seriation is shown in Fig. 2(b). The test value equals 0.644, an intermediate figure for the data sets examined. Samples are seriated from youngest at the top of the matrix to oldest at the base. The upper 8 samples (11–42) are assigned to zone VII of Guex (1978) and contain some or all of species 1–5. The only form present in three of the next four samples (43, 24, 14), in zone VI or at the base of zone VII, is species 5, a long-ranging ammonite. Samples recognized at the top of zone V include 44–47 and 31; the last of these, containing species 10, is out of position in the seriation. The typical ammonites of these samples are species 4, 10, and 11, all of which are short-ranged, and the ubiquitous species 5. The next block of samples occurs in the center of zone V and is characterized by the genera *Meekoceras* and *Anasibrites* (species 6 and 7). Samples in this interval include 48–410, 32, 33, 15, and 16. Species 5 is the only taxon listed in the oldest three samples in zone V (411–413) which are out of place in the seriation as they are placed within a group with *Meekoceras* and *Anasibrites* . Because of the averaging nature of the technique, this is a frequent problem with samples having a few long-ranged species and/or species with erratic distribution patterns. The oldest samples are in zones II and III in which species 8 and/or 9 are invariably present. These samples (414, 415, 25, 17, and 18) are classified correctly by the seriation. In general, a high degree of agreement exists between the correlation chart and the strati-graphically-constrained seriation.

A chart of the ranges of the species can be read from the seriation (Fig. 2(c)). Construction of the chart assumes that adjacent samples, with equal means and the same faunal composition are of equal age; consequently, the range chart differs from a direct overlay of the constrained seriation. It can be recognized that the youngest forms in the chart are represented by species 1, 2, and 3 whereas 4 is slightly older. These are followed by species 10, 11, and the zonal general *Meekoceras* and *Anasibrites* (species 6 and 7). The oldest taxa consist of species 8 and 9. Also, the long range of species 5 is obvious. The only inconsistency between the range chart and original data involves species 10. According to the seriation, the base of the species 10 range zone is below the upper limits of species 6 and 7. Although not impossible, the relationship is not observed in the original data and does not seem probable. The ambiguity is produced by a rare taxon, species 10, which only occurs in three samples. Such a situation is not unexpected with a data set containing only a few species.

Figure 2(d) pictures the matrix for unconstrained seriation. A test value of 0.866 is obtained versus 0.644 for the constrained seriation and eliminating the stratigraphic restriction reduces the number of embedded absences from 32 to 9. However, the assemblages of taxa which can be reconstructed from the seriated matrix are only partially in stratigraphic order. The youngest ammonites, species 1–4, are recorded in the left columns. The wide distribution of the protean species 5 is evident. Proceeding from left to right, the next two groups consist of *Meekoceras* and *Anasibrites* (species 6 and 7) and species 10 and 11. However, the stratigraphic order of the two groups is reversed. The correlation chart and unconstrained seriation indicate that the entire range zone of species 10 and upper limit of species 11 are younger than species 6 and 7. The right columns contain the oldest taxa, species 8 and 9. Although the unconstrained seriation does group taxa with similar distribution patterns, the groups are not arranged in exact stratigraphic order. Study of the sample order in the unconstrained seriation matrix reveals the same pattern with samples out of stratigraphic order. For example, most samples with species 10 and 11 are seriated below those with species 6 and 7. Another case in point is where sample 44 overlies 41 in the rearranged matrix although the stratigraphic order is opposite. In addition, all samples containing only species 5 are blocked together although these are recognized at widely separated horizons. Thus, it is clear that unconstrained seriation reduces the number of

ZONE	SECTION			
	LANDA	NARMIA	ZALUCH	NAMMAL
VII	11 12 13 14	21 22 23		41 42
VI		24		43
V	15 16		31 32 33	44 45 46 47 48 49 410 411 412 413
IV				
III				414
II	17 18	25		415

(A)

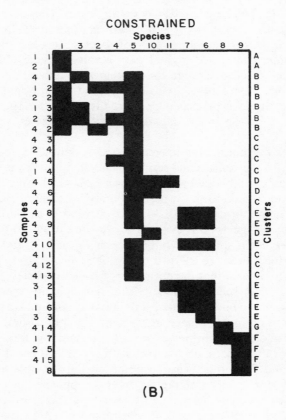

(B)

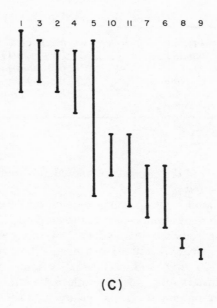

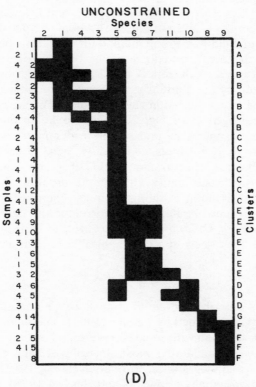

Fig. 2. Data for Triassic ammonites. (A) Correlation chart from Kummel (1966) and Guex (1978); (B) Constrained seriation; (C) Range chart of taxa (see Table 1 for list of species); (D) Unconstrained seriation

Table 1.   Ammonite species used in Case Study 3

1. *Tozericeras pakistanum*	7. *Meekoceras* sp. A
2. *Prohungarites submiddlemissi*	8. *Proptychites* sp.
3. *P. Landuensis*	9. *Gyronites frequens*
4. *Nordophiceras* cf. *planorbis*	10. *Wasatchites spinatus*
5. *Pseudosageceras*	11. *Hemiprionites* aff *typus*
6. *Anasibrites pleuriformis*	

embedded absences, but only at the expense of known information about the stratigraphic relationships of taxa and samples.

Figure 3(a) is a dendrogram for the species, where clusters were extracted from a matrix of Dice coefficients using the unweighted-pair-group method (UPGM). Cluster A includes the 4 stratigraphically youngest ammonites (species 1–4) along with the long-ranged species 5. The two oldest taxa (species 8 and 9) are grouped in cluster D. *Anasibrites* and *Meekoceras* (species 6 and 7) are found in cluster B whereas species 10 and 11 fall into cluster C. However, the clusters only partially reflect the stratigraphy because clusters B and C are reversed and long-ranging species 5 clusters with short-ranged forms recognized only in the younger samples. These results are typical of experience with cluster analysis of species (Hazel, 1970, 1971, 1977; Millendorf, Brower, and Dyman, 1978).

The dendrogram for samples is given in Fig. 3(b), where computations are based on Dice coefficients and the unweighted-pair-group method for clustering. Seven clusters, each with one to 8 samples, can be recognized at a similarity level of approximately 0.5. The youngest samples are assigned to clusters A and B whereas clusters F and G contain the oldest. Samples dominated by the ubiquitous species 5 are contained in cluster E whereas those with species 10 and 11 are placed in cluster D. For convenience, cluster assignments are listed in the seriated matrices of Fig. 2. An approximate one-to-one correlation is observed between cluster assignments and sample order in the seriated matrix for the unconstrained solution; only two samples, 44 and 41, have to be interchanged to obtain perfect correspondence between clusters and position of seriated samples (Fig. 2(d)). However, comparison of cluster assignments with sample order in the constrained seriated matrix discloses a different situation. Although there is general correlation between cluster assignments and seriated order of samples, the correlation is lower and there is overlap between samples and clusters, C, D, and E with respect to their locations in the seriated matrix (Fig. 2(b)). Here, the effects of stratigraphical constraint are obvious.

Case Study 4

The final example to be described in detail consists of Cambrian fossils from the Riley Formation of central Texas described by Palmer (1954). The main purpose is to compare the results obtained from seriation with zonations presented by other workers. The methods employed range from classical qualitative biostratigraphy of Palmer (1954) to complicated quantitative techniques including graphical correlation (Shaw, 1964), principal components (Hohn, 1978), trinomial probability (Edwards and Beaver, 1978), an algorithm based on set theory and seriation of a matrix of association coefficients (Guex, 1977) and an unpublished binomial probability technique being developed by Gradstein and Agterberg (manuscript in preparation). In addition, cluster analyses have been determined for both taxa and samples using matrices of correlation coefficients and the unweighted-pair-group method to calculate the dendrogram.

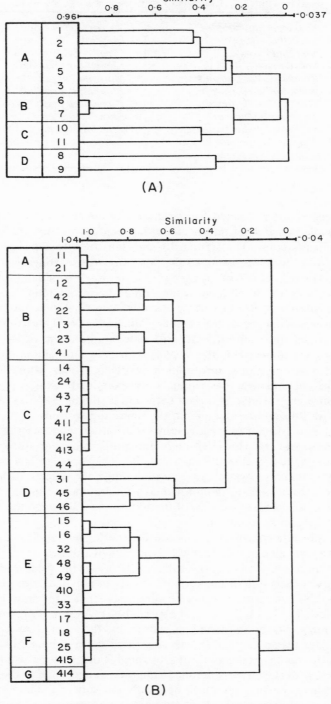

Fig. 3. Dendrograms for Triassic ammonites. Clusters were extracted from matrix of Dice coefficients with unweighted-pair-group method. (A) Analysis for taxa; (B) Analysis for samples

Table 2.   Cambrian fossils used in Case Study 4

1. *Kormagnostus simplex*	12. *Meteoraspis metra*
2. *Coosella beltensis*	13. *Coosia* cf. *albertensis*
3. *Kinsabia variegata*	14. *Crepicephalus australis*
4. *Spicule* B	15. *Llaonaspis undulata*
5. *Pseudagnostus? nordicus*	16. *Maryvillia* cf. *ariston*
6. *Arcuolimbus convexus*	17. *Llaonaspis peculiaris*
7. *Opisthotreta depressa*	18. *Dictyonina perforata*
8. *Tricrepicephalus coria*	19. *Raaschella ornata*
9. *Tricrepicephalus texanus*	20. *Angulotreta triangularis digitalis*
10. *Kingstonia pontotocensis*	21. *Labiostria conveximarginata*
11. *Llaonaspis modesta*	

21 of the species were selected for seriation (Table 2), because they occur in most of the sections studied by Palmer and because the majority of the taxa have been employed in other numerical studies. The standard Cambrian trilobite zones recognized by Palmer (1954) range from the *Bolaspidella* to the post-*Aphelaspsis* zones.

Palmer (1954) worked with 8 stratigraphic sections, of which one (Threadgill Creek) is faulted. The sections are distributed over an area of approximately 1800 square miles and lithologies recognized consist of a variety of sandstones, siltstones, and limestones deposited in a series of shallow marine environments (Palmer, 1954). Unfortunately, boundaries of paleontologic zones and rock units do not coincide and some of the lithologic units are time-transgressive (Palmer 1954, Shaw 1964). Thus, the data set is moderately complex with respect to the biostratigraphic information involved. Some workers treated data from all stratigraphic sections whereas others only examined taxa from unfaulted sections. Accordingly, we have seriated two data sets, one including 117 samples from all sections and another with 98 samples taken from the unfaulted sections.

In general, the majority of biostratigraphy seriations are concerned with both highest and lowest occurrences of taxa. However, many biostratigraphers are forced to study borehole data where only highest occurrences are reliable. Therefore, to test seriation methods on this type of data, a data set was created by recording each species as present in all samples below its highest occurrence in the individual stratigraphic sections. The seriated matrices developed in Fig. 4 represent analysis of the 98 samples, although the results derived from both data sets are almost identical.

The stratigraphically-constrained solution for the highest and lowest occurrences in 98 samples yields a test value of 0.555 with 272 embedded absences. This represents one of the lowest figures observed, with only the value of 0.506 for the Cambrian fossil data set with 117 samples lower. Scrutiny of the seriated matrix in Fig. 4(a) indicates that the majority of embedded absences are caused by rare taxa that are only known from a few samples (species 17, 15, 14, 11, 6, and especially 5). As noted previously, the effect is a regular occurrence in seriation and is caused by the overall averaging nature of the algorithm. Cluster assignments of species and samples are tabulated in the margins of Fig. 4(a) although the dendrograms are not illustrated. Four major clusters can be recognized for the samples, with cluster A generally being the youngest and D the oldest. The samples labelled 's' are 'strays' which do not fit definitely into any of the large clusters. However, a close correlation exists generally between cluster assigments and sample positions in the seriated matrix. Some overlap can be noted between clusters A and B and between cluster D and older samples of cluster C. The species also group into four clusters. The taxa of the various clusters are in distinct blocks of columns of the seriated matrix although taxa of clusters C and D intergrade.

Figure 4(b) pictures the unconstrained seriation with a test criterion of 0.571 and 255 embedded absences. Removal of stratigraphic constraint only eliminates 17 embedded absences. Samples of the seriated matrix are arranged in approximately inverse order of stratigraphic position such that the oldest samples are at the top of the matrix and youngest at the base. The overlap between clusters with respect to sample position in the seriated matrix is slightly less than for the constrained solution. Clusters of the taxa also are reversed relative to stratigraphic position.

The main purpose of this exercise is to compare zonations of taxa calculated from constrained seriation with those postulated by other authors. In order to compare results among various techniques, the following procedure was employed. The data are treated as point events where a single event represents highest or lowest occurrence of a particular taxon. For seriation, the position of an event is taken as the row or sample position where that event occurs in the final matrix for the constrained solution (Fig. 4(a)). Numbers for rows or samples increase from top to bottom (youngest to oldest) in the matrix; for example, highest and lowest occurrences of species 20 are in rows 6 and 12, respectively. Ties are ignored because the data will be subjected to nonparametric statistics. Range charts for the 21 species are illustrated in Fig. 5.

Comparative data abstracted from other authors are:

(1) Palmer (1954, Table III). Zonation by qualitative techniques for 117 samples. Palmer's zones (1954, Table III) are numbered 1 to 11 from oldest to youngest.
(2) Shaw (1964, Table A-18). Composite standard by graphical correlation for 117 samples.
(3) Guex (1977, Figs. 4, 5). Zonations for 117 samples. The technique first forms an association matrix between all possible pairs of species, where 1 is recorded if two taxa are ever noted together; other relationships are recorded by 0. The association matrix is seriated to compile associations and a range chart. The levels are numbered from 1 to 7 in which the last is youngest.
(4) Edwards and Beaver (1978). Zonations based on trinomial probability for 98 samples. The authors proposed several zonations including lowest occurrences (Table 3), highest occurrences, and both highest and lowest occurrences for some taxa (Table 5).
(5) Gradstein and Agterberg (manuscript in preparation). Zonation of lowest occurrences for 15 species from 98 samples using a relative time scale based on binomial probability.
(6) Hohn (1978, Table 1). Zonation based on scores for first principal component extracted from a correlation matrix calculated between 6 of the stratigraphic sections.

Correlation coefficients were computed for data from the seriation versus zonations from other methods. As the data are curvilinear and nonnormally distributed, coefficients employed are Spearman rank correlations. Only absolute values are provided in Table 3; signs are immaterial and reflect whether numbers increase from youngest to oldest or vice versa. The majority of resulting correlations are extremely high; values for constrained seriation versus the zonations of Palmer (1954), Shaw (1964), Geux (1977), Gradstein and Agterberg (in preparation), and the lowest occurrences of Edwards and Beaver (1978), are 0.96 or higher. The lowest correlations, 0.87 to 0.90, are for seriation versus zonations involving highest occurrences of Edwards and Beaver (1978) and for seriation versus principal com-

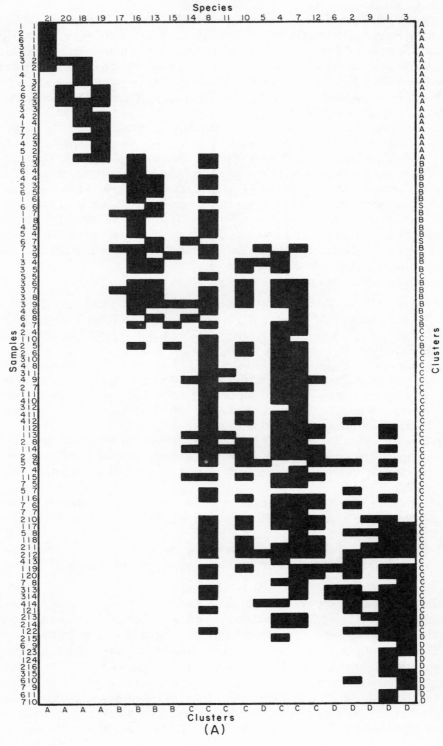

(A)

77

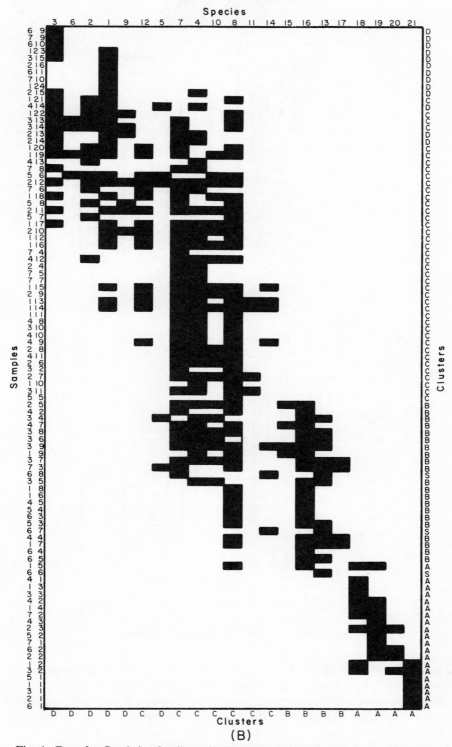

Fig. 4. Data for Cambrian fossils studied by Palmer (1954). (A) Constrained seriation; (B) Unconstrained seriation

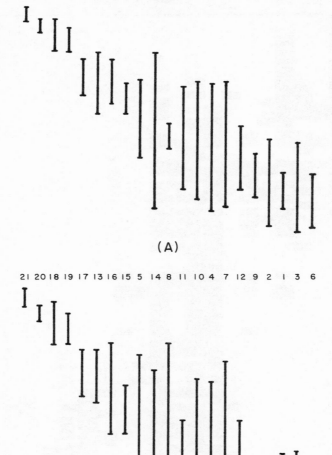

21 20 18 19 17 16 13 15 14 8 11 10 5 4 7 12 6 2 9 1 3

(A)

21 20 18 19 17 13 16 15 5 14 8 11 10 4 7 12 9 2 1 3 6

(B)

Fig. 5. Range charts for species of Cambrian fossils (see
Table 2 for list of taxa). (A) Chart for data on 98
samples; (B) Chart for data on 117 samples

ponent scores of Hohn (1978). Reasons for these discrepancies in computation are not
known completely. The most striking contrast between seriation zonation and zonation in
Edwards and Beaver (1978) is the highest occurrence of species 9 which is placed lower by
the seriation technique. Data from the principal components tend to estimate shorter
ranges for taxa in the higher and lower parts of the stratigraphic sections than does seriation. As mentioned earlier, two range charts for highest occurrences were produced by

Table 3. Magnitudes of Spearman's rank correlation coefficients for comparison of zonations of biostratigraphic events obtained with constrained seriation with those of other techniques. Student's *t* values are for comparison of observed correlation with population value of nil

Number of samples in zonation for constrained seriation	Technique and data matrix for comparison	Magnitude of Spearman's rank correlation coefficient	Number of events for comparison	Student's *t* value
98	Edwards and Beaver (1978), trinomial probability, lowest occurrences only	0.9812	15	18.35
98	Edwards and Beaver (1978), trinomial probability, highest occurrences only	0.8702	15	6.368
98	Edwards and Beaver (1978), trinomial probability, highest and lowest occurrences	0.8895	22	8.704
98	Gradstein and Agterberg (in preparation), binomial probability, lowest occurrences only	0.9777	15	16.77
117	Guex (1977) technique, highest and lowest occurrences	0.9630	34	20.20
117	Shaw (1964), graphical correlation, highest and lowest occurrences	0.9877	42	39.96
117	Hohn (1978), principal components, highest and lowest occurrences	0.9031	42	11.90
117	Palmer (1954), qualitative, highest and lowest occurrences	0.9615	42	22.13
117	Constrained seriation for highest occurrences only	0.9782	21	20.53

constrained seriation for the 117 samples; one was obtained from analysis of both highest and lowest occurrences and the second from highest occurrences only. The high correlation of 0.98 between the two range charts was not unexpected. From the previous discussion, it can be concluded that stratigraphically-constrained seriation produces similar results to those derived from qualitative techniques and other algorithms used frequently in numerical biostratigraphy.

Other case studies

In addition to the two examples discussed previously in detail, the following data sets have been seriated successfully.

(1) Smith and Fewtrell (1979) presented a qualitative seriation approach to two sets of microfossil data. The scheme minimizes taxa range zones across the samples which must remain in stratigraphic order within individual sections and is essentially a qualitative analogue of our constrained seriation. The first example includes 15 samples from 6 stratigraphic sections of Late Triassic and Early Jurassic age containing marginal marine to nonmarine clastics. Presences and absences were tabulated for 39 palynological taxa. As expected, results of the qualitative analysis are similar to our constrained seriation. For example, sample orderings from the two techniques are identical, except for three samples, and rank correlation of the two orderings is 0.975. 49 embedded absences can be counted in their seriated matrix versus 48 for our solution.

(2) The second data set studied by Smith and Fewtrell (1979) treats 27 samples from seven stratigraphic sections of the Carboniferous in Yorkshire, England. 43 microfossil taxa occurring in more than one sample, were selected for stratigraphically-constrained seriation. Again, the seriated matrices are similar with their solution having 143 embedded absences reduced to 136 by our technique. Rank correlation between the sample sequence in the seriated matrices equals 0.963.

(3) Grant (1962) described an Upper Cambrian trilobite fauna from Minnesota consisting of 24 species and 65 samples. Hazel (1977) used cluster analysis and principal components on the data, with four clusters of species and samples resulting; the clusters are tightly structured with major discontinuities between the adjacent clusters. The assemblage zones noted by Hazel correlate closely with those of Grant and only two samples were misplaced. This well-behaved data set was seriated with constrained and unconstrained methods and yielded relatively high test values of 0.806 and 0.823 respectively. The groupings of taxa and samples observed in the seriations are approximately the same as those of Grant (1962) and Hazel (1977).

(4) Numerous individuals (see, for example, the volume edited by Kummel and Teichert, 1970) have recently considered the Permian-Triassic boundary beds of Pakistan. Sweet (1970) outlined results from 89 samples of conodonts from 11 stratigraphic sections. 9 conodont assemblage zones were determined from the associations and relative abundances of the 28 taxa. Guex (1977) chose 17 of these species for analysis by a technique which combines aspects of set theory with archaeological seriation of a matrix of associations between taxa. The range chart of Guex contains 6 levels. The data set for seriation involves all 89 samples and 26 of the 28 taxa (two rare conodonts were deleted from the data). The test criteria for the constrained and unconstrained seriations are 0.717 and 0.586, respectively. Rank correlation coefficients between the seriation zonation of taxa versus zonations postulated by Guex (1977) and Sweet (1970) equal 0.925 and 0.988, respectively, reflecting the near identical results obtained from the different techniques. Correspondences between the unconstrained seriation, cluster analysis, and the conodont zones of Sweet (1970) are also high.

(5) Deboo (1965) worked on the biostratigraphy of Eocene–Oligocene boundary beds of Mississippi and Alabama based on the distribution of approximately 200 Foraminifera and Ostracoda species in 62 samples from 5 stratigraphic sections. Hazel (1970) and Brower, Millendorf and Dyman (1978) employed cluster analysis on various subsets of these data. Cluster analysis produces three or four assemblage zones depending on which subset of the data is involved. The zones are essentially time-parallel and similar to those calculated by Deboo (1965) with classical methods of biostratigraphy. Several of the data sets were subjected to seriation with test values ranging from 0.644 to 0.655. Correlations derived from seriation echo the time-stratigraphic diagrams presented by Hazel (1970) and Millendorf, Brower, and Dyman (1978).

Summary

Archaeological seriation represents a simple and powerful technique for quantitative biostratigraphy. The data matrix consists of presence-absences of m species from n samples in p stratigraphic sections. The basic procedure is to arrange the data in the form of a range chart with taxa in the columns and samples in the rows. This is accomplished by grouping the presences along the matrix diagonal so that the number of embedded absences on and near the diagonal is minimized. This also provides the shortest possible range zones of taxa. Seriation, simultaneously, calculates a range chart of taxa and places samples with the most similar faunal composition in adjacent rows of the seriated matrix. A simple test criterion measures the degree of concentration of presences along the diagonal.

Two types of seriation solution are available. Data on superposition of samples are explicitly ignored in the unconstrained solution. Consequently, samples can group in any order, which generally results in assemblages of samples and taxa that are similar to those derived from multivariate analysis of presence-absence data. However, groupings of taxa and samples are not usually in stratigraphic order. Stratigraphic control is introduced for constrained seriation in that samples are retained in order of superposition for each stratigraphic section in the seriated matrix. The stratigraphic constraint is useful in correlation.

Computational details of seriation are straightforword. Calculations are performed in a series of iterations, each of which consists of two main operations. (1) Mean location of presences is measured for each row or sample and rows are arrayed in ascending order with respect to the means. At this point, unconstrained seriation begins the next operation. For constrained seriation, the program checks samples are in stratigraphic order in each stratigraphic section. If not, samples out of stratigraphic order are interchanged in the seriated matrix to place them in stratigraphic sequence within each stratigraphic section. (2) Mean position of presences is determined for each column or species and columns are sequenced in ascending order of means. The two stages constitute one iteration and additional iterations are performed until the data converge.

Seriation techniques have been tested on 7 examples. The fossils involved include a variety of micro- and macrofossils ranging from Cambrian to Tertiary in age; the number of samples varies from 33 to 117 and the number of species ranges from 11 to 205. Reliable time-stratigraphic diagrams illustrating distributions of samples and taxa have been published for the majority of these data sets. In addition, the case studies have been subjected to one or more numerical techniques other than seriation.

Rank correlation coefficients compared results of seriation with those yielded by other qualitative and quantitative methods. All of the zonations are similar and the correlation magnitudes range from 0.87 to 0.99. It is concluded that seriation represents an effective method for numerical biostratigraphy. In essence, seriation is an averaging scheme which provides a mean or most likely zonation of taxa and samples. The logic underlying the technique is simple and similar to that employed by many nonquantitative biostratigraphers. In addition, seriation algorithms should be applicable to any data set for which an average zonation is appropriate. Hopefully, seriation will be useful to some of the biostratigraphers who have successfully resisted numerical methods to date.

References

Brower, J. C., 1981, Quantitative biostratigraphy 1830–1980, in Merriam, D. F. (ed.), *Computer Applications in the Earth Sciences—an Update of the 70's.* Plenum Press, New York, pp. 63–103.

Brower, J. C., and Millendorf, S. A., 1978, Biostratigraphic correlation within IGCP project 148, *Computers & Geosciences,* **4** (3), 217–220.

Brower, J. C., Millendorf, S. A., and Dyman, T. S., 1978, Quantification of assemblage zones based on multivariate analysis of weighted and unweighted data, *Computers & Geosciences,* **4** (3), 221–227.

Burroughs, W. A., and Brower, J. C., in press, SER, a FORTRAN program for the analysis of biostratigraphic data, *Computers & Geosciences.*

Cowgill, G. L., 1972, Models, methods and techniques for seriation, in Clarke, D. L. (ed.), *Models in Archaeology.* Methuen & Co. Ltd., London, pp. 381–424.

Deboo, P. B., 1965, Biostratigraphic correlation of the type Shubuta Member of the Yazoo Clay and Red Bluff Clay with their equivalents in southwestern Alabama, *Alabama Geol. Surv. Bull.,* **80**, 84 pp.

Doran, J. E., and Hodson, F. R., 1975, *Mathematics and Computers in Archaeology.* Harvard Univ. Press, Cambridge, Massachusetts, 381 pp.

Doveton, J. H., Gill, D., and Tipper, J. C., 1976, Conodont distributions in the Upper Pennsylvanian of Eastern Kansas; binary pattern analyses and their paleoecological implications, *Geol. Soc. America, Abstracts with Programs,* **8** (7), 842.

Edwards, L. E., and Beaver, R. J., 1978, The use of a paired comparison model in ordering stratigraphic events, *Jour. Math. Geology,* **10** (3), 261–272.

Gelfand, A. E., 1971, Rapid seriation methods with archaeological applications, in Hodson, F. R., Kendall, D. G., and Tăutu, P. (eds.), *Mathematics in the Archaeological and Historical Sciences.* Edinburgh Univ. Press, Edinburgh, pp. 186–201.

Grant, R. E., 1962, Trilobite distribution, Upper Franconia Formation (Upper Cambrian), southeastern Minnesota, *Jour. Paleont.,* **36** (5), 965–998.

Guex, J., 1977, Une nouvelle Méthode d'analyse biochronologique: note préliminaire, *Bull. Soc. Vaud. Sci. Nat.,* **73** (351), 309–322.

Geux, J., 1978, Les Trias inférieur des Salt Ranges (Pakistan): problèmes biochronologiques, *Eclogae géol. Helv.,* **71** (1), 105–141

Hay, W. W., and Southam, J. R., 1978, Quantifying biostratigraphic correlation, *Ann. Rev. Earth Planet, Sci.* **6**, 353–375.

Hazel, J. E., 1970, Binary coefficients and clustering in biostratigraphy, *Geol. Soc. America Bull.,* **81** (11), 3237–3252.

Hazel, J. E., 1971, Ostracode biostratigraphy of the Yorktown Formation (upper Miocene and lower Pliocene) of Virginia and North Carolina, *United States Geol. Surv. Prof. Paper 204,* 13 pp.

Hazel, J. E., 1977, Use of certain multivariate and other techniques in assemblage zonal biostratigraphy: examples utilizing Cambrian, Cretaceous, and Tertiary benthic invertebrates, in Kauffman, E. G., and Hazel, J. E. (eds.), *Concepts and Methods of Biostratigraphy.* Dowden, Hutchinson & Ross, Inc., Stroudsburg, Pa., pp. 187–212.

Hodson, F. R., Kendall, D. G., and Tăutu, P., 1971, *Mathematics in the Archaeological and Historical Sciences.* Edinburgh Univ. Press, Edinburgh, 565 pp.

Hohn, M. E., 1978, Stratigraphic correlation by principal components: effects of missing data, *Jour. Geol.,* **86**, 524–532.

Johnson, L., 1972, Introduction to imaginary models for achaeological scaling and clustering, in Clarke, D. L., (ed.), *Models in Archaeology.* Methuen & Co. Ltd., London, pp. 309–379.

Kendall, D. G., 1963, A statistical approach to Flinders Petrie's sequence dating, *Int. Stat. Inst. Bull.* **40**, 657–680.

Kendall, D. G., 1971, Seriation from abundance matrices, in Hodson, F. R., Kendall, D. G., and

Tăutu, P. (eds.), *Mathematics in the Archaeological and Historical Sciences*. Edinburgh Univ. Press, Edinburgh, pp. 215–252.

Kummel, B., 1966, The Lower Triassic Formations of the Salt Range and Trans-Indus Ranges, West Pakistan, *Mus. Comp. Zoology Bull.*, **134** (10), 361–429.

Kummel, B., and Teichert, C., (eds.), 1970, *Stratigraphic Boundary Problems, Permian and Triassic of West Pakistan*. Univ. of Kansas, Dept. of Geol., Special Pub. **4**, 474 pp.

Marquardt, W. H., 1978, Advances in archaeological seriation, in Schiffer, M. B., (ed.), *Advances in Archaeological Method and Theory*, **1**. Academic Press, pp. 257–314.

Millendorf, S. A., Brower, J. C., and Dyman, T. S., 1978, A comparison of methods for the quantification of assemblage zones, *Computers & Geosciences*, **4** (3), 229–242.

Miller, F. X., 1977, The graphic correlation method in biostratigraphy, in Kauffman, E. G., and Hazel, J. E. (eds.), *Concepts and Methods of Biostratigraphy*. Dowden, Hutchinson & Ross, Inc., Stroudsburg, Pa., pp. 165–186.

Palmer, A. R., 1954, The faunas of the Riley Formation in central Texas, *Jour. Paleont.*, **28** (6), 709–790.

Petrie, W. M. F., 1899, Sequences in prehistoric remains: *Jour. Anthropol. Inst.*, **29**, 295–301.

Scott, G. H., 1974, Essay review: stratigraphy and seriation: *Newsl. Stratigr.*, **3**, 93–100.

Shaw, A. B., 1964, *Time in Stratigraphy*. McGraw-Hill Book Co., New York, 365 pp.

Smith, D. G., and Fewtrell, M. D., 1979, A use of network diagrams in depicting stratigraphic time-correlation, *Jour. Geol. Soc. London*, **136**, 21–28.

Sweet, W. C., 1970, *Uppermost Permian and Lower Triassic conodonts of the Salt Range and Trans-Indus Ranges*, West Pakistan. Univ. of Kansas, Dept. of Geol., Special Publ., **4**, 207–275.

Tipper, J. C., 1977, Some distributional models for fossil marine animals, *Geol. Soc. America, Abstracts with Programs*, **9**, 1202.

Tipper, J. C., 1980, Some distributional models for fossil animals, *Paleobiology*, **6** (1), 77–95.

Wilkinson, E. M., 1974, Techniques of data analysis-seriation theory, *Archaeo-Physika*, **5**, 1–142.

Quantitative Stratigraphic Correlation
Edited by J. M. Cubitt and R. A. Reyment
© 1982 John Wiley & Sons Ltd.

The Automation of Biochronological Correlation

ERIC DAVAUD

Institut des Sciences de la Terre, 13 rue des Maraichers, 1203 Geneva, Switzerland

Abstract

Biochronological correlations between different sections can be established only with a reference scale showing the chronological sequence of biological events (range chart). This scale has to be synthesized from all local biostratigraphic data available. This fundamental operation, which is tedious when performed manually, can be characterized by a series of logical statements and automated. The principal problems of biochronological correlation are discussed and a method that establishes reliable chronological scale from raw stratigraphic data is described in detail.

Introduction

Biochronology has been a proven method for over 100 years. It has been the subject of innumerable descriptive works which established, slowly and through trial and error, the main outlines of biological evolution and, therefore, a relative chronological framework.

However, in contrast to the large number of practical applications, it is surprising that few studies have been devoted to the method itself. Perhaps, stratigraphers have assumed that the use of biochronology was self-evident and that there was no need to analyse the principles involved, as if one should discover necessarily and miraculously a chronological truth through the meticulous accumulation of paleontological data. At present, there is confusion in stratigraphic terminology, although efforts have been made recently by the International Committee of Stratigraphy (Hedberg, 1976) to clarify the situation.

The numerous controversies which continue to divide biostratigraphers into irreconcilable groups are symptoms of the malaise affecting the subject and demonstrate clearly the dangers of employing a method whose principles are poorly defined. This situation is a result of a general underestimation of the complexity of biochronological problems as condemned by Harrington (1965) in a scathing article:

'The basic concepts of stratigraphy (. . .) seem to become increasingly blurred and distorted as stratigraphic information increases. It seems to me that this hazy picture, this confused understanding of the fundamentals of stratigraphy, is the result of the unfortunate conjunction of several adverse factors: superficial thinking on the part of some stratigraphers regarding the scope and limits of their science, a scarcely philosophical approach to the troublesome problems of space and time, the peculiar tendency of the human brain toward symmetrical thinking, and last, but unfortunately certainly not least, the rather pedestrian attitude of many so-called "practical geologists" toward certain scientific problems which they class as "academic" and which they regard, therefore, as not worth the time and effort of serious and mature analysis.'

Since then, biostratigraphers have agreed on the necessity to proceed methodically and to rethink the basic concepts of traditional stratigraphy.

The scientific improvement and rationalization (illustrated by the work of Shaw (1964), Hedberg (1976), Salin (1976), Rubel (1976, 1978), Hay (1972), Hay and Southam (1978), Guex (1977, 1978, 1979), Edwards (1978), Dienes (1978), Brower (1979), and Agterberg and Gradstein (1980)) has occurred simultaneously with the introduction of computerized processing in biostratigraphy. Thus, it would seem that we are arriving at the definition of a rigorous methodology which should quickly overcome the failings of traditional empiricism. This is despite the reservations of many stratigraphers who are, as Holland (1978) suggests, 'irritated by the effusion of stratigraphical philosophers'!

The subject of our paper is to illustrate:

(1) the complexity of biochronological problems;
(2) the danger of employing out-dated concepts;
(3) the use of computers in biochronology (the use of a rigorous language).

Complexity of the Sedimentary Record

The recording of biological events in sediments is the result of superposition of two partially independent phenomena: sedimentation and evolution. These two phenomena vary in nature and intensity, both in time and space.

Let us, for the moment, ignore effects of variations in sedimentation and study closely the results of biological evolution by recalling basic principles.

The life time of a group of organisms able to multiply (species, according to the biological definition) is limited. In the period of time from first appearance of a species to eventual extinction, the group spatial distribution and its demographic importance are tied closely to environmental changes.

The appearance of the first individuals is generally limited and is the result, according to the theory of Mayr (1970), of constant genetic changes within an isolated gene pool.

Taking this into consideration, the idea of the living range of a species is defined in time as well as space and describes, in four-dimensional space, the places and moments where the species under consideration has existed. This entity, apparently abstract, can be thought of as a volume (or 'hypervolume') outside of which the probability of the species occurring is nil (either in an environment excluding its presence or outside its living period). Within this volume, the probability of the species occurring is not nil but varies considerably from one point to another depending on the demographic fluctuations of the species in time and space.

The living range of a species occurs in sediments, but shape, range, and primary population density will be changed and distorted entirely (Fig. 1). Sedimentary deposits in which that particular species lived or died are termed the 'recording range', a three-dimensional body whose geometry varies and in most situations (especially nectonic and planctonic species) has connection with the geometry of sedimentary deposits themselves. The stratigrapher has access only to certain points of this biosedimentary entity. The probability of discovering a particular species varies from one point to another, depending on primary population density, sedimentation rate, and degree of preservation of the tests. Unfortunately, none of these factors are known to us. If investigations at a given point are unsuccessful, it should be tempting to declare that absence of species is due to ecological or chronological reasons. However, if that point is within the recording range, it will be considered that this absence only reflects the lack of investigation. The ambiguity created by nonobservation of fauna is

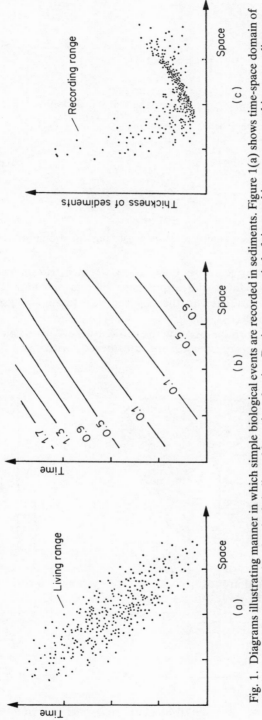

Fig. 1. Diagrams illustrating manner in which simple biological events are recorded in sediments. Figure 1(a) shows time-space domain of particular species. Population density is indicated by points density. During same period of time and in same geographic area, sedimentation rate changed regularly (Fig. 1(b)). If $S(x, y, t)$ is continuous function showing local and temporal variations of sedimentary rate, individual having lived at time t_0 and at point x_0 will be located in sedimentary record at point (x_0, y_0, z) where z is equal to $\int_0^{t_0} S(x_0, y_0, t)$ dt. Applying equation to individuals in Fig 1(a), Fig. 1(c) is obtained and illustrates the distorted information provided by sediments. Fig. 1(c) is high at base of recording range but low higher up. Therefore, chronological range of this species could be considerably underestimated.

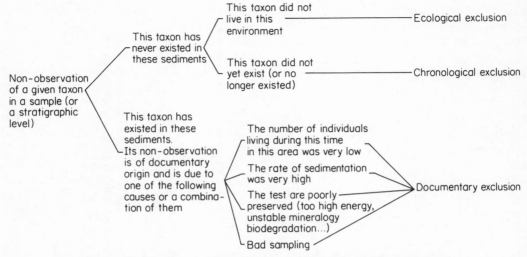

Fig. 2. Ambiguous meaning of absence of fauna in stratigraphic level

more embarrassing because it leads to uncertainties which, often, cannot be resolved. Stratigraphers, occasionally, ignore the importance of this problem (Fig. 2) and undue chronological significance is attached to the absence of a fauna (misuse of 'partial range zone' and 'interval zone' concepts).

Having discussed the basic principles, now the difficulties involved can be illustrated with an example. Suppose 7 different species are considered, with no evident evolutionary

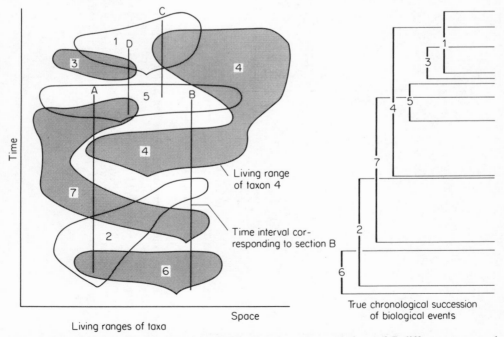

Fig. 3. Theoretical example showing distribution in space and time of 7 different taxa and true chronological succession

relationships and having a partly overlapping spatial and chronological living ranges (Fig. 3). Assume further, to simplify the analysis, that population density is even within each range but sedimentation rates vary in time and space. The result provided by the sedimentary record of these biological events will be distorted but the relative and local chronological sequence will be unchanged for all points of the space. Consider, now, the situation where only four local sections covering varying periods of time, are available (Figs 3 and 4). Forgetting for the moment that the solution is known, let us place ourselves into the shoes of a stratigrapher who studies with perplexity the four conflicting and fragmentary sections, trying to relate them chronologically. It will be seen that unless the one chronological sequence is known, it is difficult to establish chronological relationships among sections from the biological events recorded (It should be noted that none of the four cross-sections of Fig. 4 shows the real chronological sequence). Now, this knowledge, if not preexisting, can be obtained only by synthesizing paleoecological data. Since there is no preexisting biochronological scale, the stratigrapher will have to reconstitute the relative chronological sequence of observed events.

This preliminary and essential operation, which is equivalent to assembling a news item from fragmentary and contradictory evidence, is not easy and is usually performed with much trial and error. The difficulties can be demonstrated by trying to establish, mentally, the relations among recorded species in the four sections of Fig. 4.

The simplicity of the problem considered here contrasts with the difficulty in solving it. This provides an indication of the problems faced by stratigraphers when biochronological scale is established with data which are more fragmentary and conflicting than usually recognized in nature.

Furthermore, after spending much time and effort in establishing a scale, the stratigrapher would be tempted to consider the scale as a definite chronological reference. However, the scale is only temporary and can be partly wrong; it could be invalidated at any time by additional biostratigraphic data.

Because of the weakness of biochronological reconstitutions, due to the fact that they are temporary and open to improvement (Gabilly, 1971), stratigraphers should be careful and

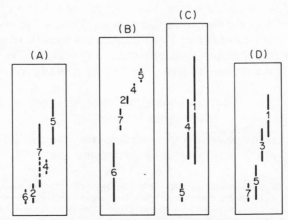

Fig. 4. Sedimentary record of biological events in four stratigraphic sections ((A), (B), (C), (D)) corresponding to theoretical example of Fig. 3. distortion due to differential rate of sedimentation has been obtained by the equation given in Fig. 1

reconsider the chronological models they use or establish. The acquisition of new and better information should result in the continual improvement of the biochronological scale. Unfortunately, it is not unusual for new evidence conflicting with existing chronological models to be dismissed on the pretext of bad data (homeomorphism or reworking).

As Hay (1978) remarks ironically: 'biostratigraphic methodology has worked well as a basic for the stratigraphic correlation as long as biostratigraphic correlations were either limited to one geologic area or carried out by one specialist. Problems and disputes began to arise when many specialists began to study larger areas and investigations began to overlap'. The complexity and vagueness of the chronological problem, and the lack of tools to solve it (hazy concepts without chronological meaning) are the main reasons for the difficulties described. The narrow pragmatism denounced by Harrington and the chaotic empiricism which is often the rule can only lead to confusion. In order to improve the situation, authors have recently reexamined the basic principles and some have attempted to treat the bio-chronology in a rigorous fashion. A method developed and 'computerized' (Davaud and Guex, 1978) is based on this rigorous way of thinking. It is derived from works of Guex (1977, 1978) and is similar to that developed by Rubel (1978).[1]

This method enables one, from an unlimited amount of stratigraphic sections, to:

(1) Determine a synthesized biostratigraphic scale (range chart);
(2) Choose from this scale the most significant chronological subdivisions;
(3) Establish chronological relations between different stratigraphical sections, which have been employed to establish the reference scale (chronostratigraphic correlations).

The concept of 'unitary associations' clearly defined by Guex is useful in biochronology. To illustrate the meaning of this concept and to demonstrate its importance, the example presented previously is examined (Fig. 3) and the real sequence (not the local one) of the biological events considered.

Theoretically, the smallest chronological subdivisions are characterized by periods of coexistence of two or more species ('concurrent range zone', 'assemblage zone'), by intervals of time defined by the appearance and extinction of a species, by the appearance of one species and the appearance of another, or by the extinction of one species and the extinction of another. It should be noted, however, that, except for periods of (co)existence, these subdivisions are defined by the lack of certain species. Now, it has been established previously that the lack (or nonobservation) of a fauna at a particular horizon is difficult to analyse because it could be the result of a number of factors most of which are unknown. Therefore, chronological subdivisions based only on the lack of certain kinds of species are useless in practice.

However, if we have a definite chronological framework, the concept of partial range zone can be employed but only under three conditions:

(1) Sedimentological and paleoecological studies show that the surroundings would have been favorable to the development of the missing species.

[1]Rubel (1976) proposes a method for organizing sets of species by employing their stratigraphical relations. Those which show contradictory relations are eliminated. When that organization is complete, Rubel uses separation intervals ('moments') between mutually-exclusive species to establish correlations. However, the correlations are usually 'crossed' and leads Hay and Southam (1978) to comment that 'Rubel's technique doesn't allow for the possibility of conflicting ranges in different sections.' Even if the first steps of Rubel's idea seem similar to Guex's (1977) method, there are fundamental differences between both approaches: Rubel does not introduce the idea of unitary associations. Consequently, his work does not allow formulation of the principle of reproducibility in the identifiability of unitary associations (discussion in Guex, 1978, 1979) and hence a chronological interpretation of biostratigraphical data.

(2) The volume of material examined is large enough to be certain that the species absence is not due to the lack of observation.

(3) This period of nonexistence is recognized systematically in most stratigraphical sections.

As, given a particular faunal assemblage, it is difficult to determine whether the missing species result from chronological, ecological, or observational reasons. Thus, the only subdivisions which can be identified locally are periods of coexistence. A series of these periods provides a reliable biochronological scale.

The traditional concepts of 'concurrent range zone' (although this does not always have a chronological value) and Oppel zone are covered by periods of coexistence, the latter corresponding exactly to the notion of unitary association defined by Guex (1977, p. 311) as 'the largest group of compatible species' (two species are classed as compatible if they have lived together at least once in the same stratigraphic horizon or have never been simultaneously observed in the same horizon but stratigraphic order is reversed from one section to another, indicating chronological coexistence). The advantage of this concept is that it can be synthesized and put into a formal language. In the light of the new terminology, a biochronological scale appears as an irreversible sequence of unitary associations.

Therefore, in order to construct a scale, unitary associations from paleontological data collected in different stratigraphic sections must be identified; those occurring in the most sections are placed together to form the scale.

Description of General Algorithm

Establishment of a biochronological scale requires segments representing species to be plotted against time; size and position of these segments are indicative of associations observed among species. As noted previously, the operation is equivalent to defining and assembling a group of unitary associations such that all local stratigraphic relations are respected.

The first step consists of assembling biostratigraphic information from as many sections as possible in a square matrix, R, in which rows and columns correspond to the various species studied.

Elements of this matrix (r_{ij}) will determine stratigraphic relationships between species i and j. By convention, (r_{ij}) will be:

$+2$ if species i has always been observed above species j;

-2 species i has always been observed below species j;

1 if species i and j have been recognized at least once in the same horizon or if they virtually coexist;

0 if relative stratigraphic positions of species i and j are unknown, never having been observed in the same section.

From this definition, it can be seen that matrix R is antisymmetric for elements r_{ij} other than 0 and 1 and that data in half the matrix below the diagonal line are redundant.

Construction of matrix R is performed by setting elements at zero and gradually inserting data from the different sections. While doing this, pairs of species whose relations differ from one section to another are frequently encountered. This apparent inconsistency, due either to incomplete data or to ecological incompatibility, implies that the species are virtually associated. Thus, if r_{ij}^{k} represents the relation between species i and j, at locality k, and r_{ij}^{n}, the relation between the same species at locality n, the inequality, $r_{ij}^{k} \neq r_{ij}^{n}$, implies that $r_{ij} = 1$.

After this preliminary operation, matrix R represents the associations between species and the stratigraphic relations observed in the sections studied. However, there remains unresolved relations ($r_{ij} = 0$) owing to the fact that some species have never been observed simultaneously in the same section.

Deductions based on transitivity of the inequality can clarify most of the unresolved relations. To illustrate the logic of this deductive step, consider two species i and j whose stratigraphic relations are unknown ($r_{ij} = 0$). If there is at least one other species (k for instance) whose stratigraphic relations with the two species are known, the relationship of i and j can be clarified, provided that the intermediate species (k) is located stratigraphically between the other two and that it does not coexist with them. The formulation of this double condition is as follows:

If

$$r_{ij} = 0, \text{ but there is, at least, one species } k \text{ such that } r_{ik} \neq 0$$

and

$$r_{kj} \neq 0, \text{ then } r_{ij} = r_{ik}, \text{ if and only if } r_{ik} = r_{kj} \neq 1.$$

The algorithm used to clarify these relationships acts in an iterative manner: firstly, unresolved relations are checked and, if possible, modified (intermediary species must show the same stratigraphic relationship between the two species considered). Secondly, those remaining with unresolved relations are double checked and changed using stratigraphic relations determined in the previous cycle. This iterative procedure is repeated as long as improvements occur from one cycle to another, and stops when further correlations are not possible. The species for which we have the least stratigraphic information are eliminated from matrix R. The process is repeated until no such species remain.

It will be noted that species excluded at this stage will be ones which appear only rarely in the different sections under study. Their stratigraphic interest is therefore limited.

After these preliminary operations, matrix R shows, synthetically, the basic stratigraphic information contained in the original data. With this information, groups of species having actually or virtually coexisted can be chosen and placed together in a synthetic list (bio-chronological scale). The fit will be optimised only if the following three conditions, for the regrouping of species having coexisted, are observed:

(1) All species belonging to a group must have shown at one time, evidence of coexistence.
(2) The number of groups must be minimal (or the coverage of groups should be maximal).
(3) All actual or virtual coexistences, previously determined, must be expressed.

Groups of species, according to the conditions, correspond to the unitary associations of Guex (1977). Each unitary association is a square submatrix of matrix R containing only 1 ($r_{ij} = 1$, whatever i and j). Research of unitary associations is performed in two steps. First, the spectrum of association[1] of species is reduced to a group of compatible species[2], by eliminating, step by step, species showing incompatibility within the spectrum.

Let us consider for instance a species (i), row or column i of matrix R represents relations

---

[1]The spectrum of association of a species contains all the species having actually or virtually coexisted at some time with that species.
[2]Group of species having coexisted at a certain moment. A unitary association is a compatible group but not vice versa: thus a compatible group can only be a subgroup of a unitary association.

between this particular species and all others. But only a few coexisted with it; these are termed the spectrum of association of species i (Guex, 1977). Let us place these species together, after species (i), in a square matrix B illustrating their stratigraphic relations (b_{kl}). There are then two possibilities:

(1) All species have coexisted $(b_{kl} = 1$ whatever k and $l)$. We have here a compatible group: in this situation, the spectrum should not be changed.

(2) Some species are anterior or posterior to others, although they coexisted with species i: the association is not compatible and some of its species must be eliminated. Criteria for elimination are based on the frequency of incompatibilities $(b_{kl \neq 1})$ observed for each species of matrix B and on the necessity to determine all observed coexistences (condition 3) within the unitary association. The species showing most incompatibilities is eliminated first unless the coexistence with species i has not been expressed in previous compatible groups. The eliminating process continues in the same manner until a consistent group is obtained, in which species i is necessarily included.

After the operations have been performed for each spectrum of association (each line of matrix R), as many compatible groups as species are derived. However, it is evident that information in many of these groups is redundant and can be eliminated. Groups remaining correspond to unitary associations; their quantity is minimal and relations of coexistence are specified. The last step, before building a schematic chart illustrating the associations and exclusions between species, is to reorganize, in time, all unitary associations using stratigraphic relations provided by the main matrix (r_{ij}). This can be performed by creating a new matrix D showing stratigraphic relations, no longer between species but between unitary associations. The relationships in time between two unitary associations is determined by the stratigraphic relations observed for species belonging to the two associations. If this group is consistent, which implies that all typical species of the first association have been systematically determined to be above (or below) typical species of the second association, the relative stratigraphic position of the two associations is absolutely determined. Alternatively, if stratigraphic relations between species of these two associations are inconsistent, the relative position of the two associations remains unknown, due to the fact that, in one of the two associations, we do not have sufficient information about the species. Therefore, it will be necessary to eliminate them.

Identification of these species is achieved by counting the number of times a species is responsible for anomalies in compiling the matrix of relations between unitary associations (D). The species responsible for the most anomalies then is eliminated and the unitary associations, to which it belonged, are redefined by incorporating species which excluded its presence. It should be noted that this procedure can lead to some redundant unitary associations. The latter are noted immediately and removed from the matrix D.

One can repeat this process until all remaining relations between unitary associations are defined unequivocally. The information contained in matrix D then can order the unitary associations to respect the stratigraphic relations observed locally and to create a synthetic range chart.

However, this chart is not necessarily a biochronological scale; a close look at each unitary association is necessary to determine the subdivisions of the chart in terms of zones and to employ them chronostratigraphically.

The principles involved in this critical examination confront us with a problem discussed previously: the absence of a fauna.

For example, consider two species, whose spectrum of association is the same but have never been observed at the same stratigraphic level (Fig. 5). The two species will provide two unitary associations whose significance will depend on the manner in which the fact that the species have never been recognized together is interpreted. There are three possibilities:

(1) Mutual exclusion of the two species is due to ecological reasons (they do not affect the same biotope).
(2) Exclusion is only apparent, due to the lack of observation.
(3) Exclusion has a chronological reason (the two species were never contemporaneous).

The first two explanations imply that the unitary associations resulting from this exclusion have no chronological meaning. By choosing these explanations, we consider implicitly that:

(1) The data (from local stratigraphic sections) are too fragmentary to provide real chronological relations between the two species considered.
(2) Further data would allow us to determine their synchronism (Fig. 5). Alternatively, if the third explanation is chosen, the resulting unitary associations must have a chronological explanation which could be confirmed by increased information.

We see here that there is a close relationship between the chronological validity of a unitary association and the quantity of information used to establish it.

However, the problem of the validity of a unitary association is to know whether there is enough information to exclude what could be an artefact without any chronological meaning. Intuitively, it is tempting to say yes if the mutually-excluding species generating these unitary associations were observed in most sedimentary sections where they were expected. However, if these characteristic species were recognized only in a few sections, then, the

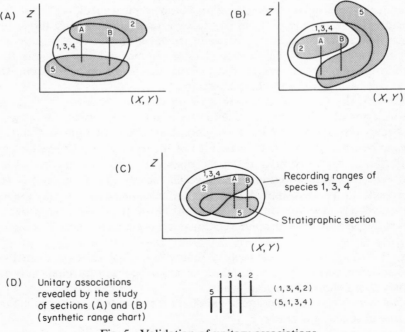

Fig. 5. Validation of unitary associations

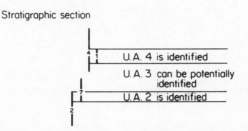

Fig. 6. Real or potential identification of unitary associations

validity would remain doubtful. This principle of spatial replicability is the only criterion by which the chronological value of subdivisions can be judged (Fig. 5). Furthermore, quantification of spatial replicability requires geological commonsense and reminds us of the complementary notions introduced by Guex (1979) of 'real or potential identification of unitary association'.

(1) A unitary association, characterized by an assemblage of species, or pairs of species, is identified in all sections where at least part of this assemblage can be observed.

(2) A unitary association can be potentially identified in all stratigraphic sections where unitary associations, surrounding the one considered, have been identified (Fig. 6).

If each time a unitary association is identified (M_i) or potentially identified (N_i) is counted the following ratio providing a chronological value of the reference scale can be obtained:

$$P_i = \frac{M_i}{N_i + M_i}$$

A sequence of unitary associations showing a high index of replicability ($P_i \rightarrow 1$) provides the best chronological environment that can be established. With this scale, chronostratigraphic correlations can be calculated easily. Furthermore, it is only necessary to reconsider, step by step, all stratigraphic sections studied, to examine their faunas, level by level, and then to identify unitary assocations present. Once a species or a characteristic group of species of a unitary association is recognized at a certain horizon, it can be stated that this horizon was formed during the period of time corresponding to the unitary association. It may be, however, that there is no characteristic group at a particular horizon but faunas common to several successive unitary associations are recognized. This horizon therefore belongs to any of the unitary associations $A_k, A_{k+1}, \ldots A_{k+n}$. Although, the position in a chronological scale will be uncertain, nevertheless, it has been formed during the period of time from the end of A_{k-1} to the start of A_{k+n+1}.

This deductive step, based on fundamental principles of stratigraphy, can be generalized and written in a rigorous manner (Fig. 7). Although transcription into a formal language is

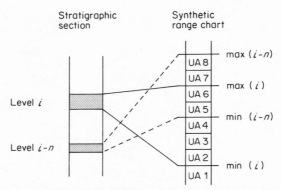

Fig. 7. Rules for improvement of chronostratigraphic definition of horizons without characteristic fauna

straightforward, it has not been performed in such a manner as to give the stratigrapher more freedom in interpretation. The algorithm is therefore limited only to establishing relations between faunas recognized in sections and subdivision of the reference scale. The information is presented as stratigraphic 'logs' and enables us to establish chronostratigraphic correlations manually (Fig. 8). Three operations are left to the stratigrapher:

(1) Choice of the unitary associations which seem most typical as far as chronology is concerned (employing indices of replicability, P_i).
(2) Improvement of the chronostratigraphic definition assigned to horizons with no characteristic fauna, achieved by the basic principle of stratigraphic superposition (geological commonsense and use of rules given in Fig. 7 will aid this operations).
(3) Chronostratigraphic interpolations between different sections.

The efficiency of this algorithm has been demonstrated in resolving particularly complex problems involving large numbers of species (Davaud and Guex, 1978).

Conclusion

The sedimentary record of biological events is the result of superposition of two different phenomena varying in quality and quantity in relation to time and space, biological evolution and sedimentary evolution. Data collected on these phenomena, are necessarily sporadic and fragmentary and, therefore, can provide an incorrect and contradictory image of evolution.

Single stratigraphic sequences only reflect the order in which a site has been colonized locally, but this order does not correspond necessarily to the true chronological succession of biological events. Therefore, a biological scale only provides a provisional idea of the chronological reality.

The choice of stratotypes, an expression of this reality, should be made only when a reference scale showing the hypothetical and virtual succession of biological events is obtained. Leaving the establishment of a reference scale until the discovery of a stratotype, is nonsense.

The absence of a certain fauna (or, more exactly, nonobservation) has an ambiguous meaning. It may be an ecological or chronological exclusion, or simply a lack of observation. Therefore, the only chronological subdivisions which can be identified without doubt

INPUT:

Title ——————————— EXAMPLE OF FIG. 4
4 strategic sections ——— 4   7
 and 7 different taxa ⸺5
5 taxa have been observed ⸺ 6   0 2   2 0 4   7 3 16   4 6 8   5 12 21
 in the first section   4
taxon 6 has been found from⸺ 7   0 4   5 0 7   3 8 15   1 13 22
 level 0 to level 2, taxon   3
 2 from level 0 to level 4   5   0 4   4 9 11   1 8 29
 . . .   5
         6   0 12   7 14 19   2 20 21   4 22 24   5 25 27

OUTPUT:

(A)

STRATIGRAPHIC RELATIONSHIP BETWEEN SPECIES 3 AND 4 REMAINS UNDETERMINED;
SPECIES 3 IS ELIMINATED.

SYNTHETIC RANGE CHART

(SPECIES OCCURENCE FREQUENCY INDICATED IN %)

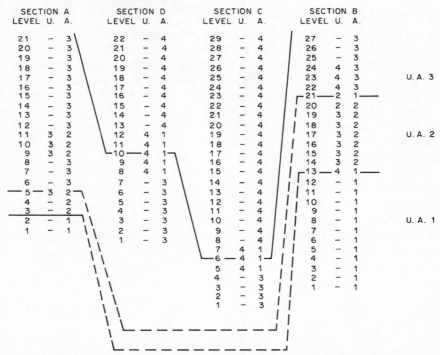

UNITARY
ASSOCIATIONS

		1	2	3	4
SPECIES 2	(50%)	===	===		
SPECIES 6	(50%)	===			
SPECIES 7	(75%)		===	===	
SPECIES 4	(75%)		===	===	
SPECIES 5	(100%)		===		
SPECIES 1	(50%)			===	

UNITARY ASSOCIATIONS	OCCURENCE FREQUENCY	REPLICABILITY INDEX
1	50%	1.0
2	25%	0.5
3	100%	1.0
4	50%	1.0

(B) CORRELATION OF STRATIGRAPHIC SECTIONS

SECTION A LEVEL U. A.	SECTION D LEVEL U. A.	SECTION C LEVEL U. A.	SECTION B LEVEL U. A.	
21 − 3	22 − 4	29 − 4	27 − 3	
20 − 3	21 − 4	28 − 4	26 − 3	
19 − 3	20 − 4	27 − 4	25 − 3	
18 − 3	19 − 4	26 − 4	24 4 3	
17 − 3	18 − 4	25 − 4	23 4 3	U. A. 3
16 − 3	17 − 4	24 − 4	22 4 3	
15 − 3	16 − 4	23 − 4	21—2 1—	
14 − 3	15 − 4	22 − 4	20 2 2	
13 − 3	14 − 4	21 − 4	19 3 2	
12 − 3	13 − 4	20 − 4	18 3 2	
11 3 2	12 4 1	19 − 4	17 3 2	U. A. 2
10 3 2	11 4 1	18 − 4	16 3 2	
9 3 2	10—4 1	17 − 4	15 3 2	
8 − 3	9 4 1	16 − 4	14 3 2	
7 − 3	8 4 1	15 − 4	13—4 1—	
6 − 3	7 − 3	14 − 4	12 − 1	
5—3 2	6 − 3	13 − 4	11 − 1	
4 − 2	5 − 3	12 − 4	10 − 1	
3 − 2	4 − 3	11 − 4	9 − 1	
2 − 1	3 − 3	10 − 4	8 − 1	U. A. 1
1 − 1	2 − 3	9 − 4	7 − 1	
	1 − 3	8 − 4	6 − 1	
		7 4 1	5 − 1	
		6—4 1	4 − 1	
		5 4 1	3 − 1	
		4 − 3	2 − 1	
		3 − 3	1 − 1	
		2 − 3		
		1 − 3		

Fig. 8. Codification of problem mentioned in Fig. 4 and solution provided by algorithm
developed in this paper

(the only ones which can be used practically) must be based on the presence of a species or a group of species. Using the absence of a fauna as a chronological fact implies that with certainty the reason for absence is not ecological or for lack of observation.

Biochronological correlations are straightforward to make as long as a reliable pre-established reference scale is created. If this is not the situation or if the stratigraphic data contradict the working scale, it is necessary then to establish a new scale. However, this operation is difficult and time consuming when performed manually and even though new data may be in conflict with the established scale, it remains frequently unchanged for this reason. Therefore, interest in computerizing (or at least formalizing) the construction of such reference scales is evident.

The concept of unitary associations which differs from current stratigraphic concepts by rigorous formulation is definitely an advance. By using this concept, the empirical methods used to construct biochronological reference scale can be analyzed and replaced by a series of unequivocal elementary operations written in a computer language. This process can be summarized as follows:

(1) Compilation of biostratigraphic data from several sections and synthetic formulation of stratigraphic relations between species.
(2) Extraction of unitary associations.
(3) Stratigraphic listing of these entities.
(4) Study of their chronological value.

These operations provide a series of reliable unitary associations, from which a reference scale can be constructed and used to establish biochronological correlations.

Characteristics and Use of the Program

The program used to construct the biochronological scales is written in FORTRAN IV and has been developed on a UNIVAC 1108. It employs 30 K-words.

With this program, one can establish biochronological scales based on range-through data from an unlimited number of sections but with no more than 110 different species. This limit can be changed by modifying dimensions of the matrix and vectors defined at the start of the program. Times of execution do not depend on the number of sections, but are proportional directly to the total number of species used and to the number of anomalous species. Input and output are illustrated in Fig. 8.

A listing can be obtained directly from the author.

Acknowledgments. This research was supported by the Swiss National Science Foundation (grant 2.766.0.77). The author expresses his thanks to Dr. M. Septfontaine, J. Charollais, J. Guex, R. Wernli, J.-M. Jaquet for numerous fruitful discussions. He also acknowledges Dr. H. De Souza for his help in translating this paper and C. Laborde for typing the various drafts.

References

Agterberg, F., and Gradstein, F., 1980, A statistical method for the clustering of biostratigraphic events, *26th Int. Geol. Congress Abs.*, **II**, 841.

Brower, J. C., 1979, Special notes on quantitative biostratigraphy, *IGCP Project 148 Newsletter*, Spring 1979, 6 pp.

Davaud, E., and Guex, J., 1978, Traitement analytique manuel et algorithmique de problèmes complexes de corrélations biochronologiques, *Eclogae Geol. Helv.*, **71**(3), 581–610.

Dienes, I., 1978, Methods of plotting temporal range charts and their application in age estimation, *Computers & Geosciences*, **4**(3), 269–272.

Edward, L. E., 1978, Range charts and no-space graphs, *Computers & Geosciences*, **4**(3), 269–272.

Gabilly, J., 1971, Méthods et modèles en stratigraphie du Jurassique, *Mem. B.R.G.M.*, **75**, 5–16.

Guex, J., 1977, Une nouvelle méthode d'analyse biochronologique. Note préliminaire, *Bull. Soc. Vaud. Sci. Nat.*, **351**(73), 309–322.

Guex, J., 1978, Influence du confinement géographique des espèces fossiles sur l'élaboration d'échelles biochronologiques et sur les corrélations, *Bull. Soc. Vaud. Sci. Nat.*, **354**(74), 115–124.

Guex, J., 1979, Terminologie et méthods de la biostratigraphie moderne: commentaires critiques et propositions, *Bull. Soc. Vaud. Sci. Nat.*, **74**(3), 169–216.

Harrington, H. J., 1965, Space, things, time and events. An essay on stratigraphy, *Bull. Amer. Assoc. Petrol Geol.*, **49**(10), 1601–1646.

Hay, W. W., 1972, Probabilistic stratigraphy, *Eclogae Geol. Helv.*, **65**(2), 255–266.

Hay, W. W., and Southam, J. R., 1978, Quantifying biostratigraphic correlation, *An. Review Earth Planet. Sci.*, **6**, 353–375.

Hedberg, H. D., 1976, *International Stratigraphic Guide: a Guide to Stratigraphic Classification, Terminology and Procedure*. J. Wiley and Sons, New York, 200 pp.

Holland, C. H., 1978, Stratigraphical classification and all that: *Lethaia*, **11**(1), 85–90.

Mayr, E., 1970, *Populations, Species and Evolution*. Harvard Univ. Press, Cambridge, Mass., 453 pp.

Rubel, M., 1976, On biological construction of time in geology, *Eesti NSV Tead. Akad. Toimetised Keemia Geologia.*, **25**(2), 136–144.

Rubel, M., 1978, Principles of construction and use of biostratigraphical scales for correlation, *Computers & Geosciences*, **4**(3), 243–246.

Salin, Y. S., 1976, Algorithm of stratigraphic correlation, *Modern Geology*, **5**(2), 191–199.

Shaw, A. B., 1964, *Time in Stratigraphy*. McGraw-Hill Book Co., New York, 365 pp.

Quantitative Stratigraphic Correlation
Edited by J. M. Cubitt and R. A. Reyment
© 1982 John Wiley & Sons Ltd.

The Conceptual Basis for Lateral Tracing of Biostratigraphic Units

STEVEN A. MILLENDORF and MARION T. MILLENDORF

617 Soniat Street, New Orleans, La., 70115, U.S.A.

Abstract

The general strategy for performing lateral tracing of biostratigraphic units is outlined. Lateral tracing being a stepwise analysis of multivariate similarity between geographically-adjacent stratigraphic sections and concomitant analysis of dissimilarity within each section utilizing binary, presence/absence data. These analyses divulge lateral continuity of biostratigraphic zones, and display vertical discontinuities between these zones, respectively. The method inludes a factor for weighing the similarity analysis to minimize crossings in correlation (that is, older to younger, and younger to older). As a generalized technique, lateral tracing can be performed on correlation problems of any scale (from detailed samplings within a small area to interbasinal and global projects), use various similarity/dissimilarity coefficients and handle either biostratigraphic or lithostratigraphic data.

Introduction

Lateral tracing was conceived by Sneath (1975a, 1975b), who presented it as an alternate method for quantified correlation of sedimentary units. Millendorf and Heffner (1978) elaborated on this technique, presented a FORTRAN computer program for its application, and, based upon successful usage by Millendorf, Brower, and Dyman (1978), touted lateral tracing as a significant refinement to the then popular approaches to quantified biostratigraphy which were based primarily upon methods of numerical taxonomy. The purpose of this paper is to reiterate and elucidate the conceptual basis of the lateral tracing technique, and thereby demonstrate its superiority over those methods borrowed from numerical taxonomy.

Lateral Tracing

In the multivariate analyses of faunal zones, the basic data are binary, denoting the presence or absence (1 or 0, respectively) of a fossil taxon in a particular sample collected from one of several stratigraphic sections to be correlated (Fig. 1). The popular methods derived from numerical taxonomy, such as cluster analysis, factor analysis, principle components, and principle coordinates, generally analyze all samples from all stratigraphic sections simultaneously, thus their designation as 'en masse' techniques. However, as stated earlier (Millendorf and Heffner, 1978), this approach is reasonable only when the result is to depict ecological zones or biofacies because such en masse analysis is insensitive to the effects of faunal gradation within an isochronous unit with respect to geographic position.

101

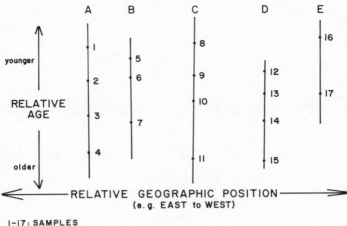

I-17: SAMPLES
A-E: STRATIGRAPHIC SECTIONS

Fig. 1. Typical geologic cross-section to be correlated

Thus, as the faunal composition of an isochronous unit changes across the study area, samples taken from distant points in the unit might be dissimilar enough not to cluster. Or, simply, as the lateral variation in faunal composition, and the distance between sample points increase, the similarity between the samples decreases. The en masse approach is also deficient in ability to define vertical zonations within individual sections. As an alternative, we propose lateral tracing which is designed to alleviate both of these difficulties, and thereby offer a truer representation of biostratigraphic zonations.

Lateral tracing makes use of the definition of a biostratigraphic zone as a faunal assemblage which has spacial continuity and temporal discontinuity, and proceeds to portray this dimensionality amongst and between the samples. The two-dimensional nature of the problem requires a two-step solution. First, lateral, geographic continuity is established by step-wise analysis of geographically-adjacent pairs of stratigraphic sections. These analyses can be performed using numerous multivariate techniques, though our preference is a simple calculation of binary similarities.

The method begins by calculating a matrix of similarity coefficients between samples from the two sections. For example, the analysis of two sections containing N and M numbers of samples yields an $N \times M$ similarity matrix (Fig. 2). Then, the greatest similarity

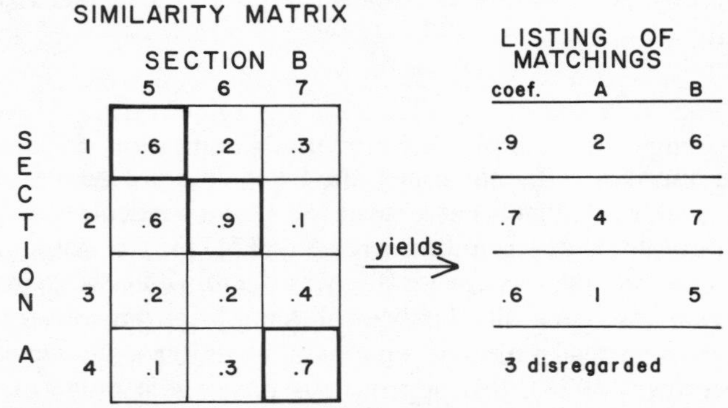

SIMILARITY MATRIX

SECTION B

		5	6	7
S E C T I O N A	1	.6	.2	.3
	2	.6	.9	.1
	3	.2	.2	.4
	4	.1	.3	.7

yields

LISTING OF MATCHINGS

coef.	A	B
.9	2	6
.7	4	7
.6	1	5
3 disregarded		

Fig. 2. Hypothetical output from first half of analysis which
establishes lateral continuity of biostratigraphic units

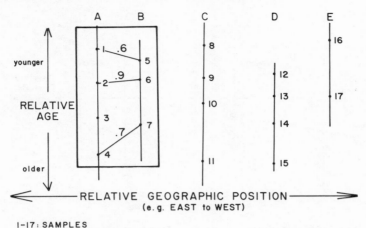

I-17: SAMPLES
A-E: STRATIGRAPHIC SECTIONS

Fig. 3. Matchings from first half of analysis applied to cross-section
with similarity coefficients given

is determined from the matrix and corresponding samples matched. These two samples are discounted and the next greatest similarity is determined and samples matched. This procedure is repeated until all samples in the section with fewer samples are matched. Any remaining unmatched samples are disregarded (Fig. 2).Presently, only a single $N \times M$ similarity matrix is calculated, and from it the largest coefficients are systematically selected. The alternative would be to recalculate the matrix subsequent to each selection, and in doing so, omit those previously-selected samples from the calculations (after selection of the first two samples, recalculation would yield an $N-1 \times M-1$ similarity matrix; then an $N-2 \times M-2$ matrix and so on). This iterative approach has not been attempted, and though computationally more complex than the current method, such recalculation might offer better results. Figure 3 shows these hypothetical matchings as applied to our cross section. This entire procedure then is repeated and correlations are developed between the second and third sections, third and fourth sections, and so on for as many sections as desired.

A graphic representation of faunal gradation within units will help to illustrate the benefits to be derived from this phase of the analysis. Figure 4 is a hypothetical plot of character distance, or multivariate descriptor for two isochronous units, R_y and R_x, versus lateral, geographic, location with vertical lines A through N denoting the positions of stratigraphic sections from which samples were taken. The displacement of each unit line relative to character distance as it progresses from section A through N represents the lateral change in faunal composition, with the slope of each segment directly proportional to the magnitude of this change. In a simple, nongradational, situation the units would be portrayed as horizontal, parallel lines. The area between the dashed lines represents a possible cluster 'window,' or area, within which samples would form a cluster in an en masse analysis and be interpreted erroneously as belonging to one specific unit. It is seen that by restricting this window to two adjacent sections at a time, as is the situation in lateral tracing, we can accommodate lateral gradation so that resulting matchings are more likely to represent true biostratigraphic units than in simple, en masse analysis encompassing all sections simultaneously. Lateral tracing also provides a list of the similarity coefficients for a ready comparison of the relative strength of each correlation.

Second, vertical discontinuities are revealed, and the biostratigraphic units defined, by

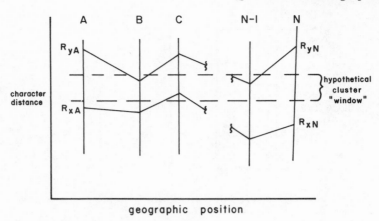

Fig. 4. Graphic representation of faunal gradation within two isochronous units, R_y and R_x, versus geographic position

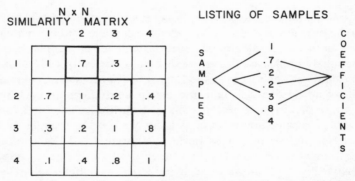

Fig. 5. Hypothetical output from second half of analysis which
reveals vertical discontinuities between biostratigraphic units

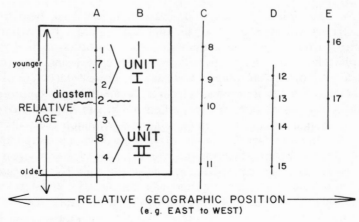

Fig. 6. Discontinuities from second half of analysis applied to
cross-section with similarity coefficients given

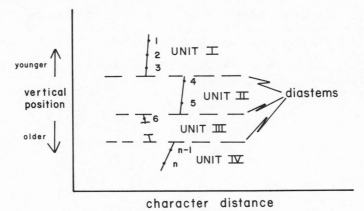

Fig. 7. Graphic representation of vertical discontinuities between four biostratigraphical units, I–IV

analysis of each section individually. This is accomplished by calculating similarities, or dissimilarities, between vertically-adjacent pairs of samples within the section (Fig. 5). Small similarities, or large dissimilarities, are interpreted as major diastems and representative of a unit boundry (Fig. 6).

Again, a graphic representation will assist. Figure 7 is a hypothetical plot of character distance of a single stratigraphic section, S, containing samples l to n, versus vertical, stratigraphic position. The horizontal, dashed lines are diastems, or unit boundries, at which points major displacements in faunal character occur. These diastems define the temporal limits of an individual unit, each of which is represented by a line segment whose length is proportional to its vertical persistence (how many feet/meters thick the unit is) and whose slope represents minor faunal gradation vertically within the unit. This second aspect of biostratigraphic analysis is completely overlooked by the *en masse* techniques.

In the process of establishing lateral continuity, crossings will frequently occur in the lines of correlation (young to old and old to young) between two adjacent sections. Though generally restricted to the later samples that join together, these crossings occur at greater similarity levels. In order to minimize the number and severity of these crossings, Millendorf and Heffner (1978) formulated the vertical weighting factor (*VWF*). The *VWF* is a coefficient which, when introduced into the analysis, will eliminate some crossings, and, if made large enough, will force all samples to link horizontally and parallel. The effectiveness of the *VWF* in enhancing biostratigraphic correlations has not been evaluated fully at this time.

Lateral tracing has been utilized in the analysis of three large data sets of varying degrees of geologic complexity (Millendorf, Brower, and Dyman 1978). Other, en masse methods were also utilized in these analyses, with lateral tracing proving to be the most useful for biostratigraphic correlation.

In closing, we offer not a panacea for the many philosophical and practical problems of time-stratigraphic analysis, but an improvement to the popular quantitative techniques currently in use.

References

Millendorf, S. A., Brower, J. C., and Dyman, T. S., 1978, A comparison of methods for the quantification of assemblage zones, *Computers & Geosciences*, **4**(3), 229–242.

Millendorf, S. A., and Heffner, T., 1978, Fortran program for lateral tracing of time-stratigraphic units based on faunal assemblage zones, *Computers & Geosciences*, **4**(3), 313–318.

Sneath, P. H. A., 1975*a*, Quantitative method for lateral tracing of sedimentary units, *Computers & Geosciences*, **1**(3), 215–220.

Sneath, P. H. A., 1975*b*, Clarification on a quantitative stratigraphic correlation technique, *Computers & Geosciences*, **1**(4), 353–354.

Quantitative Stratigraphic Correlation
Edited by J. M. Cubitt and R. A. Reyment
© 1982 John Wiley & Sons Ltd.

Properties of Composite Sections Constructed by Least-Squares

MICHAEL ED. HOHN

West Virginia Geological and Economic Survey, Morgantown, WV 26505, USA

Abstract

Constructing a composite section through least-squares fitting provides an ordering on an interval scale rather than the seriation of probabilistic methods. Given a large number of sections, taxa, and samples within sections, building an interval or ordinal scale should yield equally-useful, and perhaps equivalent results; these requirements are the same as saying that events must be spaced more closely than the desired stratigraphic resolution.

The events—appearances and disappearances of taxa—might not be dispersed uniformly in the stratigraphic column because of bursts in taxonomic diversity. Similarly, not all sections represent the same rate of sedimentation throughout, or offer equally-favorable chances of sampling. The steps required in making a composite on an interval scale provide the mechanism for detecting inequities in event dispersal without the need to inspect large numbers of bivariate plots amongst sections or the composite and sections. Just as the first principal component represents a composite section in an eigenvector solution to the fit, less-important components indicate major deviations from the main trends, and provide the mechanism for piece-wise fitting of a composite. Some suggestions for fitting this piece-wise composite exist in the statistical literature.

Introduction

Uses of quantitative methods in biostratigraphy have centered on two foci: the establishment of assemblage zones and construction of composite sections. The resulting diversity of methods can treat both qualitative and quantitative data. Shaw (1964) proposed an approach to the creation of composite sections using regression by least-squares; Hohn (1978) modified Shaw's technique through principal components analysis. Although retaining the least-squares criterion, Hohn's method computes a composite section from the observed sections simultaneously, whereas Shaw works through pair-wise comparisons of sections.

The principal components solution requires estimation of missing values in the original data array. Missing values occur because not all events—the upper and lower ranges of fossil taxa—are observed in all sections. In addition, some sections can represent a more restricted stratigraphic range than others in a data set. Procedures exist for estimating missing values for multivariate analysis, using multiple regression (Buck, 1960) or principal components analysis (Dear, 1959) as part of the estimation process. Unfortunately, these methods may be sensitive to systematic patterns of missing entries present in biostratigraphic data. Allowance for missing entries thus requires an algorithm specific to the biostratigraphic problem. The earlier paper (Hohn, 1978) proposed a naive method of estimation; this paper includes a revised procedure.

The present work also considers problems not treated in the earlier paper such as differential rates of sedimentation within sections, inequities in sampling, and relationships with other methods. Accounting for depositional and sampling effects should lead to sophisticated methods in creating composite sections, and improve the stratigraphic resolution of the result.

Procedure

The revised algorithm for the principal component solution was programmed using SAS (Statistical Analysis System: Barr, and others 1976). In brief, the procedure runs as follows:

(1) Input data, using the SAS default feature of designating missing values with a full stop (.). Throughout the calculations, events are 'observations' and sections are 'variables'. Nonmissing values in each section can be centered to a zero mean and scaled to a constant standard deviation.

(2) Compute a correlation matrix between the sections; the CORR procedure of SAS employed all available data for computation of correlation coefficients, that is the coefficient between two variables is calculated from observations for which there are values present for both variables.

(3) Enter the SAS procedure MATRIX to extract eigenvectors (E) and eigenvalues, a matrix containing means (MEAN (I, J)) of Section I when compared with Section J; and a matrix containing standard deviations (SD (I, J)) of Section I when compared with Section J. Note that in general, MEAN (I, J) $\neq$ MEAN (J, I) and SD (I, J) $\neq$ SD (J, I). The I, J entry of MEAN is the mean stratigraphic level in Section I of observations present in both Sections I and J.

Because MATRIX treats missing values as zeros, care must be taken to flag missing values with some large number, for example 10^8.

(4) Assume M sections. For each range bottom I missing in Section J, loop through all Sections K, where K = 1, M and $k \neq J$, to find the lowest value of $(X (I, K) - \text{MEAN} (K, J))/(SD(K, J) \cdot E(K, I))$, where X(I, K) is the entry for event I in Section K, K = 1, M; the entry for Section K is not missing; and E(K, 1) is the loading of Section K on the first principal component. Employing calculated means and standard deviations for each pair-wise comparison of sections, enables a sliding and shifting of sections with respect to each other. This procedure provides an estimate of missing values in situations where given pairs of sections represent different, but overlapping stratigraphic ranges. Once a minimum value (MIN) is recognized, then, the missing entry X (I, J) is computed as: MEAN (J, K) + MIN/(SD (J, K) $\cdot$ E (J, 1)).

(5) Follow the same procedure with the upper ranges, but calculate maximum values rather than the previous minimum values. The procedure differs from that in Hohn (1978), by differentiating between upper and lower range limits in the estimation process.

(6) Compute the composite as X$\cdot$E1, where E1 is the first column of E.

The result of applying this algorithm to the data used in Hohn (1978) compares favourably with the composite standard of Shaw (1964), shown in Table 1 and Fig. 1. In general, lower range limits are estimated higher than in Shaw's composite, and upper range limits are lower. The difference in results stems probably from the fact that Shaw's procedure attempts to mitigate the effects of collection terraces on observed ranges.

Table 1. Scores of events on first principal component

Taxon	Event	Score	Taxon	Event	Score
Aphelaspis walcotti	T	0.369	*Syspacheilus* cf. *S. camurus*	T	−0.351
	B	0.275		B	−0.422
Labiostria conveximarginata	T	0.384	*Raaschella ornata*	T	0.290
	B	0.315		B	0.273
Coosia cf. *C. albertensis*	T	0.222	*Cheilocephalus breviloba*	T	0.368
	B	0.195		B	0.260
Kingstonia pontocenis	T	0.136	*Meteoraspis metra*	T	0.072
	B	−0.036		B	−0.018
Maryvillia cf. *M. ariston*	T	0.228	*Cedarina cordillarae*	T	−0.757
	B	0.160		B	−0.823
Pseudagnostus ?nordicus	T	0.160	*Bolaspidella burnetensis*	T	−1.006
	B	−0.040		B	−1.070
Tricrepicephalus coria	T	0.195	*Angulotreta triangularis*	T	0.356
	B	−0.036		B	0.304
Llanaspis peculiaris	T	0.234	*Dictyonina perforata*	T	0.295
	B	0.213		B	0.278
Crepicephalus australis	T	0.111	Spicule B	T	0.162
	B	0.064		B	−0.173
Llanoaspis undulata	T	0.142	*Opisthotreta depressa*	T	0.168
	B	0.131		B	−0.092
Coosella beltensis	T	−0.078	*Diaphora sp.*	T	0.114
	B	−0.150		B	−0.070
Kormagnostus simplex	T	−0.250	*Kingsabia variegata*	T	−0.037
	B	−0.601		B	−0.310
			Paterina sp.	T	−0.186
				B	−0.332

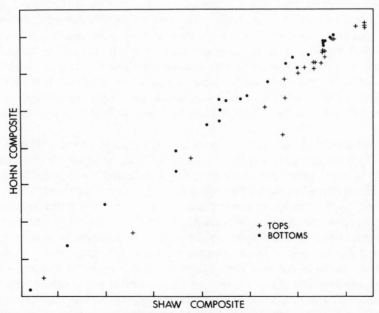

Fig. 1. Comparison of Shaw's composite section of events (1964) with that obtained from revised principal component algorithm

Other Methods: Assemblage Zones

Hazel (1970, 1977) advocates the use of multivariate statistics in delineation of assemblage zones for biostratigraphy; although differing in end product from composite-standard methods, the assemblage-zone approach shares with the principal components composite, the use of multivariate statistics. It is instructive to draw further comparisons in order to clarify aspects of the computations, and the extent to which the two methods might yield similar results.

Both approaches employ the same type of data, viz. at any locality, presence or absence of fossil taxa recorded at a level within the stratigraphic column. The 'locality' may in fact be a core. One could observe the fossil content in a continuous traverse across an outcrop or core, but the inequities of fossil diversity and abundance ensure that a relatively small number of discrete samples contain the majority of information, that is upper and lower stratigraphic limits of occurrences.

Following the range-through procedure (Hazel, 1977), the data comprise a discrete representation of continuous fossil ranges. The table may be reduced by eliminating samples in which no upper or lower range limits are observed; the reduction does not involve any loss of information useful in subsequent analysis. Note that the range-through procedure are not only justified, but necessary if the sampling effects of varying paleoenvironments and lithologies are to be lessened.

The composite-standard approach collapses the presence-absence table of samples versus taxa into a table of sections versus events, using rank order or footage of each sample. As outlined previously and in Shaw (1964), subsequent computations treat the sections as variables to be stretched, squeezed, and slid next to each other. Stratigraphic level in the method of Shaw (1964) or Edwards and Beaver (1978) performs a role intrinsic to the analysis. Assemblage zones can be delineated from positions of upper and lower range limits on the composite. Individual sections may be zoned by plotting the composite against a section, estimating or computing a line of best fit, and projecting zone limits from the composite to this line and to the section. Building a composite-standard through principal components parallels the traditional methods of the biostratigrapher; the resulting composite escapes the criticism of *petitio principii* leveled by Hazel (1977) through the need for the composite to provide reasonable correlations with new sections. A plot of composite versus new section will reveal a large scatter about a line of best fit if the composite has no geologic information. Certainly, assemblage zones are subject to the same test; addition of new data should not change delineated assemblages, although new assemblages could be added.

In the construction of assemblage zones through cluster analysis (Hazel, 1970) or variants of principal components analysis (Hazel, 1977), cooccurrences of taxa are intrinsic to the analysis. From the pattern of cooccurrences, stratigraphic ordering of samples in a given section can be recovered in part; however, true rank order or stratigraphic levels of samples and, therefore, upper and lower range limits, remain extrinsic to the calculations.

Data employed in the creation of assemblage zones can have a geographic component in addition to a stratigraphic component. In fact, raw data entered into computation of similarities do not include explicit information on the assignment of individual samples among sections. Resulting ordinations of taxa or samples may reflect geographic as much as stratigraphic relationships. This can be a useful feature if geographic range of assemblage zones is of interest. The composite standard does not provide information about geographic relationships, and indeed, the procedure of estimating missing values should eliminate many of the geographic effects.

Properties of the Composite

An advantage in the principal component solution lies in the use of variances and covariances among sections rather than events or taxa; the result is frequently a small array for extraction of eigenvectors (Hohn, 1978).

One possible drawback is the need to estimate missing entries. As discussed previously, this feature may have advantages in mitigating paleoenvironmental effects on sampling.

Because the composite standard yields an ordering of events, new sections can be compared directly with the composite. The principal component solution for the definition of assemblage zones can provide a stratigraphically-meaningful ordering of taxa, particularly if geographic and sampling effects are minimal and the first principal component accounts for a high percentage of the total variance. The ranking does not represent relative stratigraphic positions of the upper and lower range limits. Therefore, assemblage-zone ordination cannot be converted to a composite section, although the resemblance to one constructed from the midpoints of ranges on a composite of range-limits should be significant.

Ancillary statistics associated with the composite section provide an indication of the severity and deviation of events from the composite. The first eigenvalue indicates the proportion of total variance explained by the composite section, although one must employ this number with caution, for statistical correlation among sections can be high because upper range limits must be observed above lower range limits. Tests of significance on eigenvalues and correlation coefficients are invalid (Hohn, 1978) except in the case of data limited to either upper or lower range limits.

Despite restrictions on significance tests, principal components indicate sections contributing significantly to the ordering, and sections showing the greatest deviation. In an analysis of six sections described by Palmer (1955) and employed by Shaw (1964), the sections yield nearly-equal weights for the first principal component (Table 2), although sections 3 and 4 are the major contributors and section 1 is the minor contributor to the composite standard. Loadings on the second to sixth components (Table 2) indicate sections that deviate from the composite and whether groups of sections deviate in the same manner. As with principal components analysis, loadings may be graphed to aid interpretation (Fig. 2). Knowledge of specific sections that fit the composite poorly could lead to greater sampling, consideration of possible errors in observed ranges, or detection of systematic errors. For instance, facies changes across a sedimentary basin could indicate restricted fossil ranges in a subset of localities; the principal components plot would reveal a grouping of points representing these sections. Restricted range limits could be noted by graphing one of these sections against the composite. In the presence of many sections, such

Table 2. Loadings of 6 sections on principal components

Section	Eigenvector					
	1	2	3	4	5	6
1	0.38	0.80	−0.17	−0.06	0.23	−0.36
2	0.40	−0.45	−0.30	−0.59	−0.15	−0.42
3	0.42	0.01	−0.50	−0.02	0.08	0.75
4	0.42	−0.30	0.46	0.10	0.72	0.00
5	0.41	0.19	0.63	−0.21	−0.54	0.24
6	0.41	−0.19	−0.14	0.77	−0.33	−0.27
Eigenvalues	5.39	0.34	0.17	0.10	0.00	−0.02

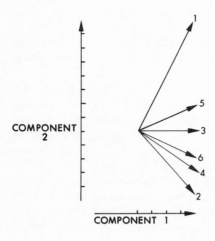

Fig. 2. Loadings of 6 stratigraphic sections on first two principal components. Sections are numbered in order: James River, Little Llano River, Morgan Creek, Pontoto, Streeter, and Threadgill Creek

a procedure removes the need to graph each section against the composite in evaluating goodness-of-fit and sources of error.

Differential Rates of Deposition

Differential rates of deposition between pairs of sections are taken into account by scaling each section. Shaw (1964) illustrates the graphical effects of changes in depositional rates within sections. Calculation of the composite standard by the method outlined does not consider such changes with concomittant foreshortening of section. However, graphing individual sections against the composite should reveal systematic deviations attributable to sedimentation slowdowns or halts. A difficulty lies in separating systematic error from sampling error, and employing information gleaned from graphs to improve the composite standard.

If gaps and foreshortening in the sections could be located, then a broken line might be fit to the data, each segment of the line representing a stratigraphic interval of constant sedimentation rate within individual sections. The resulting composite would have different scales for each segment, but scales of adjacent segments might be related through sections that appear to maintain constant rates of deposition.

A necessary first step in such a procedure is the piece-wise fitting of a composite standard. One must partition the events into subsets that possess an optimum property in fitting a straight line. One property is an F-type criterion, that is, a ratio between explained and error sums of squares. A method described by McGee and Carleton (1970) employs a clustering algorithm to group points possessing a common regression equation. Hawkins and ten Krooden (1979) published an algorithm for partitioning an ordered set of points into groups with homogeneous variance and covariance. In extending this algorithm to the present application, one must decide on an ordering of points; the first principal component can serve this purpose. The algorithm of Hawkins and ten Krooden (1979) considers variance and covariance in testing for goodness-of-fit, whereas fitting of a piece-wise composite needs only to consider differential covariances among segments. However, heterogeniety in variance should be noted in that it can reveal parts of the data set characterized by poor or insufficient sampling.

Data employed to demonstrate the algorithm comprised observed and estimated data on

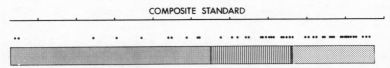

Fig. 3. Positions of events on composite standard. Segmentation of composite gave three sequences of events indicated by adjacent patterns

6 sections described previously; the estimates were calculated by a variant of the method in Hohn (1978). Input to the program includes maximal number of segments to be fit, and minimal number of points per segment, which in the present analysis is three.

A three-segment fit produced results that are reasonable from the distribution of events in the composite (Fig. 3), and bivariate plots of sections with the composite (Fig. 4). Segments I and II possess similar density of events on the composite, whereas segment III has a sparse sampling. Of the three segments, number II conforms least with the composite (Fig. 4). Performing principal component analyses on data representing each segment, segments I and III are seen to exhibit a single large eigenvalue each (Fig. 5). However, the middle segment does not yield a single large eigenvalue; in fact, neither of the first two components have positive loadings for all sections (Fig. 6). The results suggest inadequacy of the composite standard in the interval represented by this segment.

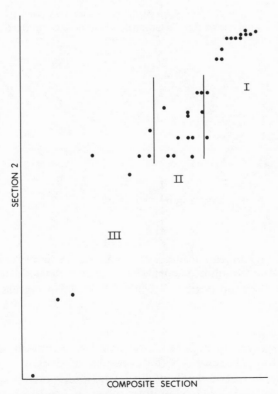

Fig. 4. Scatter plot of Section 2 against composite standard. Three areas delineated correspond to the segments in Fig. 3

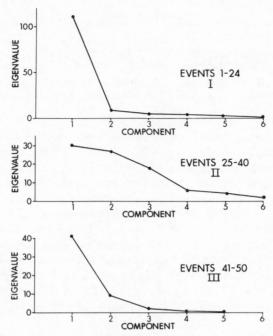

Fig. 5. Eigenvalues from principal components
analysis of three segments of composite standard

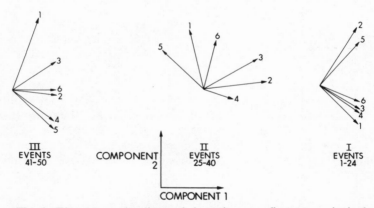

Fig. 6. Eigenvector loadings of 6 sections on first two principal
components for three segments. Events are numbered from top to
bottom in composite standard

Ranking

Advantages that the composite-standard constructed on a continuous scale possesses compared to a simple ranking of events, are that measures of diversity could be calculated from the result, and the patchiness of events might reveal intervals of poor sampling. If a stratigraphic interval represents a time of uniform diversity, and diversity is high enough to guarantee fine division of the interval, then, the composite based on depths and that based on ranks should produce similar stratigraphic resolution. An algorithm described by

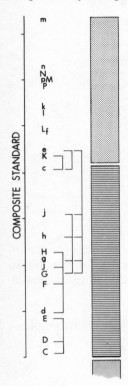

Fig. 7. Plot of selected events studied by Edwards and Beaver (1978) on Hohn composite standard, comparing segmentation with paired-comparisons model

Edwards and Beaver (1978) indicates events having inconsistent ordering relative to each other; an interval characterized by several such events might reflect a time of rapid environmental change with corresponding turnover in fauna, or a time of slow deposition. Either process leads to a large number of events per unit of rock.

The composite standard employed in the previous section compares favourably with the paired-comparisons ranking of Edwards and Beaver (1978) in the sense that events with unclear ordering possess similar scores on the principal component (Fig. 7). Taken from the paper on paired comparisons, the brackets define groups of events judged to have similar relative ordering, that is, show no consistent ordering. Note that events of low ordering fall generally within the second zone as defined by the segmentation procedure. The fact that approximately half of the events studied by Edwards and Beaver (1978) fall within this zone of inconsistent range limits, could account for the poor correspondence observed (p. 271) between their results and Shaws' (1964) composite.

The individuals's choice between the two kinds of composite-standard might rest upon the computer software available. If depths are replaced by ranks of events within each section, then, applying the principal components method offers an alternative to the paired-comparisons approach of Edwards and Beaver (1978). Although the method might lack the statistical rigor of the paired-comparisons model, it can yield useful results. The approach shares with computations based on absolute depths, partitioning of data into a composite and subsidiary eigenvectors from which sections contributing to the composite, and those contributing to the error can be determined.

Comparison of eigenvalues and eigenvectors between analyses of ranked and unranked data (Table 3) indicates a degree of similarity. In particular, the order of importance of

Table 3. Eigenvalues and first eigenvector loadings of principal components analysis of ranked and unranked data

	Eigenvalues			First eigenvector	
	Unranked	Ranked		Unranked	Ranked
1	5.39	5.46	1	0.379	0.390
2	0.34	0.33	2	0.404	0.396
3	0.17	0.13	3	0.423	0.420
4	0.10	0.07	4	0.415	0.418
5	0.00	0.06	5	0.412	0.411
6	0.02	0.05	6	0.414	0.413

variables on the first principal components remains unchanged. Furthermore, ordering of events on the composite standard (Fig. 8) does not differ greatly from the previous result (Fig. 7). Principal components analysis of the ranked events does offer an easily-calculated ordination of events.

Acknowledgment. This work is published by permission of the Director and State Geologist, West Virginia Geological and Economic Survey.

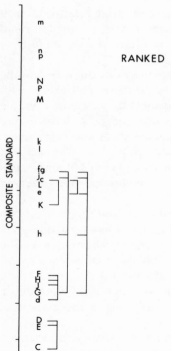

Fig. 8. Plot of selected events on composite standard calculated from ranked data

References

Barr, A. J., Goodnight, J. H., Sall, J. P., and Helwig, J. T., 1976, *A User's Guide to SAS 76*. SAS Institute, Raleigh, North Carolina, 329 pp.

Buck, S. F., 1960, A method of estimation of missing values in multivariate data suitable for use with an electronic computer, *Jour. Roy. Stat. Soc. Ser. B*, **22**(2), 302–307.

Dear, R. E., 1959, *A Principal-component Missing-data Method for Multiple Regression Models*. SP-86, System Development Corporation, Santa Monica, California.

Edwards, L. E., and Beaver, R. J., 1978, The use of a paired comparison model in ordering stratigraphic events, *Jour. Math. Geology*, **10**(3), 261–272.

Hawkins, D. M., and ten Krooden, J. A., 1979, Zonation of sequences of heteroscedastic multivariate data, *Computers & Geosciences*, **5**(2), 189–194.

Hazel, J. E., 1970, Binary coefficients and clustering biostratigraphy, *Geol. Soc. Amer. Bull.*, **81**(11), 3237–3252.

Hazel, J. E., 1977, Use of certain multivariate and other techniques in assemblage zonal biostratigraphy: examples utilizing Cambrian, Cretaceous, and Tertiary benthic invertebrates, in Kauffman, E. G., and Hazel, J. E. (eds.), *Concepts and Methods of Biostratigraphy*. Dowdon, Hutchison and Ross, Stroudsburg, Pennsylvania, pp. 187–212.

Hohn, M. E., 1978, Stratigraphic correlation by principal components: Effects of missing data, *Jour. Geol.*, **86**(4), 524–532.

McGee, V. E., and Carleton, W. T., 1970, Piecewise regression, *Jour Amer. Stat. Assoc.*, **65**(33), 1109–1124.

Palmer, A. R., 1955, The faunas of the Riley Formation in central Texas, *Jour. Paleontol.*, **28**(6), 709–786.

Shaw, A. B., 1964, *Time in Stratigraphy*. McGraw-Hill Book Co., New York, 365 pp.

Quantitative Stratigraphic Correlation
Edited by J. M. Cubitt and R. A. Reyment
© 1982 John Wiley & Sons Ltd.

Models of Cenozoic Foraminiferal Stratigraphy—Northwestern

Atlantic Margin

F. M. GRADSTEIN AND F. P. AGTERBERG

Geological Survey of Canada, P.O. Box 1006, Dartmouth, N.S., Canada

Geological Survey of Canada, 601 Booth St., Ottawa, Ontario, Canada

Abstract

Stratigraphic events defined by the highest or lowest occurrences of fossils in a group of wells or natural sections can be ordered into an optimum sequence. In the statistical model used the relative position of events in the most likely sequence is an 'average' of all the relative positions encountered. Frequency of crossover (mismatch) of events in the sections correlated has been used to estimate average distances between successive events in the optimum sequence. The events are clustered by the estimated distances between them, which gives results similar to those of the assemblage zone approach in biostratigraphy. The statistical model has been employed to erect a zonation for Cenozoic benthonic and planktonic foraminiferal records (exits of 206 taxa) in 22 wells on the Canadian Atlantic continental margin between 43° and 60°. Northern and southern optimum sequences occur, containing respectively 41 and 60 taxa, 14 of which are in common. The southern sequence contains 32, mostly Eocene and Miocene index planktonics and the northern sequence 6, essentially Eocene ones. The difference reflects pronounced post-Middle Eocene latitudinal water mass heterogeneity and differential post-Eocene shallowing across the continental margin.

The preferred northern and southern 'optimum' zonations consist of up to 9 clusters of Paleocene through Pliocene/Pleistocene age; distortion of the taxa stratigraphic order in the clusters is negligible. The latitudinally-differentiated, 9-fold 'optimum' zonation is related to a conventionally-constructed, planktonic biostratigraphic scheme which has 7 (northern area) to 12 (southern area) successive assemblages. This combined scheme is useful to 'optimum' middle and higher latitude Cenozoic biostratigraphy for the Northwest Atlantic continental margin.

Introduction

Stratigraphy, like most geological sciences, is essentially empirical, which implies that it is rooted in a body of organized, cumulative observations. The two main attributes of this historical philosophy are:

(1) the irreversible flow of time (the arrow of time),
(2) events as fossilized in the earth record and their spatial and temporal relations.

The reconstruction of the likely order and geographic extent of events (concrete) and their placement within the geological time scale (abstract) provide a framework called earth geological history. There are many categories of events, each with special properties and values to geological history. Lithostratigraphy, magnetostratigraphy, and biostratigraphy are frequent classification systems of such categories; reasonable rules for these systems are provided in the International Stratigraphic Guide (International Subcommission on Stratigraphic Classification, 1976).

The properties of the paleontological record form the basis of biostratigraphy. The record includes entry, range, peak occurrence, and exit of fossil taxa; these paleontological events are results of the continuing evolutionary trends of life on Earth. They differ from physical events in that they are unique, nonrecurrent, and that their order is irreversible.

Paleontological correlation depends on comparing similar fossil occurrences in or between regions and is referred to frequently as a paleontological zonation. We assume that correlation lines established through correlation of zones over an area correspond to time lines.

Partly under the influence of the development of a paleomagnetic reversal scale, which promises virtually isochronous correlations for specific horizons, efforts have been made to arrive at detailed sequences of evolutionary fossil datums particularly in the marine planktonic record. These are based on First and Last Appearance Datums (FADs and LADs in biochronology; Berggren and Van Couvering, 1978) in ocean and land sections. In theory, this allows for more or less reliable point correlations in time.

When correlating discrete events or zones reflecting short time increments, the error in time becomes relatively important. Unfortunately, paleontological correlation is fraught with uncertainties which are hard to quantify. Uncertainty factors which bear on the accuracy in time of correlations are a function both of personal bias and of inherent paleontological/biological characteristics, including:

(a) sampling and the frequency of fossil taxa,
(b) confidence of taxonomic identifications,
(c) influence of facies (change) on stratigraphic range of fossils,
(d) differential rate of taxa evolution in different parts of the world.

Careful and innovative field and laboratory techniques lead to a more objective biostratigraphy and a better understanding of limitations. Such techniques include:

(1) the application of multiple correlation criteria,
(2) replicate sampling and replicate taxonomic and stratigraphic procedures by different students,
(3) quantification of conventional biostratigraphic analysis,
(4) quantitative taxonomy.

A wealth of information is accumulating in each of the four techniques mentioned. Rather arbitrarily, we refer to a few of the studies in micropaleontology. Zachariasse and others (1978) pay much attention to replicate sampling and multiple correlation techniques with emphasis on the statistical evaluation of errors. Quantification of biostratigraphical interpretations with simple and complex statistical methods is examined in the *Computers & Geosciences* issue on Quantitative Stratigraphic Correlation (Cubitt, 1978), Hazel (1970), Hay and Southam (1978), and by Brower and Burroughs (this volume) who appraise many techniques and systems. A thorough review of the potential and problems of quantitative taxonomy in micropaleontology is performed by Scott (1974).

Early studies by Shaw (1964, 1969) and Miller (1977) focus on observational and intrinsic problems in paleontology and biostratigraphy and the authors introduce a simple bivariate graphical technique for correlation of sections based on the entries and exits of taxa in sections and aimed at a regional composite standard, that combines the local stratigraphic ranges of taxa in total stratigraphic ranges.

Recently, attention has been focused on the theoretical, mathematical solution of correlation type problems which view biostratigraphic sequences as random deviations from a true solution. The solution faces four sources of uncertainty:

(1) The uncertainty that the optimum, likely sequence of fossil events has been established. This, composite standard problem, is discussed in detail by W. W. Hay and colleagues (Hay, 1972; Worsley and Jorgens, 1977; Hay and Southam, 1978).

(2) The uncertainty that distance of fossil events along a relative time scale is known (spacing problem). In conventional biostratigraphy, extensive use is made of distances in time between events or (non) overlap of ranges to produce assemblage zones. In the composite standard technique, distance between successive events is a function of the relative dispersion of each event in the sections considered; first occurrence levels are minimized and last occurrence levels are maximized but no direct standard errors are available on the composite positions.

(3) The uncertainty that the geographic distribution of an event is known (its traceability; Drooger, 1974). Linkage of events through second or nth order of correlation from type sections to the area under study, although proportionally less accurate in time, must be considered also.

(4) The error in fixation of biostratigraphic events at the scale in a well, or outcrop section. This is basically a technical error which calls for an understanding and mathematical expression of the errors in field and laboratory techniques.

Expressing correlation as a numerical probability will require considerable quantitative insight into these four parameters.

In this paper, we consider the construction and application of most likely sequences and groupings based on estimates of relative order and distance of joint events. The Cenozoic foraminiferal record which consists of extinction points of 206 taxa, as recorded by one of us (FMG) in 22 Canadian Atlantic margin wells is employed. For this purpose, a computer-based statistical system has been developed (by FPA) which produces two types of answers:

(1) the optimum, most likely sequence of extinction points along a relative time scale;
(2) the optimum clustering of extinction points along a relative time scale, using the optimum sequence.

The first method (1) identifies regional correlation markers, helps in evaluating their stratigraphic positions, and provides a standard to which individual (new) sequences can be compared. The second method (2) approximates the assemblage zone technique in biostratigraphy. Because of provincialism in the fossil record, attention is given to the determination of optimum standards for parts of the region studied using two geographically-separate groups of wells.

Because the 'probabilistic' zonation does not give weight to those known zonal markers (mostly planktonic index fossils) which are too rare to be locally correlative, the 'probabilistic' zonation is compared to a subjective one. Regional zonal schemes are presented for the planktonic record and a conventionally-derived planktonic zonation complements the optimum clusters of the mixed benthonic and planktonic record. No attempt has been made to define zones formally and proliferate formal zonal nomenclature based on a regional data base. Our record also suffers from a lack of insight in the regional lower stratigraphical limits of the marker taxa.

7 plates of scanning electron microscope illustrations and few camera-lucida drawings (Geological Survey of Canada catalogue numbers 64168 to 64262) show the external morphology of key taxa in the optimum clusters.

Acknowledgments. During the accumulation of micropaleontological identifications in the wells the senior author has benefited from counsel given by several colleagues and from comparative studies. He particularly likes to mention discussions with W. A. Berggren (WHOI, Woods Hole), J. H. Lamb (Exxon, Houston), R. C. Tjalsma (Cities Services, Tulsa), G. Vilks (GSC, Dartmouth), and the help and hospitality at the Utrecht University, Netherlands, to examine collections of D. A. J. Batjes and J. P. H. Kaaschieter, and at the U.S. National Museum, Washington, to study the J. A. Cushman collections. Frank Thoma (GSC, Dartmouth) made the many scanning electron microscope illustrations and provided other valuable technical assistance.

We are indebted to Jim Brower and Bill Burroughs (Syracuse University), P. Doeven (GSC, Dartmouth), C. W. Drooger (Utrecht University), and Lucy E. Edwards (USGS, Reston) for helpful discussions regarding the development and implementation of the methodology used. The technical assistance, particularly of Louis Nel (Ottawa) and Art Jackson, and of G. Grant, G. Cooke, K.-G. Shih and A. Johnston (GSC, Dartmouth) is much appreciated.

Why Should 'Probabilistic' Biostratigraphy be Applied?

It is well known that the sequence of first and last occurrences of planktonic foraminiferal species in open marine Cenozoic sediments in the low-latitude regions of the world is closely spaced and shows a regular order. As a result, standard planktonic zonations provide a stratigraphic resolution of 30 to 45 zones over a time span of $65 \times 10^6 y$ (Blow, 1969; Postuma, 1971; Berggren, 1972a; Stainforth and others, 1975). Although several Cenozoic taxa are indigenous to mid latitudes, the absence of many lower-latitude forms and the longer stratigraphic ranges of mid-latitude taxa causes stratigraphic resolution to decrease away from the lower-latitude belt. In high latitudes (65°N and S), the virtual absence of planktonic foraminiferal taxa makes standard zonations inapplicable.

The northwestern Atlantic margin, offshore eastern Canada, spans the mid- to high-latitudinal realm (north of 42°) and although there were temporal northward incursions of lower-latitudinal taxa in Early or Middle Eocene times, there is a drastic overall diminution of the number of biostratigraphically-useful Cenozoic planktonic species (from about 75 to 30 as determined to date) from the Scotian Shelf to the Grand Banks to the Labrador Shelf. A change from a deeper, open marine facies in the Paleogene to nearshore, shallower conditions in the Oligocene to Neogene (Gradstein and others, 1975; Gradstein and Srivastava, 1980) also curtails the number of taxa present in the younger Cenozoic section.

As a consequence the construction of a planktonic zonation is mainly applicable to the southern Grand Banks and Scotian shelf where 12 zones have been recognized based on species of standard zonations which are not too rare locally to be of practical value in correlation. Similarly, on the northern Grand Banks and Labrador Shelf a 7-fold planktonics subdivision of the Cenozoic sedimentary strata is possible; the regional application is limited but the zonal markers and associated planktonic species improve chronostratigraphic calibration for the benthonic zones.

Independently, the Cenozoic benthonic foraminiferal record also shows temporal and spatial trends in taxonomic diversity and number of specimens. Calcareous benthonic species diversity and number of specimens decreases northward from the Scotian Shelf to the Grand Banks to the Labrador Shelf whereas the Early Cenozoic agglutinated species diversity and numbers of specimen drastically increases on the Labrador Shelf. This benthonic provincialism is complicated by incoherent geographic distribution of some taxa, which in part is due to sampling.

Few of the agglutinated taxa, only a dozen out of more than 50 determined, are of biostratigraphic value (Gradstein and Berggren, 1981), but among the hundreds of calcareous benthonic forms determined, more potentially locally-useful or widely-known index species occur. As a consequence of the ecological sensitivity of these bottom dwellers, and because of the long stratigraphic ranges, facies changes can be expected to modify stratigraphic ranges. This is known as the problem of total versus local stratigraphic range. As a result, the benthonic stratigraphic correlation framework based on exits forms the appearance of a weaving pattern with numerous small and few large scale cross-correlations. Considerable mismatch in correlation is the result of misidentification, reworking, or large differences between local stratigraphic ranges of a taxon. In addition, some correlation lines only transverse part of the combined shelves area.

The previous summary provides insight into some of the constraints on a regional foraminiferal zonation. The most important additional one is sampling methods. Only samples of cuttings obtained dominantly over 30-ft. intervals, are available generally from the wells, inferring that instead of entry, relative range, peak occurrence, and exit, only the exit of taxa is known. Furthermore, downhole contamination in cuttings hinders recognition of stratigraphically-separate benthonic or planktonic homeomorphs. Other limiting factors are that species occur frequently in low numbers and that tests are reworked normally in the younger Neogene section of the Labrador Shelf.

In summary, the data base shows the following properties, ranked according to their importance towards stratigraphic resolution:

(1) Samples are predominantly cuttings, which forces use of the highest part of stratigraphic ranges or of the highest occurrences (tops, exits), and restricts the number of stratigraphically-useful taxa.

(2) There is limited application of standard planktonic zonations, due to the mid- to high-latitude setting of the study area and the presence of locally-unfavourable facies.

(3) There are minor and major inconsistencies in relative extinction levels of benthonic taxa.

(4) Many of the samples are small which limits the detection of species represented by few specimens; this contributes to factor (3) and to the erratic, incoherent geographic distribution pattern of some taxa.

(5) There is geographic and stratigraphic provincialism in the benthonic record from the Labrador to Scotian Shelves which makes detail in a general zonation difficult.

Despite the limiting factors, it is possible to erect a zonation based on a partial data base which uses few taxa. Gradstein and Williams (1976) used four Labrador Shelf–northern Grand Banks wells to produce an 8-fold (benthonics) subdivision of the Cenozoic section. Similar stratigraphic resolution and improved zone delineation was obtained by Gradstein (unpublished) using 9 wells on the Labrador Shelf and northern Grand Banks. Some of the zones were tentative and their ages not well defined.

Increase of the Cenozoic data base through incorporation of more wells has clarified the broader correlation pattern and increased the number of chronostratigraphic calibration points based on planktonic foraminifera occurrences. It also increased noise in the stratigraphic signal (factors 3 and 4) due to more stratigraphic inconsistencies and geographic incoherences of exits. In an attempt to optimize stratigraphic resolution based on all observations that could be employed for a zonation, the 'optimum' biostratigraphy methods of this study were developed.

Statistical Models

Methods for paired comparison

A basic statistical approach consists of defining a 'population' with parameters that can be estimated from 'samples' drawn from this population. Two types of population can be defined for biostratigraphic events occurring along a relative time axis. The first type of model aims to describe the order of event occurrence and results in methods of estimating the optimum sequence of events from a set of observed sequences. In the other type of model, each event is assumed to describe a probability distribution along a relative time scale. These probability distributions can overlap with one another. From the observed amounts of overlap, it is attempted to estimate the distance between mean values of the probability distributions for the events. The second method serves to describe the clustering of events along the relative time axis.

Practical methods resulting from the preceding two statistical models will be described in the next two sections. The optimum sequence derived by means of the first model may be employed as a starting point for the calculations required for the second model.

The approach advocated in this paper builds on the work by Hay (1972) and other workers in the field of probabilistic stratigraphy including Worsley and Jorgens (1977). Hay (1972) was primarily concerned with recognizing the 'most likely sequence' of stratigraphic events. However, it can also be useful to order events along a scale with variable intervals between them. For example, suppose two microfossils have observed extinction points (A and B) in 10 sections with A occurring 5 times above B and 5 times below B. Then, the distance between A and B along a relative time scale is likely to be close to zero and it is useful to distinguish this from the situation in which one event occurs above the other event with certainty. For the clustering of events along a scale, extensive use was made of methods for paired comparison developed by mathematical statisticians including David (1963) and Kendall (1975) and by quantitative psychologists (Torgerson, 1958).

Statistical models for paired comparison are concerned not only with the problem of ranking objects according to an optimum sequence but also with the problem of locating the objects along a scale. Interesting examples of these methods are the Thurstone–Mosteller model which uses Gaussian probability distributions and the Bradley–Terry model which uses logistic probability distributions for the occurrence of events along a line. Both models have been reviewed by David (1963). In the Bradley–Terry model, each object is compared to all other objects in order to estimate the probability $P(A \rightarrow B)$ that A ranks above B in the optimum sequence. It is implied that $P(A \rightarrow B) = 1 - P(B \rightarrow A)$ denotes the probability that B occurs above A in the optimum sequence. Davidson (1970) has extended this model to consider the situation of simultaneous events (ties) and Edwards and Beaver have applied Davidson's method to stratigraphic events. Although Edwards and Beaver (1978) have pointed out that the purpose of their application is not to estimate amounts of spacing between events, their estimated probabilities provide some measure of relative closeness of events along a relative time scale.

Generally, the applicability of statistical models for paired comparison in probabilistic stratigraphy is seriously impaired by the following factors. The presence or absence of stratigraphic events within a given interval along a section is dictated by nature and stratigraphic events cannot be subjected to the statistically-designed sampling experiments frequently applied in statistical models for paired comparison. This point can be illustrated as follows. A usual procedure for paired comparison consists of asking a number of judges to rank a number of objects pairwise with respect to specific characteristics of the objects, for example their quality. It usually is feasible not only to change the number of judges but also to ask each judge to make a specific number of comparisons. In probabilistic stratigraphy, the judges are replaced by wells or sections and the objects are stratigraphic events. In a large data base such as the one shown in the appendix, many events are absent in each individual well and this has a significant influence on the frequencies and pairs of events. The application of standard statistical methods for paired comparison is impaired because there may be many missing data. A further drawback is that the total time-span of events compared to one another in problems of probabilistic stratigraphy usually is wider than the total range of characteristic features used for rating objects in experiments of paired comparison. In the Bradley–Terry model (cf. David, 1963), each object is compared with all other objects. However, in biostratigraphic applications, it may be certain that the events near the top of a section occur above those near the bottom of this section and therefore, it is not useful to compare all events with one another.

Because of the preceding reasons, the approach developed in this paper, like the method of Hay (1972), differs significantly from most other statistical models for paired comparison.

Construction of the optimum sequence

This method was originally proposed by Hay (1972) and applied in the example reproduced in Table 1. This example will be used in this section and later in order to illustrate the newly-developed method applied in this paper. There are 9 sections ($n = 9$) labelled A to I and 10 stratigraphic events ($p = 10$) labelled 1 to 10 in Table 1A. Hay determined the number of times each event occurs above or below all other events and this yields a cumulative order matrix. Simultaneous events are indicated by minus signs in Table 1A. Hay did not count simultaneous events occurring within the same section. It is preferable (see also Edwards and Beaver, 1978) to score simultaneous events within the same section as 0.5 on both sides of the diagonal indicated by x-es in Tables 1B and 1C. This gives the cumulative order matrix of Table 1B which differs from the original cumulative order matrix in Hay (1972).

When an element A_{ij} in row i and column j above the diagonal in Table 1B is greater than or equal to its corresponding element A_{ji} in the lower triangle, it remains unchanged. It is convenient to move from left to right in each successive row i. If $A_{ji} > A_{ij}$, row i is interchanged with row j and column i with column j. After this, all elements in row i are again subjected to the same test and another interchange is made whenever $A_{ji} > A_{ji}$. The procedure can be repeated until $A_{ij} \geq A_{ji}$ for all elements in the upper triangle with $i = 1, 2, \ldots, p$ and $j = i + 1, i + 2, \ldots, p$. The events are ranked according to the optimum or most-likely sequence. It should be kept in mind, however, that this method of successive iterations may not provide a solution when events in subgroups cluster along the time axis, as will be explained later.

In the computer algorithm developed for this method, successive elements ($j = i + 1, \ldots, p$) are tested for successive rows ($i = 1, 2, \ldots, p$) starting at the beginning ($j = i + 1$) of row i when there has occurred an iteration for any element within that row. Application of the method to Table 1B yields the final order relation matrix shown in Table 1C, which represents a unique solution to the problem. The corresponding optimum sequence is shown in the margin of Table 1C.

Table 1. A. 10 stratigraphic events (numbered 1–10) in 9 sections (A to I). First event in each section is written as a positive number. Negative numbers indicate events which are simultaneous to the events preceding them in the series for a section. Positive numbers indicate events which are distinct from the events preceding them (for original data, see Hay, 1972); 1B. Cumulative order matrix. For example, events 1 and 2 occur as a pair in 7 sections. They are simultaneous in sections A and G; 1 precedes 2 in F; 2 precedes 1 in C, D, E, and I. This gives $1 + 2 \times 0.5 = 2$ for the second element in the first row, and $4 + 2 \times 0.5 = 5$ for the first element in the second row; 1C. Final optimum sequence matrix where the 10 events have been ordered so that none of the elements in the lower triangle is greater than its counterpart in the upper triangle

1A. Section — Observed events

A	1	−2	−3	−4	−5	−6	7	8	9
B	2	−3	−7	−4	−5	−6	−10	9	
C	2	5	1	9					
D	2	1	7	5	8	9	10		
E	2	−5	1	3	7	8	4	6	9
F	1	−3	4	−5	2	7	−8	9	10
G	7	3	−4	1	−2	−5	10	−8	9
H	7	10	−1	−5	9	4			
I	2	3	−1	5	4	6	9	10	

1B. Events (original order)

	1	2	3	4	5	6	7	8	9	10
1	×	2	$2\frac{1}{2}$	$4\frac{1}{2}$	$4\frac{1}{2}$	$2\frac{1}{2}$	4	5	8	$4\frac{1}{2}$
2	5	×	3	3	5	3	$4\frac{1}{2}$	5	8	$4\frac{1}{2}$
3	$2\frac{1}{2}$	3	×	$4\frac{1}{2}$	4	3	$3\frac{1}{2}$	4	6	$3\frac{1}{2}$
4	$1\frac{1}{2}$	3	$1\frac{1}{2}$	×	$2\frac{1}{2}$	3	$2\frac{1}{2}$	3	6	$3\frac{1}{2}$
5	$3\frac{1}{2}$	3	2	$4\frac{1}{2}$	×	3	$3\frac{1}{2}$	5	9	5
6	$\frac{1}{2}$	1	1	1	1	×	$1\frac{1}{2}$	1	4	$1\frac{1}{2}$
7	2	$1\frac{1}{2}$	$1\frac{1}{2}$	$3\frac{1}{2}$	$3\frac{1}{2}$	$1\frac{1}{2}$	×	$4\frac{1}{2}$	7	$4\frac{1}{2}$
8	0	0	0	1	0	1	$\frac{1}{2}$	×	5	$2\frac{1}{2}$
9	0	0	0	1	0	0	0	0	×	3
10	$\frac{1}{2}$	$\frac{1}{2}$	$\frac{1}{2}$	$1\frac{1}{2}$	1	$\frac{1}{2}$	$\frac{1}{2}$	$\frac{1}{2}$	3	×

1C. Events (optimum sequence)

	2	1	3	5	7	4	6	8	9	10
2	×	5	3	5	$4\frac{1}{2}$	3	3	5	8	$4\frac{1}{2}$
1	2	×	$2\frac{1}{2}$	$4\frac{1}{2}$	4	$4\frac{1}{2}$	$2\frac{1}{2}$	5	8	$4\frac{1}{2}$
3	3	$2\frac{1}{2}$	×	4	$3\frac{1}{2}$	$4\frac{1}{2}$	3	4	6	$3\frac{1}{2}$
5	3	$3\frac{1}{2}$	2	×	$3\frac{1}{2}$	$4\frac{1}{2}$	3	5	9	5
7	$1\frac{1}{2}$	2	$1\frac{1}{2}$	$3\frac{1}{2}$	×	$3\frac{1}{2}$	$1\frac{1}{2}$	$4\frac{1}{2}$	7	$4\frac{1}{2}$
4	3	$1\frac{1}{2}$	$1\frac{1}{2}$	$2\frac{1}{2}$	$2\frac{1}{2}$	×	3	3	6	$3\frac{1}{2}$
6	1	$\frac{1}{2}$	1	1	$1\frac{1}{2}$	1	×	1	4	$1\frac{1}{2}$
8	0	0	0	0	$\frac{1}{2}$	1	1	×	5	$2\frac{1}{2}$
9	0	0	0	0	0	1	0	0	×	3
10	$\frac{1}{2}$	$\frac{1}{2}$	$\frac{1}{2}$	1	$\frac{1}{2}$	$1\frac{1}{2}$	$1\frac{1}{2}$	$1\frac{1}{2}$	3	×

Table 2. Example of cycling events (Initial matrix from Worsley and Jorgens, 1977). Unlike the example of Table 1, the algorithm for ordering does not yield an optimum sequence because the initial matrix returns after 6 iterations. Note that event *D* does not participate in the cycling (see also Fig. 1)

(a) ABCD	(b) CBAD	(c) BCAD	(d) ACBD	(e) CABD	(f) BACD	(g) ABCD
× 232	× 243	× 511	× 322	× 423	× 151	× 232
1 × 51	5 × 11	2 × 43	4 × 23	3 × 22	2 × 32	1 × 51
42 × 3	32 × 2	23 × 2	15 × 1	51 × 1	24× 3	42× 3
074 ×	470 ×	740 ×	047 ×	407 ×	704 ×	074 ×

Worsley and Jorgens (1977) have discovered that the preceding algorithm does not always yield an optimum sequence. Their example of cycling events is shown in Table 2. Application of the preceding algorithm results in cycling of the three events *A*, *B*, and *C*. This cycling would be continued indefinitely because the original matrix returns after each set of 6 consecutive iterations. The same situation is shown in Fig. 1 by a graphical method originally used by Kendall (1975). The cyclicity of the elements *A*, *B*, and *C* is immediately apparent in Fig. 1 because of the arrows in the triangle *ABC* are pointing in a consistent direction at both sides of each corner point. Kendall termed the occurrence of a set of three cycling events a 'circular triad'.

Worsley and Jorgens (1977) proposed the following method for obtaining an optimum sequence when cycling events occur during successive iterations. Hay (1972) has developed a binomial test by which the observed frequency, $F_{ij} = A_{ij}/(A_{ij} + A_{ji})$ for a pair of elements A_{ij} and A_{ji}, is compared to the theoretical frequency 0.5 corresponding to the situation that the two events are simultaneous. Each observed frequency F_{ij} can be tested for the significance of departure from randomness after choosing an appropriate level of significance. Worsley and Jorgens (1977) subjected all elements of the cumulative order matrix to this test before starting the iterative process. Pairs of elements which do not pass the binomial test were replaced by zeros. They recognized that the modified, cumulative order matrix usually produces a unique optimum sequence.

The method of Worsley and Jorgens has two drawbacks which can be serious. The binomial test does not allow for the occurrence of more than two events per comparison. Suppose that the two events *A* and *Z* occur in only three sections with *A* three times above *Z*. Because little can be said on the basis of a small sample consisting of three pairs, the relative order of *A* and *Z* remains undecided when the binomial test is applied. However, if *A* occurs above many events which, in turn, occur above *Z*, then the relative positions of *A*

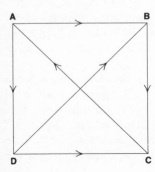

Fig. 1. Three-event cycle (ABC) in set of four events is characterized by successive arrows pointing in same direction at both sides of corner points (A, B and C). Arrow between two events indicates that one event precedes other event

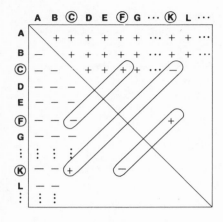

Fig. 2. Graphical illustration of algorithm developed to locate three-event cycle. Elements in successive rows of upper triangle are tested proceeding from left to right. Row and column interchanges only take place when element is less than its counterpart in lower triangle. In example, element circled in margin Ⓒ then, will be replaced by Ⓚ which, in turn, will be followed by Ⓕ. Cycle ⒸⓀⒻ will repeat indefinitely

and Z in the optimum sequence are well determined. In that situation, it is not useful to test the hypothesis that A and Z are simultaneous events because the binomial test would suggest that they were. A second drawback of the binomial test or any other test which is applied to all elements of the cumulative order matrix is that the number of cycles of events may be relatively small. In a typical situation involving a large data base, thousands of elements may have to be set equal to zero when each pair of elements is tested separately whereas the number of cycles may be less than 10. For this reason, it can be advantageous to locate the cycles in the cumulative order matrix and to set only the pair (or pairs) with the smallest difference $|A_{ij} - A_{ji}|$ in each identified cycle equal to zero. Figure 2 illustrates the computer algorithm developed for this purpose. Suppose that the event C is the first event in the original sequence which is part of a cycle, for example, along with events F and K. The elements in the first two rows (for A and B) can be tested using the previous method. However, iterations would be repeated indefinitely for the elements in the third row. The event in the margin of the third column of Fig. 2 can be scanned. For the three-event cycle of C, F, and K, it will begin showing the cycle $CKFCKF$ . . . when all elments before the column of F have passed the test that they are not less than their counterparts in the lower triangle. Once the events involved in a cycle have been identified, their smallest difference $|A_{ij} - A_{ji}|$ can be set equal to zero.

Kendall (1975) has shown that not more than three events can be involved in a cycle when ties are not permitted. In probabilistic stratigraphy, a tie corresponds to the simultaneous occurrence of two events in a section or well. In Kendall's theory, the circular triad is of primary importance. However, when ties are permitted, four-event cycles and higher-order cycles are possible as illustrated in Fig. 3. For a four-event cycle $ABCD$, it is required that the elements for the pairs (A, C) and (B, D) are equal to one another. In practical applications, this condition is fulfilled rather frequently, and four-event cycles do occur regularly although less frequently than three-event cycles. 5-event cycles are encountered occasionally. The largest number of events involved in a single cycle in the large number of experiments performed by the authors was equal to 6.

A disadvantage of the preceding algorithm is that the resulting optimum sequence is not necessarily unique, when there is more than one cycle in the system. Suppose that a pair of elements, which are set equal to zero for having the smallest difference in one cycle, also belong to another cycle. This other cycle, then, cannot be detected by the algorithm. Situations of this type may arise when the number of cycles is large. Its effect can be assessed by reversing the observed sequences of events in all sections and repeating the

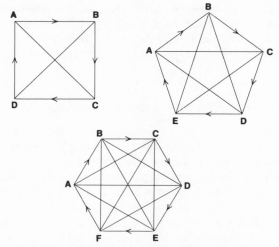

Fig. 3. Cycles of more than three events can occur when all events, except those involved in cycle, are pairwise simultaneous (frequency F_{IJ} is equal to 0.5). Pair of events that are simultaneous on average have connecting lines without arrows in examples for 4-, 5-, and 6-event cycles shown

application of the algorithm. It is also noted that during the application of the algorithm illustrated in Fig. 2, the event that is being scanned can initially show a pseudocycle which is unstable and will disappear after continued iterations. This feature has been considered in the algorithm developed.

The number of cycles increases rapidly when events are clustering strongly along the relative time scale and when the samples (numbers of pairs) are small. Apart from the problem of increased number of cycles, use of small samples causes uncertainties in the optimum sequences and subsequent results. For this reason, it is useful to require that each event occurs in at least k_c sections and each pair of events in at least m_c sections where k_c and m_c are threshold numbers to be selected for each application of the algorithm. Unless $k_c \geq m_c$ there may be events whose positions in the optimum sequence are completely undetermined.

The data base shown in the appendix was subdivided into two groups and analyzed with $k_c = 5$ (at least 5 occurrences of each event) and $m_c = 3$ (at least three pairs of events). This gave the optimum sequences shown in Fig. 8 which will be discussed later in more detail. During the iterative process used for deriving the result for 16 wells shown on the right side of Fig. 8 and 9 three-event cycles and one four-event cycle were eliminated. One three-event cycle and one four-event cycle had to be eliminated for obtaining the result shown on the left side of Fig. 8.

Because of the occurrence of zeros in the final order relation matrix, each event in an optimum sequence is shown with its range in Fig. 8. The range (i, k) for event j indicates that j occurs between events i and k $(i < j < k)$. If $i = j - 1$ and $k = j + 1$, as for most events in Fig. 8, the position of event j in an optimum sequence is fully determined by its two neighbours. Otherwise, it can occur anywhere between events i and k. If $i > j - 1$, there occur $(j - i - 1)$ zero elements above the diagonal in the jth column before a nonzero element is recognized. Likewise, if $k > j + 1$, there occur $(k - j - 1)$ zero elements to the right of the diagonal in the *j*th row. Then, the fact that an event assumes a specific position within its range in the optimum sequence is caused by its position along the margin of the cumulative order matrix. This, in turn, reflects the arbitrary rank initially assigned to the event for coding purposes. For this reason it is useful to consider the ranges of events in the optimum sequences.

Estimation of distances between events along a relative time scale

Biostratigraphic events are clustering along a relative time scale when they frequently change positions with respect to one another in the wells or sections. This feature can be quantified by calculating crossover frequencies for pairs of events.

The situation of events which interchange places with one another in different wells can be modeled by assuming that each event is described by a different probability distribution with its own expected value. The model adopted for the results described in this paper is shown in Fig. 4. The event *A* would take position x_{Ai} in section *i* where x_{Ai} is the distance to *A* from an origin with an arbitrary location along the relative time scale. The distance x_{Ai} is the realization of a random variable X_A whose probability distribution is shown in Fig. 4. Similar random variables can be defined for the other events *B*, *C*.

The random variable X_A satisfies the normal (Gaussian) probability distribution $N(EX_A, \sigma^2)$ with expected (or mean) value EX_A and variance σ^2. Mean values of events differ from one another but standard deviations of all events are equal to σ in the model of Fig. 4.

The normal populations for events *A* and *B* are also shown in Fig. 5. In practice, it is not possible to estimate the arbitrary location of the origin where $x = 0$. Neither will it be possible to estimate the standard deviation σ. Consequently, the time scale will remain relative. However, it is feasible to estimate the ratio Δ_{AB}/σ for the distance between the population means $\Delta_{AB} = EX_B - EX_A$ from the observed proportion F_{AB} by which *A* occurs above *B* in a number of sections. The matrix of crossover frequencies corresponding to the example of Table 1 is given in Table 3A. When the observed frequency F_{AB} provides an approximation of the true probability $P(X_B - X_A > 0) = P(D_{AB} > 0)$, use can be made of the formula:

$$P(D_{AB} > 0) = \frac{1}{\sqrt{2\pi}} \int_{-\Delta_{AB}/\sigma\sqrt{2}}^{\infty} e^{-z^2/2} \, dz \tag{1}$$

Equation (1) follows from the fact that the difference $D_{AB} = X_B - X_A$ has a normal

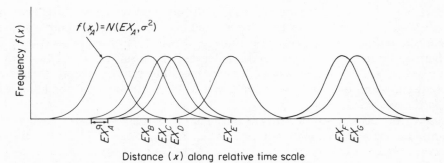

Fig. 4. Probabilistic model for clustering of biostratigraphic events (A, B, C . . .) along relative time scale (*x*-axis). Relative position of event (for example, A) in section or well is random variable (X_A) which is distributed normally around average location (EX_A) with standard deviation σ

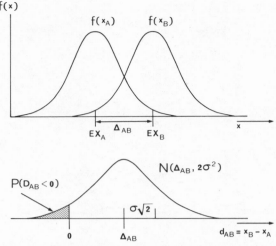

Fig. 5. Direct estimation of distance Δ_{AB} between events A and B from cross-over frequency $P(D_{AB} < 0)$. Random variable $D_{AB}(= X_B - X_A)$ is negative only when order of A and B in section is reverse of order of EX_A and EX_B. Variance of D_{AB} is twice as large as variance σ^2 of individual events A and B

distribution written as $N(\Delta_{AB}, 2\sigma^2)$ with expected value Δ_{AB} and a variance which is twice as large as the variance σ^2 of X_A and X_B. This relationship is shown graphically in Fig. 5 where $P(D_{AB} < 0)$ is equal to the shaded portion of $N(\Delta_{AB}, 2\sigma^2)$ which has unity area. It follows that $P(D_{AB} > 0) = 1 - P(D_{AB} < 0)$.

The distance between events A and B in a specific section can be written as $d_{AB} = x_B - x_A$. In practice, d_{AB} is not measured directly but use is made of the relative frequency F_{AB} for the number of times that A occurs above B in a number of sections. It is noted that F_{AB} can be less than 0.5. In this situation, setting $P(D_{AB} > 0)$ equal to F_{AB} yields negative value of $\Delta_{AB}/\sigma\sqrt{2}$ when equation (1) is applied. The advantage of using F_{AB} instead of d_{AB} is that in this manner, it is possible to counteract differences in the rate of sedimentation for different sections. Such differences can be significant, especially when the sections are distant from one another.

A precise estimate of F_{AB} is seldom available in practical applications because a large number of sections with both A and B would be required in order to achieve this. However, it is possible to estimate distances such as Δ_{AB} using indirect methods. In this manner, use can be made of the fact that the number of events can be large although samples remain small. This indirect approach is illustrated in Fig. 6. The probability distributions of the differences d_{AC} and d_{BC} are shown together with the probability distribution of their difference $d_{AB.C} = d_{AC} - d_{BC}$ which is normal with $N(\Delta_{AB}, 4\sigma^2)$.

Suppose that an event A has a variable amount of overlap with N other events B, C, D, ..., then, it is possible to obtain one direct estimate $\Delta_{AB}/\sigma\sqrt{2}$ and $(N-2)$ indirect estimates by comparing A and B with the $(N-2)$ other events. For convenience, σ, which determines the scale along the relative time axis (see Fig. 6), can be set equal to $\frac{1}{\sqrt{2}}$. It follows that Δ_{AB} can be estimated as:

$$\bar{\Delta}_{ABu} = \{z_{AB} + (z_{AC} - z_{BC}) + (z_{AD} - z_{BD}) + \ldots\}/N \tag{2}$$

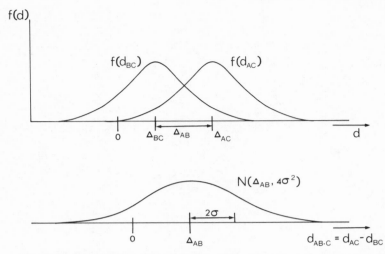

Fig. 6. Indirect estimation of distance Δ_{AB} between events A and B from cross-over frequencies with event C. Indirect distance $D_{AB\cdot C} = D_{AB} - D_{BC}$ has variance which is four times as large as variance of individual events A, B and C

Table 3.   A. Matrix of frequencies F_{IJ} corresponding to matrix with scores in Table 1C. 3B. Matrix of z-values corresponding to frequencies of Table 3A

3A	(2)	(1)	(3)	(5)	(7)	(4)	(6)	(8)	(9)	(10)
(2)	×	5/7	3/6	5/8	4½/6	3/6	3/4	5/5	8/8	4½/5
(1)		×	2½/5	4½/8	4/6	4½/6	2½/3	5/5	8/8	4½/5
(3)			×	4/6	3½/5	4½/6	3/4	4/4	6/6	3½/4
(5)				×	3½/7	4½/7	3/4	5/5	9/9	5/6
(7)					×	3½/6	1½/3	4½/5	7/7	4½/5
(4)						×	3/4	3/4	6/7	3½/5
(6)							×	1/2	4/4	1½/2
(8)								×	5/5	2½/3
(9)									×	3/6
(10)										×

3B	(2)	(1)	(3)	(5)	(7)	(4)	(6)	(8)	(9)	(10)
(2)	×	0.566	0.000	0.319	0.674	0.000	0.674	1.645	1.645	1.282
(1)		×	0.000	0.157	0.431	0.674	0.967	1.645	1.645	1.282
(3)			×	0.431	0.524	0.674	0.674	1.645	1.645	1.150
(5)				×	0.000	0.366	0.674	1.645	1.645	0.967
(7)					×	0.210	0.000	1.282	1.645	1.282
(4)						×	0.674	0.674	1.068	0.524
(6)							×	ND*	1.645	ND*
(8)								×	1.645	0.967
(9)									×	0.000
(10)										×

*ND, not determined because $m_c \geq 3$ for this example

where each z_{IJ}-value denotes the value of the cumulative standard normal distribution corresponding to the observed relative frequency F_{IJ} ($I = A$ or $B; J = B, C, D, \ldots ; J \neq I$). The z-values also can be obtained from a statistical table of probits ($z =$ probit-5). The subscript u in $\bar{\Delta}_{ABu}$ of equation (2) indicates that this estimate is unweighted contrary to the weighted estimate $\bar{\Delta}_{ABw}$ that will be derived later (see equation 3).

Use of the resulting statistical method is illustrated in Tables 3 to 5. Table 3B contains z-values corresponding to pairs of elements at both sides of the diagonal of Table 1C. For example, $F_{AB} = 5/(5 + 2) = 5/7$ which is the first element in Table 3A. From tables of the cumulative normal distribution in standard form, it is derived that the corresponding z-value is 0.566 (the corresponding probit value is 5.566). Likewise, $F_{AC} = 3/6 = 0.5$ gives $z = 0.000$. When an element in the lower triangle of Table 1C is equal to zero, its z-value is set equal to 1.645 which corresponds to $F = 19/20$ instead of $F = 1$.

Application of equation (2) to estimate the distance between events originally labelled 7 and 4 yields $1.563/9 = 0.174$ because the sum of 9 unweighted estimates is equal to $0.210 + (0.000–0.674) + (0.674–0.431) + (0.674–0.524) + (0.366–0.000) + (0.000–0.674) + (1.282–0.674) + (1.645–1.068) + (1.282–0.524) = 1.563$.

Table 4 shows estimates of all 9 distances between 10 successive events of the optimum sequence. This includes a value of 0.174 for the distance between events 7 and 4. Not all distances of Table 4 are based on a complete set of 9 values ($N = 9$), for example, distance between events 8 and 9 in Table 4 is based on 4 instead of 9 unweighted estimates, because the four zero differences resulting from the sequence of four successive pairs of 1.645 values were not used for estimation. In the computer program developed for this method, zeros for pairs of 1.645 values are used ony if they occur between nonzero differences. Also, if one or both values of a pair are not determined (*ND* in Table 3B), the corresponding distance cannot be estimated. This accounts for another missing unweighted estimate for the distance between events 8 and 9 which is equal to 0.861 as derived from $N = 9 - 4 - 1 = 4$ values.

The cumulative distance of each event from the first event (2) of the optimum sequence is also shown in Table 4. The events can be reordered on the basis of this cumulative distance giving a set of distances between successive events in the reordered optimum sequence. However, the z-values of Table 3 are not equally precise because they are derived from observed proportions F_{IJ} for small samples of variable size n_{IJ}. The effect of unequal

Table 4. Calculation of distances between successive events in optimum sequence from z-values of Table 3B (see text for example of detailed calculation). Later the events may be reordered on the basis of cumulative distance from the first event

Pair of events	Distance	Cumulative distance	Sum of (N) values
2–1	0.000	0.000	0.00(9)
1–3	−0.057	−0.056	−0.51(9)
3–5	0.214	0.157	1.92(9)
5–7	0.178	0.335	1.60(9)
7–4	0.174	0.509	1.56(9)
4–6	0.166	0.675	1.16(7)
6–8	0.812	1.487	4.87(6)
8–9	0.842	2.329	3.37(4)
9–10	−0.435	1.894	−3.48(8)

sampling size can be accounted for by replacing equation (2) by equation (3) with:

$$\overline{\Delta}_{ABw} = \frac{(w_{AB}z_{AB} + w_{AB\cdot C}(z_{AC} - z_{BC}) + w_{AB\cdot D}(z_{AD} - z_{BD}) + \dots}{(w_{AB}) + w_{AB\cdot C} + w_{AB\cdot D} + \dots} \qquad (3)$$

where weights w_{AB} and $w_{AB\cdot K}$ satisfy:

$$w_{AB} = \frac{n_{AB}\,e^{-z^2}AB}{2\pi F_{AB}(1 - F_{AB})}\,; \qquad w_{AB\cdot K} = \frac{w_{AK}\cdot w_{BK}}{w_{AK} + w_{BK}}\,; \qquad K = C, D, \dots$$

Weights of the z-values were obtained by calculating the propagation of variance of a proportion with the binomial distribution when this proportion is converted into a z-value. Equation (3) can be used to calculate the distance $\overline{\Delta}_{IJw}$ between any pair of events (I, J) by replacing A by I and B by J $(I \neq J, K \neq I, J)$.

Results of using equation (3) instead of equation (2), are shown in Table 5A. These values were reordered on the basis of their cumulative distance from the first position yielding the reordered optimum sequence of Table 5B. The distances between successive events can be plotted and clustered as in the dendrogram commonly used in cluster analysis (cf. Davis, 1973). In a dendrogram for successive interfossil distances (for example, Fig. 9),

Table 5. Calculation of distances between successive events in optimum sequence repeated after weighting each z-value according to its sample size (= number of pairs used for comparison). 5B. The events have been reordered on the basis of their cumulative distance from the first event. The new interfossil distances can be represented by a dendrogram

5A

Pair of events	Distance	Standard deviation	Cumulative distance	Sum and weight (W)
2–1	0.111	0.134	0.111	1.63(14.8)
1–3	−0.110	0.082	0.000	−1.47(13.3)
3–5	0.287	0.060	0.287	4.19(14.6)
5–7	0.148	0.079	0.435	2.32(15.7)
7–4	0.157	0.153	0.592	2.34(15.0)
4–6	0.266	0.163	0.858	2.44(9.2)
6–8	0.770	0.203	1.628	3.62(4.7)
8–9	0.861	0.306	2.488	3.59(4.2)
9–10	−0.317	0.100	2.171	−3.27(10.3)

5B

Reordered sequence	Distance from origin	Pair of events	Recalculated distance	Standard deviation
3	0.000	3–2	0.006	0.124
2	0.006	2–1	0.117	0.147
1	0.123	1–5	0.195	0.082
5	0.318	5–7	0.157	0.085
7	0.475	7–4	0.157	0.153
4	0.632	4–6	0.266	0.163
6	0.898	6–8	0.770	0.203
8	1.668	8–10	0.176	0.289
10	1.844	10–9	0.317	0.100
9	2.161			

the pair of events with the shortest distance between them are interconnected first and this procedure is repeatedly applied to sets of unconnected pairs of events until all pairs have been connected to other pairs or groups of pairs.

The occurrence of pairs of 1.645 values as previously discussed in the context of Table 4 becomes more frequent when the total time-span of all events is significantly wider than the range of the probability distributions of single events (cf. Fig. 4). As before, these pairs were not used for the distance calculation unless they are followed by one or more pairs of ordinary z-values above or to the left of the diagonal (cf. Table 3). Likewise, missing z-values result in a decrease in the number of differences $(z_{IK} - z_{JK})$ that can be used for the distance calculation of equation (3).

Finally, it is possible to calculate standard deviations of the estimated distances $\bar{\Delta}_{IJw}$. In order to explain this, it is useful to introduce the following simplified notation. Equation (3) can be rewritten as

$$\bar{x} = W^{-1} \sum_{i=1}^{N} w_i x_i$$

with

$$\bar{x} = \bar{\Delta}_{ABw}; \quad W = \sum_{i=1}^{N} w_i;$$

and

$$x_1 = z_{AB}, \qquad w_1 = w_{AB}$$
$$x_2 = z_{AC} - z_{BC}, \quad w_2 = w_{AB \cdot C}$$
$$x_3 = z_{AD} - z_{BD}, \quad w_3 = w_{AB \cdot D}$$
$$\dots \dots \dots \dots \dots \dots \dots$$

The standard deviation $s(\bar{x})$ then is the positive square root of the variance

$$s^2(\bar{x}) = W^{-1}(N-1)^{-1} \sum_{i=1}^{N} w_i (x_i - \bar{x})^2 \tag{4}$$

Standard deviations $s(\bar{x})$ are given for the distances in the second column of Table 5A. Although the distances listed in the fourth column of Table 5B are calculated from the information provided in Table 5A, it is not possible to estimate from this information standard deviations of new distance values resulting from the reordering, because the estimated distances should not be regarded as realizations of independent random variables. For example, the variance for the sum of two successive estimated distances is not equal to the sum of the variances for these distances. However, it is possible to replace Table 3 by a table in which the order of the rows and columns corresponds to that given in the first column of Table 5B. Application of equations (4) and (5) to this new table yields the recalculated distances and corresponding standard deviations shown in the last two columns of Table 5B.

In general, standard deviations provide useful information on relative positions of events along the relative time scale. For example, the standard deviation of an estimated distance can be used for comparison with zero reflecting the null hypothesis that the two events connected to each other can be regarded as simultaneous on the average. In order to be significantly different from zero, an estimated distance should be at least twice as large as its standard deviation when the level of significance is set equal to $\alpha = 0.05$. In Table 5A, 5 of the 9 estimated distances differ significantly from zero when this test is applied. This confirms that it is meaningful to separate these events along the relative time scale.

Finally, the model presented in this section permits an evaluation of the relative positions of separate events in the original sections. For example, it is noted that event 10 occurs near the beginning (stratigraphically-lowest point) of section H (see Table 1), whereas it occurs at or near the end of the sequences B, D, F, and I. In section H, event 10 is simultaneous to events 1 and 5. According to Table 5B, its distance from event 1 should be equal to $(1.844 - 0.123 =)$ 1.721, and its distance from 5 equal to $(1.844 - 0.318 =)$ 1.526. Approximately, the sum of these two values $(1.721 + 1.526 =)$ 3.247 would be a realization of a random variable with $N(0, 4\sigma^2)$. Because $\sigma^2 = \frac{1}{2}$ (see previous discussion), it follows that the value 3.247 for event 10 in section H can be directly compared to a normal distribution in standard form. Setting $\alpha = 0.05$, this gives a significance limit equal to 2.326 which is less than 3.247, indicating that event 10 occurs significantly lower in section H than in the other sections. Similar normality tests can be devised for all events in all sections. Application of these tests has shown that the model of Fig. 4 is satisfied probably for all observations of Table 1 with the exception of event 10 in section H.

The newly-developed method of clustering biostratigraphic events along a relative time axis described in this paper shows similarity to cluster analysis as practiced by biometricians for the classification of taxa. Both techniques are based on the principle of calculating distances between points representing observations. These distances may be subjected to simple transformations which are kept the same for all comparisons made. Although this type of statistical treatment of the data is simple admittedly its purpose is to provide the scientist with a tool for performing objective classifications. Cluster analysis helps the biometrician to classify n objects in p-dimensional space on the basis of p characteristics. The scaling technique used in this paper is simpler in that it clusters n events along an axis which is equivalent to a one-dimensional space. This simplicity has the advantage that it allows the application of statistical significance tests to check the validity of the hypothesis of employing the same normal distribution for locating different events along an axis (Fig. 4).

The curves shown in Fig. 4 should not be confused with frequency curves for the changing abundance of a taxon in the course of time. They represent the position along an axis at which the highest occurrence of a taxon is observed. This position generally occurs below the true time of exit of the taxon considered and its range generally is less than the total time span of the taxon.

The assumption of normality or near-normality for locations along an axis has been made previously in the Thurstone–Mosteller and Bradley–Terry models for paired comparison (cf. David, 1963). In reality, the frequency curves of a biostratigraphic event may not be normal but skewed either in the stratigraphically-upward or downward direction. The central-limit theorem of statistical theory implies that the addition (or subtraction) of nonnormal random variables yields new random variables which in shape are closer to the normal (Gaussian) distribution than the original, nonnormal random variables. Thus, the curves used for distance calculations (Figs. 5 and 6) are closer to normality than the possibly nonnormal frequency curves for the occurrence of the events.

It can be argued, however, that the assumption of equal variance for all events is not realistic in some applications. A possible remedy in a situation of that type would be (1) to identify the events which, in one or more wells or sections, occur in positions that are either too high or too low according to the normality assumption, and (2) to repeat the statistical analysis deleting these events from the wells or sections in which they occupy anomalous positions.

The Stratigraphical Meaning of the Models

In order to appreciate the stratigraphical and paleontological meaning of the optimum sequences and the optimum clusters, we will discuss briefly their characteristics.

The optimum sequence shows the most likely order of fossil events. An event is defined as the stratigraphically-highest occurrence of a taxon. It is required that the events occur in at least k_c sections and each pair of events in at least m_c sections ($k_c \geq m_c$). The relative position of the events in the most likely sequence is an 'average' of all the relative positions encountered. By combining the results based on lowest and highest stratigraphic occurrences, charts of optimum stratigraphic ranges can be constructed.

As discussed in the next sections, the method provides information on regional markers. A comparison of the optimum or most likely sequence in the region to the sequence of events in the individual sections may provide insight into reworking, misidentifications, and other sources of mismatch. Rigorous solutions based on high k_c and high m_c values serve in point correlations. This is promising and simple for geological studies which require local, detailed well-to-well correlations.

Optimum clusters based on interfossil distances in time resemble assemblage zones; the clusters employ the same events as used in the optimum sequence. A simple example will serve to clarify the basis for the clusters. We consider the extreme situation of all fossil events occurring in the same order in all sections examined. There are no crossovers of events. Here, the optimum, most likely sequence of events is fully determined. Since the optimum clustering is a function of crossover frequency of events no distances between events can be determined and there will be no clustering.

The real situation is that some events show a more regular, consistent order than others in the sections studied. The more adjacent events crossover, the better they cluster in the optimum clustering diagram; vice versa, the tighter events cluster in this diagram, the less the exact order is determined. Large breaks in the clustering sequence agree with stratigraphic horizons separating assemblages of events, which is exactly as accomplished in a subjective zonation.

There is one virtue of zonation based on optimum clusters over conventional assemblage zonations. A regular, subjective zonation frequently makes no distinction between relatively abundant and relatively rare species. It will give weight to index forms, however rare, whose stratigraphic value is understood. Such a procedure will lead to an idealistic zonation which reflects the best stratigraphic resolution to be obtained in the area studied. The zonation based on optimum clustering uses only events frequent enough to occur in the most likely sequence. As a result, the objective zonation scores high in practical value.

The Data

As outlined in the previous section, the basis for this Cenozoic stratigraphy consists of the stratigraphically-highest occurrence of benthonic and planktonic foraminiferal taxa in 22 exploratory wells between latitude 42° and 60°N on the Scotian shelf, Grand Banks, and Labrador Shelf, eastern Canada. The main concern in well selection has been a broad regional coverage. The wells are numbered from north to south (Fig. 7). The footages in parentheses refer to the Cenozoic intervals studied; the measurements are below rotary table.

The foraminiferal record was artificially truncated at the Cretaceous–Tertiary boundary as determined from the tops of Cretaceous taxa. The next step was to prepare a dictionary of species recorded in the wells which are of potential biostratigraphical significance. The present dictionary (Appendices 1–3) consists of 206 foraminifera and few miscellaneous

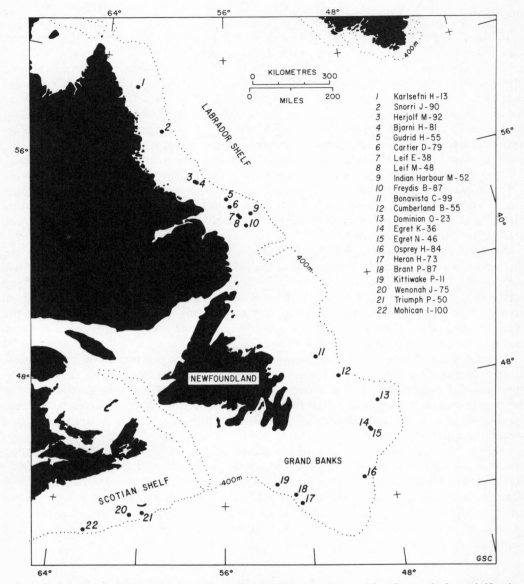

Fig. 7. Location of 22 wells along Eastern Canadian margin used for Cenozoic foraminiferal
stratigraphy

microfossil taxa (including a megaspore, a pteropod, a scaphopod, and '*Coscinodiscus*'
diatom species), numbered from 1 to 257.

Subsequently, the record was coded using the dictionary and three simple rules:

(1) a local identification with the qualifier aff. (affinis) is the same as the nominate taxon,
(2) a local identification with the qualifier cf. (confer) for a regular dictionary taxon has
    not been taken into account,

Labrador Shelf	Eastcan and others	Karlsefni H-13	(1760–12 990')
		Snorri J-90	(1260–9950')
		Herjolf M-92	(3030–7800')
		Bjarni H-81	(2760–6060')
		Gudrid H-55	(1660–8580')
		Cartier D-79	(1950–6070')
	Tenneco and others	Leif E-38	(1210–3557')
	Eastcan and others	Leif M-48	(1300–5620')
	BP Columbia and others	Indian Harbour M-52	(1740–10 480')
	Eastcan and others	Freydis B-87	(1000–5260')
Northern Grand Banks	BP Columbia and others	Bonavista C-99	(1860–11 940')
	Mobil Gulf	Cumberland B-55	(920–11 830')
		Dominion O-23	(1380–10 260')
	Amoco Imp Skelly	Egret N-64	(1060–2070')
		Egret K-36	(860–2270')
		Osprey H-84	(1190–2660')
Southern Grand Banks	Amoco Imp	Heron H-73	(970–5800')
	Amoco Imp Skelly	Brant P-8	(1050–6270')
	Amoco Imp	Kittiwake P-11	(970–5560')
Scotian Shelf	Petrocanada Shell	Wenonah J-75	(1000–4750')
	Shell	Triumph P-50	(990–5490')
		Mohican I-100	(1276–5320')

Fig. 7—*continued*

(3) obviously reworked highest occurrences of taxa have not been considered. Reworking is apparent from anomalous, poor preservation of tests relative to the rest of the assemblage and from highly-erratic stratigraphic position.

The complete, coded data file is listed in Appendix 4.

Taxonomic Concepts

The planktonic species determinations are broadly compatible with the systematics presented by Stainforth and others (1975), Tjalsma (1976), Zachariasse (1975), and Berggren (1977); use was also made of Subbotina's (1953) monograph for comparison of midlatitude acarininids morphology. The generic plankton nomenclature cautiously follows evolutionary classification schemes as eloquently reviewed by Steineck and Fleisher (1978). In this respect, it is clear that the present status of the genera *Subbotina* (Paleogene) and *Globigerina* and *Globorotalia* (Oligocene–Neogene) remains unsatisfactory; use of these three genera in the next sections is overly comprehensive, akin to the concepts in the spirit of the anatomical, inventory classification approach.

The taxonomy for benthonics, far less stable in the literature than the planktonics one, is based on many sources, including for the Paleogene, Aubert and Berggren (1976), Batjes (1958), Bettenstaedt and others (1962), Cushman (1951), Cushman and Renz (1946), Gradstein and Berggren (1981) for *Rzehakina, Glomospira, Spiroplectammina* and *Cyclammina* and some other agglutinants, Hansen (1972) for *Turrilina,* Hillebrand (1962), Kaasschieter (1961), Lamb (1964) for some *Uvigerina* spp., and Tjalsma (in prep.). Neogene–Recent taxonomic nomenclature has been compared to concepts in Batjes (1958) for *Asterigerina*, Barker (1960), Bettenstaedt and others (1962), Cushman (1918), Feyling-

Hansen and others (1971), Indans (1962), and Sen Gupta (1971). Use was also made of the Catalogue of Index Smaller Foraminifera (Ellis and others, 1969). Amendations to the taxonomy made after completion of the computer runs are listed with the dictionary in Appendix 3.

Optimum Cenozoic Sequences

Optimum sequences for the Cenozoic foraminiferal events in the 22 selected wells on the Labrador Shelf–Grand Banks–Scotian Shelf were constructed using different combinations of wells and different minimum conditions for the occurrences of the events. After a number of trials, solutions of two geographically-separate groups of wells were chosen: Group 1 consists of 16 wells on the Labrador Shelf and northern Grand Banks, and Group 2 consists of 6 wells on the Scotian Shelf and southern Grand Banks. These solutions best reflect provincialism in the record. The wells are listed previously. From the total of 206 events, 150 occur in Group 1 and 157 in Group 2. The cumulative distribution of the number of events present in 1, 2, 3, or more wells of Groups 1 and 2 is as follows:

Group 1

Number of wells	1	2	3	4	5	6	7	8	9	10	11	12	13	14	15	16
Number of events	150	94	68	55	41	30	26	21	19	17	14	10	5	3	2	0

Group 2

Number of wells	1	2	3	4	5	6
Number of events	157	101	60	31	16	6

The cumulative distributions provide a simple guide to select the referred requirements for the optimum sequences. The distributions indicate that much information is lost when it is required that the taxa of the optimum sequences must occur in most of the wells. The data base is not homogeneous enough to set such high standards.

The preferred requirements for the optimum sequence of Group 1 are that each event occurs in at least 5 wells and each pair of events in at least three wells (further on abbreviated as $k_c \geq 5/m_c \geq 3$). This solution retains 41 (including 6 planktonic ones) out of a total of 150 events. The $k_c \geq 3/m_c \geq 3$ solutions which retain 68, including 15 planktonic, events is useful where information is desirable on the probable stratigraphic position of rarer taxa.

The preferred requirements for the optimum sequences of the Group 2 wells are $k_c \geq 3/m_c \geq 3$, that is both each exit and each pair of exits occur in at least three out of a total of 6 wells. This solution retains 60 out of a total of 157 events.

Figures 8 and 10 show the preferred optimum sequences of events. Only 14 taxa are common to Group 1 and 2 wells; the southern one has about 5 times more planktonics (32 in Group 1 versus 6 in Group 2). In the southern sequence, the planktonics are uniformly distributed throughout the stratigraphical column; in the northern one, planktonic taxa occur largely in the stratigraphically-lower part.

The benthonics also show provincialism with *Asterigerina gurichi, Turrilina alsatica, Uvigerina batjesi, Cibicidoides alleni,* various agglutinants like *Cyclammina amplectens, Spiroplectammina spectabilis,* and *Glomospira corona,* and the megaspore sp. 1 and pteropod sp. 1 being less abundant or absent in the southern region. Personal observation in more wells indicates that the latter fact is not influenced by the lower number of wells in the southern sequence.

It is important to realize that the statistical range of an event, which refers to the (stratig-raphic) interval in which the position of an exit is not determined, is generally narrow in the optimum sequences. Most events are determined within one or two immediately-adjacent positions (see columns under Range in Fig. 8). Exceptions are at the top of each sequence where several events are undetermined between positions 0 and 5. This is the result of sample truncation (no stratigraphically-higher samples available), reworking and low number of specimens. Another exception are several events (positions 37, 38, 39, 40, 41, 42, 44, 45, 46, 47, 48, 53, and 54) in the lower part of the southern sequence; this exception may disappear when more wells are used.

A correlation of the events in common between both groups of wells (Fig. 8) shows considerable crossover of lines (mismatch) for the species *Alabamina wolterstorffi* and *Ceratobulimina contraria*. Both forms only occur in three out of the 6 wells of the southern region; we suspect taxonomic problems to underlie the mismatch of relative exits of *Alabamina*.

The optimum exit sequences constructed serve as standards to which individual, new well records in each region can be compared. The sequences provide useful age calibration for the likely (local) exits of many benthonic markers. The solutions can be used for point correlation of exits from well to well. The 40 to 60 rungs on the stratigraphic ladder, each rung corresponding to an exit in the optimum sequence, although not rigidly fixed in time, provide measure of resolution not previously available in the region.

The Optimum Cenozoic Clusters

Group 1 wells–Labrador Shelf, northern Grand Banks

The optimum clustering in Group 1 (16 wells) on the Labrador shelf, northern Grand Banks, based on the optimum sequence with $k_c \geq 5/m_c \geq 3$, consists of 41 taxa in 9 clearly-separated clusters (Fig. 9). The optimum clustering uses the same taxa as shown in the optimum sequence of Fig. 8, but the relative order of taxa shows minor changes. These are a function of the fact that the clusters express both interfossil distances (derived from the crossover frequencies of taxa exits) and the optimum rank in the stratigraphical sequence. Each interfossil distance is weighted also (ranked) according to the number of pairs of events (m_c) from which it was calculated, $(m_c \geq 3)$. The clusters resemble biostratigraphical assemblages (assemblage zones) in which the order of taxa in the clusters expresses the arrow of time; the upper events in a cluster are younger than the lower ones.

The preferred 9 assemblages of Fig. 9 show low interfossil distances except in parts of the upper (Neogene) sequence where sample truncation, lower fossil content, and reworking account for much 'noise.' Analysis of the clusters based on the presence of index taxa with known stratigraphical value shows that each cluster can be given a progressively younger age. A shading pattern was used to enhance the stratigraphically most useful part of the individual clusters.

The $k_c \geq 3/m_c \geq 3$ solution based on the 16 wells of Group 1 (Fig. 10) retains 68 taxon exits in distinct and progressively younger clusters. As in Fig. 9 a shading pattern was used to enhance the stratigraphically most useful part of the individual clusters. Again, 9 preferred assemblages are shown resembling those of the $k_c \geq 5/m_c \geq 3$ solution but incorporating less frequent events.

The basal cluster contains *Gavelinella beccariiformis* which tops in P5 (Tjalsma, in prep.; Aubert and Berggren, 1976) of Late Paleocene age. This cluster consists only of two taxa because Paleocene beds are relatively unfossiliferous and not always present.

In the next two clusters, planktonic taxa occur—*Acarinina soldadoensis* and *Subbotina patagonica*—accompanied in the $k_c \geq 3/m_c \geq 3$ solution by (in ascending order)

SEQUENCE POSITION	FOSSIL NUMBER	RANGE	FOSSIL NAME
1	1	0 - 5	NEOGLOBOQUADRINA PACHYDERMA*
2	5	0 - 5	GLOBOROTALIA CRASSAFORMIS*
3	77	0 - 5	ELPHIDIUM SP
4	109	0 - 5	CASSIDULINA CURVATA
5	9	4 - 6	FURSENKOINA GRACILIS
6	11	5 - 7	NONIONELLA PIZARRENSE
7	112	6 - 8	MARGINULINA BACHEI
8	14	7 - 9	TEXTULARIA AGGLUTINANS
9	78	8 - 10	UVIGERINA PEREGRINA
10	8	9 - 11	ORBULINA UNIVERSA*
11	113	10 - 12	GLOBOROTALIA MENARDII GROUP*
12	115	11 - 13	GLOBOROTALIA OBESA*
13	18	12 - 14	SPIROPLECTAMMINA CARINATA
14	219	13 - 15	MARTINOTIELLA COMMUNIS
15	117	14 - 16	SPHAEROIDINA BULLOIDES
16	120	15 - 17	GLOBOROTALIA SIAKENSIS*
17	71	16 - 18	EPISTOMINA ELEGANS
18	67	17 - 19	SCAPHOPOD SP1
19	15	18 - 20	GLOBIGERINA PRAEBULLOIDES*
20	128	19 - 21	GLOBOROTALIA PRAEMENARDII*
21	122	20 - 22	SPHAEROIDINELLOPSIS SEMINULINA*
22	124	21 - 23	GLOBOQUADRINA DEHISCENS*
23	123	22 - 24	GLOBIGERINOIDES TRILOBUS*
24	222	23 - 25	GLOBOQUADRINA ALTISPIRA*
25	242	24 - 26	GLOBOROTALIA FOHSI GROUP*
26	127	25 - 28	GLOBIGERINITA NAPARIMAENSIS*
27	217	25 - 28	GLOBOROTALIA SCITULA*
28	20	27 - 29	GYROIDINA GIRARDANA
29	125	28 - 31	"GLOBOROTALIA" CONTINUOSA*
30	231	28 - 31	UVIGERINA RUSTICA
31	220	30 - 32	CIBICIDOIDES WUELLERSTORFFI
32	21	31 - 33	GUTTULINA PROBLEMA
33	26	32 - 34	UVIGERINA DUMBLEI
34	16	33 - 36	CERATOBULIMINA CONTRARIA
35	25	32 - 36	COARSE ARENACEOUS SPP.
36	139	35 - 37	"GLOBOROTALIA" OPIMA OPIMA*
37	29	36 - 41	CYCLAMMINA AMPLECTENS
38	223	32 - 42	GLOBIGERINA CIPEROENSIS*
39	142	32 - 49	GYROIDINA SOLDANII MAMILLIGERA
40	143	36 - 43	UVIGERINA GALLOWAY
41	148	37 - 44	GLOBIGERINATHEKA INDEX*
42	81	38 - 43	GLOBIGERINA VENEZUELANA*
43	147	42 - 46	CATAPSYDRAX AFF. DISSIMILIS*
44	82	41 - 47	SUBBOTINA LINAPERTA*
45	70	42 - 48	ALABAMINA WOLTERSTORFFI
46	79	43 - 49	GLOBIGERINA TRIPARTITA*
47	33	44 - 49	TURBOROTALIA POMEROLI*
48	190	45 - 49	ANOMALINOIDES ACUTA
49	90	48 - 50	ACARININA DENSA*
50	37	49 - 51	ACARININA AFF PENTACAMERATA*
51	93	50 - 52	ACARININA AFF. BROEDERMANNI*
52	154	51 - 55	ANOMALINOIDES MIDWAYENSIS
53	57	43 - 60	SPIROPLECTAMMINA SPECTABILIS
54	52	43 - 57	ACARININA SOLDADOENSIS*
55	157	52 - 56	TRITAXIA SP3
56	158	55 - 58	SUBBOTINA INAEQUISPIRA*
57	96	54 - 60	ACARININA INTERMEDIA WILCOXENSIS*
58	92	56 - 59	MOROZOVELLA CAUCASICA*
59	159	58 - 60	MOROZOVELLA ARAGONENSIS*
60	166	59 - 61	MOROZOVELLA SUBBOTINAE*

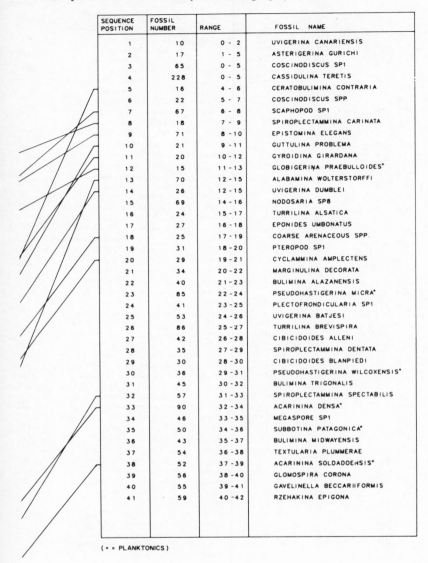

SEQUENCE POSITION	FOSSIL NUMBER	RANGE	FOSSIL NAME
1	10	0 - 2	UVIGERINA CANARIENSIS
2	17	1 - 5	ASTERIGERINA GURICHI
3	65	0 - 5	COSCINODISCUS SP1
4	228	0 - 5	CASSIDULINA TERETIS
5	16	4 - 6	CERATOBULIMINA CONTRARIA
6	22	5 - 7	COSCINODISCUS SPP
7	67	6 - 8	SCAPHOPOD SP1
8	18	7 - 9	SPIROPLECTAMMINA CARINATA
9	71	8 - 10	EPISTOMINA ELEGANS
10	21	9 - 11	GUTTULINA PROBLEMA
11	20	10 - 12	GYROIDINA GIRARDANA
12	15	11 - 13	GLOBIGERINA PRAEBULLOIDES*
13	70	12 - 15	ALABAMINA WOLTERSTORFFI
14	26	12 - 15	UVIGERINA DUMBLEI
15	69	14 - 16	NODOSARIA SP8
16	24	15 - 17	TURRILINA ALSATICA
17	27	16 - 18	EPONIDES UMBONATUS
18	25	17 - 19	COARSE ARENACEOUS SPP.
19	31	18 - 20	PTEROPOD SP1
20	29	19 - 21	CYCLAMMINA AMPLECTENS
21	34	20 - 22	MARGINULINA DECORATA
22	40	21 - 23	BULIMINA ALAZANENSIS
23	85	22 - 24	PSEUDOHASTIGERINA MICRA*
24	41	23 - 25	PLECTOFRONDICULARIA SP1
25	53	24 - 26	UVIGERINA BATJESI
26	86	25 - 27	TURRILINA BREVISPIRA
27	42	26 - 28	CIBICIDOIDES ALLENI
28	35	27 - 29	SPIROPLECTAMMINA DENTATA
29	30	28 - 30	CIBICIDOIDES BLANPIEDI
30	36	29 - 31	PSEUDOHASTIGERINA WILCOXENSIS*
31	45	30 - 32	BULIMINA TRIGONALIS
32	57	31 - 33	SPIROPLECTAMMINA SPECTABILIS
33	90	32 - 34	ACARININA DENSA*
34	46	33 - 35	MEGASPORE SP1
35	50	34 - 36	SUBBOTINA PATAGONICA*
36	43	35 - 37	BULIMINA MIDWAYENSIS
37	54	36 - 38	TEXTULARIA PLUMMERAE
38	52	37 - 39	ACARININA SOLDADOENSIS*
39	56	38 - 40	GLOMOSPIRA CORONA
40	55	39 - 41	GAVELINELLA BECCARIIFORMIS
41	59	40 - 42	RZEHAKINA EPIGONA

(* = PLANKTONICS)

Fig. 8. Correlation of stratigraphically-optimum sequences for 16 wells of Group 1 (right), Labrador Shelf-northern Grand Banks and 6 wells of Group 2 (left), Scotian Shelf-southern Grand Banks. Relative position of events in most likely sequence is average of relative positions encountered. Requirements for Group 1 sequence are that each event (exit) occurs in at least 5 wells and each pair of events in at least three wells ($k_c \geq 5/m_c \geq 3$). For Group 2 solution it was required that each event and each pair of events occur in three or more wells ($k_c \geq 3/m_c \geq 3$). Fossil number is Dictionary number shown in Appendix; Range refers to (stratigraphic) interval in which position of exit is not determined due to insufficient information. 14 taxa are common to two optimum sequences; southern one has about 5 times more planktonics. For further explanations see text

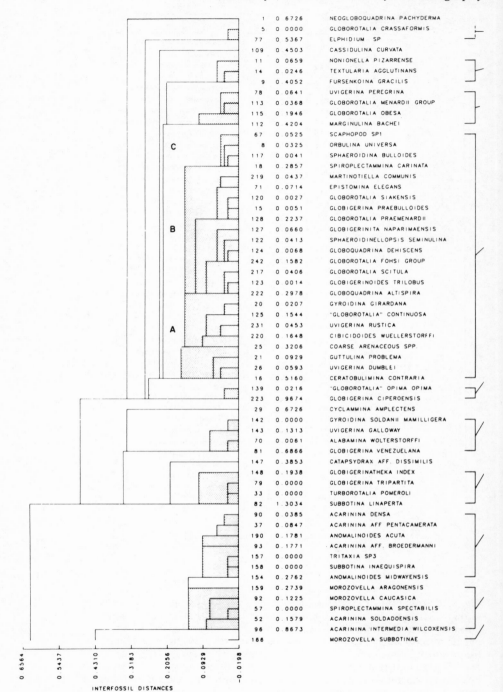

1	0 6726	NEOGLOBOQUADRINA PACHYDERMA
5	0 0000	GLOBOROTALIA CRASSAFORMIS
77	0 5367	ELPHIDIUM SP
109	0 4503	CASSIDULINA CURVATA
11	0 0659	NONIONELLA PIZARRENSE
14	0 0246	TEXTULARIA AGGLUTINANS
9	0 4052	FURSENKOINA GRACILIS
78	0 0641	UVIGERINA PEREGRINA
113	0 0368	GLOBOROTALIA MENARDII GROUP
115	0 1946	GLOBOROTALIA OBESA
112	0 4204	MARGINULINA BACHEI
67	0 0525	SCAPHOPOD SP1
8	0 0325	ORBULINA UNIVERSA
117	0 0041	SPHAEROIDINA BULLOIDES
18	0 2857	SPIROPLECTAMMINA CARINATA
219	0 0437	MARTINOTIELLA COMMUNIS
71	0 0714	EPISTOMINA ELEGANS
120	0 0027	GLOBOROTALIA SIAKENSIS
15	0 0051	GLOBIGERINA PRAEBULLOIDES
128	0 2237	GLOBOROTALIA PRAEMENARDII
127	0 0660	GLOBIGERINITA NAPARIMAENSIS
122	0 0413	SPHAEROIDINELLOPSIS SEMINULINA
124	0 0068	GLOBOQUADRINA DEHISCENS
242	0 1582	GLOBOROTALIA FOHSI GROUP
217	0 0406	GLOBOROTALIA SCITULA
123	0 0014	GLOBIGERINOIDES TRILOBUS
222	0 2978	GLOBOQUADRINA ALTISPIRA
20	0 0207	GYROIDINA GIRARDANA
125	0 1544	"GLOBOROTALIA" CONTINUOSA
231	0 0453	UVIGERINA RUSTICA
220	0 1648	CIBICIDOIDES WUELLERSTORFFI
25	0 3206	COARSE ARENACEOUS SPP.
21	0 0929	GUTTULINA PROBLEMA
26	0 0593	UVIGERINA DUMBLEI
16	0 5160	CERATOBULIMINA CONTRARIA
139	0 0216	"GLOBOROTALIA" OPIMA OPIMA
223	0 9674	GLOBIGERINA CIPEROENSIS
29	0 6726	CYCLAMMINA AMPLECTENS
142	0 0000	GYROIDINA SOLDANII MAMILLIGERA
143	0 1313	UVIGERINA GALLOWAY
70	0 0061	ALABAMINA WOLTERSTORFFI
81	0 6866	GLOBIGERINA VENEZUELANA
147	0 3853	CATAPSYDRAX AFF. DISSIMILIS
148	0 1938	GLOBIGERINATHEKA INDEX
79	0 0000	GLOBIGERINA TRIPARTITA
33	0 0000	TURBOROTALIA POMEROLI
82	1 3034	SUBBOTINA LINAPERTA
90	0 0385	ACARININA DENSA
37	0 0847	ACARININA AFF PENTACAMERATA
190	0 1781	ANOMALINOIDES ACUTA
93	0 1771	ACARININA AFF. BROEDERMANNI
157	0 0000	TRITAXIA SP3
158	0 0000	SUBBOTINA INAEQUISPIRA
154	0 2762	ANOMALINOIDES MIDWAYENSIS
159	0 2739	MOROZOVELLA ARAGONENSIS
92	0 1225	MOROZOVELLA CAUCASICA
57	0 0000	SPIROPLECTAMMINA SPECTABILIS
52	0 1579	ACARININA SOLDADOENSIS
96	0 8673	ACARININA INTERMEDIA WILCOXENSIS
166		MOROZOVELLA SUBBOTINAE

0 6564   0 5437   0 4310   0 3183   0 2056   0 0929   -0 0198

INTERFOSSIL DISTANCES

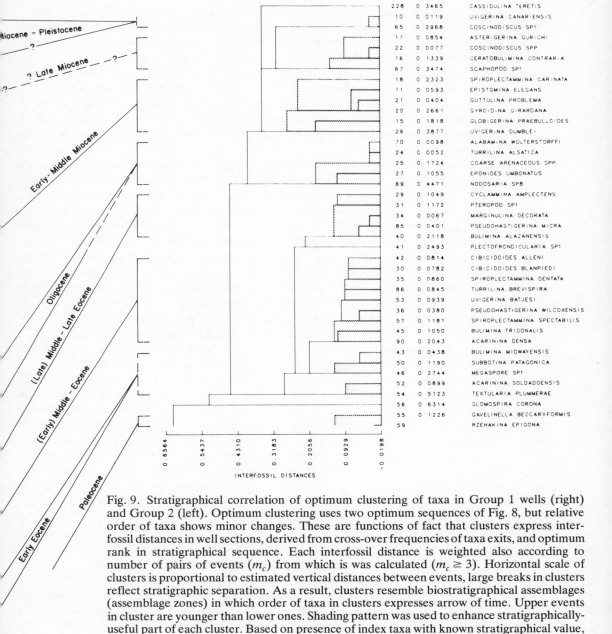

Fig. 9. Stratigraphical correlation of optimum clustering of taxa in Group 1 wells (right) and Group 2 (left). Optimum clustering uses two optimum sequences of Fig. 8, but relative order of taxa shows minor changes. These are functions of fact that clusters express interfossil distances in well sections, derived from cross-over frequencies of taxa exits, and optimum rank in stratigraphical sequence. Each interfossil distance is weighted also according to number of pairs of events (m_c) from which is was calculated ($m_c \geq 3$). Horizontal scale of clusters is proportional to estimated vertical distances between events, large breaks in clusters reflect stratigraphic separation. As a result, clusters resemble biostratigraphical assemblages (assemblage zones) in which order of taxa in clusters expresses arrow of time. Upper events in cluster are younger than lower ones. Shading pattern was used to enhance stratigraphically-useful part of each cluster. Based on presence of index taxa with known stratigraphical value, each cluster can be given a progressively younger age, and can be correlated

Morozovella aragonensis, Planorotalites planoconicus, Acarinina aff. *broedermanni* and *A. intermedia-wilcoxensis*. The age is Early Eocene. The (frequent) occurrence of *Subbotina patagonica*, as observed in several of the wells, correlates to the Rosnaes Clay of Denmark, which is a lithostratigraphical equivalent of the London Clay and the Ypresian in the type-area (= *Morozovella formosa–M. aragonensis* Zones; Hardenbol and Berggren 1978).

The upward following large and tight cluster includes the planktonics *Acarinina densa* and *Pseudohastigerina wilcoxensis* known to disappear in the Middle Eocene and *Uvigerina batjesi, Turrilina brevispira, Spiroplectammina spectabilis, Cibicidoides alleni*, and *C. blanpiedi*. These benthonics are either restricted to, or known to disappear in, the Eocene (Kaasschieter, 1961; Hansen, 1972; Hiltermann, 1972; Berggren and Aubert, 1975; Gradstein and Williams, 1976). In the $k_c \geq 5/m_c \geq 3$, the next higher cluster is separated from the immediately underlying one; in the $k_c \geq 3/m_c \geq 3$ solution, the exact level of separation is arbitrary. The presence of the exits of *Pseudohastigerina micra* and *Cyclammina amplectens* and less frequently of *Turborotalia pomeroli* correlates to (Late) Middle to Late Eocene beds.

The cluster with *Turrilina alsatica* is of Oligocene age. According to Batjes (1958), this taxon occurs in the Boom Clay of Belgium, and is restricted to the 'Middle' Oligocene; Hansen (1972) reports that it is abundant in Middle and Upper Oligocene clays of Denmark. Tjalsma (in prep.) reports *Turrilina*, resembling *T. alsatica* in all Oligocene planktonic standard zones (P18–P22) in few DSDP sites but curiously enough in our wells *T. alsatica* was never found with standard Oligocene planktonics (see Fig. 9, Group 2-Oligocene). The Labrador well record shows a coincidence of shallowing, a local Oligocene hiatus and of coarser clastics influx in the Eocene to Miocene transitional interval, events probably related to a large scale Oligocene sea-level change. Zonal resolution is insufficient to determine if part of Oligocene beds are missing.

The next higher cluster with *Globigerina praebulloides, Gyroidina girardana*, and *Spiroplectammina carinata* is likely of Early-Middle Miocene age (Bettenstaedt and others, 1962; Gradstein and Williams, 1976) in agreement with the occurrence of these events in the Group 2 well cluster with Early-Middle Miocene planktonics, discussed later.

The immediately higher cluster with *Ceratobulimina contraria* and *Asterigerina gurichi* has also a Miocene connotation and, by virtue of local superposition, may be of Middle or Late Miocene age. Unfortunately, there is no firm tie in our wells of Miocene planktonics (as in Group 2 wells) and the occurrence of *Asterigerina* (exclusively present in Group 1 wells). The occurrence of *A. gurichi* at the base of the Pliocene/Pleistocene cluster of Fig. 10 and not in the one below is a function of (a) subjective separation of the clusters, which express the arrow of time, in 'disjunct' chronostratigraphic unit (le question d'accolade), (b) insufficient information on clustering in the stratigraphically-youngest part of the section and reworkings of taxa. It would have been more appropriate to label (provisionally) the youngest cluster Late Miocene to Quaternary.

An interesting question is how much Upper Miocene strata are present in the wells; few wells (in Group 2) have Late Miocene planktonics (Figs. 11 and 12) which may be evidence of a widespread hiatus, the result of erosion due to drastic sea-level lowering.

The cluster at the top of the sequence contains taxa known from local recent deposits, including *Cassidulina teretis*, used as a post-Miocene regional marker (Gradstein and Williams, 1976). This agrees with the finding in some of the wells of Pliocene planktonics (Figs. 11 and 12).

Group 2 wells–Scotian shelf, southern Grand Banks

The optimum clustering in Group 2, consisting of 6 wells on the Scotian Shelf–southern Grand Banks, is based on the exits of 60 taxa ($k_c \geq 3/m_c \geq 3$, Fig. 9). Condensation in the lower part of the Paleogene record and the low number of wells produce less detailed clustering than in Group 1. The fossiliferous record with few cross-overs of exits in Neogene strata of Group 2 wells accounts for the expanded Miocene cluster.

The optimum clusters of Group 2 can be ranked according to progressively-younger age from Early Eocene to Pliocene-Pleistocene. One only has to match the planktonic record as present in most of the clusters to the standard zonal scheme of Stainforth and others (1975).

The basal listing of *Morozovella subbotinae* corresponds to the *Morozovella subbotinae* and *M. formosa formosa* Zones of early Eocene age. In the next higher cluster occur *Acarinina soldadoensis*, *Morozovella caucasica*, and *M. aragonensis* which equate with the *Acarinina aragonensis* and *A. pentacamerata* Zones of late Early Eocene age. *Subbotina inaequispira*, *Acarinina* aff. *broedermanni*, *A.* aff. *pentacamerata*, and *A. densa* in the next higher cluster correlate to the *Hantkenina aragonensis–Morozovella lehneri* Zones of early Middle Eocene age. The stratigraphically-following cluster, which is a distinct entity, contains the average tops of *Subbotina linaperta*, *Turborotalia pomeroli*, *Globigerinatheka index*, and *Globigerina tripartita*. It corresponds to the zones of *Truncorotaloides rohri–Turborotalia cerroazulensis* s.l. of the youngest part of the Middle Eocene and of Late Eocene age.

The two small clusters following stratigraphically-upwards contain the benthonic *Uvigerina galloway* of Oligocene connotation and the planktonic zone markers *Globorotalia opima opima* and *Globigerina ciperoensis* which together match the Late Oligocene Standard Zone of '*Globorotalia*' *opima opima*.

The next higher cluster is large. The use of many more wells might have split it into stratigraphically more useful separate assemblages. If the well record is continuous, the basal part probably straddles the Oligocene–Miocene boundary. The presence upwards of '*Globorotalia*' *continuosa* (probably including *G. acrostoma*), *Globigerinoides trilobus*, *Globoquadrina altispira*, *G. dehiscens* (near A) is reminiscent of Berggren's (1972b) and Poore's (1979) Early Miocene zones with *Globoquadrina*. The presence halfway up the cluster of the *Globorotalia fohsi* group (mostly *G. fohsi peripheroronda*) firmly correlates to the Zones of *Catapsydrax stainforthi* to *Globorotalia fohsi* s.l. which range from Early to the Middle Miocene.

The average tops (near B) of *Globorotalia praemenardii*, *G. siakensis* and *Globigerina praebulloides* probably correspond to the *Globorotalia siakensis* Zone of late Middle Miocene age. The exit of *Martinotiella communis* also clusters well with these taxa. Schrader and others (1976) report this deeper water form from Middle Miocene levels in the Norwegian Sea.

The tight cluster of taxa near C composed of Scaphopod sp. 1, *Orbulina universa* and *Spiroplectammina carinata* correlate to the *S. carinata* and *Asterigerina gurichi* clusters of the Group 1 wells. The age may be Middle Miocene or slightly younger.

The next three small clusters stratigraphically consist of taxa known in Recent Atlantic continental margin sediments. The upper cluster with *Globorotalia crassaformis* together with the topmost one at the list, *Neogloboquadrina pachyderma*, is of post Miocene age. This taxon in one or two wells also includes some *N. atlantica* type forms (Poore, 1979) but poor quality of the samples prohibits a stratigraphically-meaningful differentiation of this group. Rare occurrences of *Globorotalia* aff. *puncticulata* and *G. inflata* testify to a Pliocene interval in the wells, but their record is too sporadic to be preserved in the optimum sequence. Inspection of the original data shows that clusters with *Globorotalia menardii* group and *G. obesa* would have shown *Neogloboquadrina acostaensis* if this species had been recognized in one more well. As a consequence, the age of this cluster is probably Late Miocene or Early Pliocene.

The differences discussed in the previous sections between the southern and northern optimum sequences also apply to the optimum clusters. Some are worth noting although no firm conclusions should be drawn because of the few wells in the southern group. In this southern group, *Spiroplectammina spectabilis* disappears in the Early rather than in the Middle Eocene cluster; this conclusion can be misleading because in the optimum sequence this taxon spans positions 43 to 60 and as a result its accuracy of position is low. The same is true for its lower neighbour *Acarinina soldadoensis*. *Cyclammina amplectens*, a typical Eocene taxon (as in the northern group wells), occurs in the southern Oligocene cluster. In the south, it only occurs in three out of 6 wells and its top may have been misidentified or it occurs in a reworked state.

Objective and subjective zonations

The fact that the clusters of the optimum-clustering solution of well-Groups 1 and 2 can be given a progressively-younger age provides one way of model verification. It shows that the statistical model which expresses crossover frequency of events in an optimum sequence produces meaningful biostratigraphic results resembling conventional assemblage zonation. It is necessary (as in every zonation) to have a sufficient number of wells which in our situation may be 10 or more.

Another way to 'test' the stratigraphic meaning of the model is to compare directly optimum clustering solutions to subjective attempts. For this purpose, a comparison and species score (Fig. 10) was made of the $k_c \geq 3/m_c \geq 3$ clustering solution for the 16 wells of Group 1 with the assemblage zonations of Gradstein and Williams (1976) using four of the wells, and of Gradstein (unpublished) using 9 of the same wells.

The subjective biostratigraphic solution for both the four wells and 9 wells produced an 8-fold subdivision of the Cenozoic section, which yields a geologically-meaningful regional correlation. Some of the zones were tentative and their ages not well defined. No good Paleocene events were recognized in these wells.

The sequence of species in the optimum clustering assemblages shows a similar sequential arrangement as in the two conventional assemblage zonations. There is a match between species in corresponding successive assemblages, with deviating scores being a minority. This is also obvious from the order of key markers in common between the probabilistic and conventional (T) zonation and indicated with black dots in Fig. 10. Each group of key markers in the succession of T zones occurs in a successively-younger and stratigraphically-corresponding probabilistic one, with the exception of *A. gurichi*, as mentioned earlier. There is no noticeable effect from the fact that the stastistical method uses average event positions whereas a subjective one would not tend to average such positions but might prefer absolute ones, that is highest (and lowest). Three cases can be made. Firstly, there are not many examples in our data where absolute and average tops would be substantially different at the level of stratigraphic resolution. Secondly, the use of many wells tends to bring the average tops closer to the absolute ones. Thirdly, in the subjective approach, no rigid use was made of the absolute top concept but rather of the highest part of the range of taxa. For a detailed evaluation of the difference between the two concepts, it is necessary to produce subjective and objective range charts, preferably in detailed, cored sections.

Conventional Planktonic Biostratigraphic Framework

In this section, we examine subjectively the Cenozoic planktonic foraminiferal record in the 22 wells on the Scotian Shelf, Grand Banks, and Labrador Shelf. Assemblages or single occurrences of index taxa are stacked in each well and correlated. The planktonic biostratigraphical framework is used to outline a latitudinally-differentiated zonal scheme, one for the Scotian Shelf–southern Grand Banks and one for the northern Grand Banks–

Labrador Shelf. The scheme is related to a low-latitude standard zonation (Stainforth and others, 1975) and to informal mid-latitude schemes established for the North Atlantic Ocean (Berggren, 1972b; Poore, 1979).

The reasons for this are threefold:

(1) to provide a 7- to 12-fold conventional planktonic zonation which complements and supports the 'probabilistic' zonation based on the benthonic and planktonic record.
(2) to provide a comprehensive overview of the regional Cenozoic planktonic assemblage complementary to that discussed in the context of the 'probabilistic zonation';
(3) to contribute to studies aimed at a definitive Cenozoic planktonic foraminiferal zonation for the middle to high-latitude Atlantic realm. For this reason, we will not formally define zones and proliferate zonal nomenclature based on one, regional, data base only. Our record also suffers from a lack of insight into the regional lower stratigraphical limits of the ranges of most markers.

Several earlier studies have involved regional planktonic record in Cenozoic strata, including Bartlett (1973), Bartlett and Hamdan (1972), and Ascoli (1977) for the Scotian Shelf and southern Grand Banks, and Gradstein and Srivastava (1980) for the Labrador Shelf and northern Grand Banks.

Pioneering studies were performed by G. A. Bartlett who had available an assortment of samples including cuttings from the earliest Grand Banks–Scotian exploratory wells. Although stratigraphically incomplete, the record was clearly recognized as of low-latitude affinity. His studies furnished a planktonic paleobiogeography as a function of regional paleoecology, Atlantic paleogeography, and paleoclimate and paleocirculation trends. Consistent tropical water (Gulf Stream) influences in Paleogene and Miocene time were inferred to wane over the continental margin in the Middle Miocene (post *Globorotalia fohsi* Zones). After that time, tropical microfauna (although north of the areal distribution of *Hantkenina* in Eocene assemblages) was replaced by a high-latitude assemblage made up of *Neogloboquadrina pachyderma* and *Globigerina bulloides*.

Ascoli (1977) in his well-illustrated report on Scotian Shelf, Mesozoic and Cenozoic micropaleontology tabulated the Cenozoic planktonic record in 7 wells. His listing of 43 taxa matches part of the list in Fig. 11. Based mainly on the stratigraphical ranges of markers in Postuma (1971), a 9-fold regional zonal scheme was proposed. The zones are *Subbotina pseudobulloides* zone—Early Paleocene; *Morozovella velascoensis, Planorotalites pseudomenardii* zone—Middle to Late Paleocene; *Morozovella subbotinae* zone—early Eocene; *Turborotalia cerroazulensis* group, *Acarinina densa* zone—Middle to Late Eocene; *Globigerina ampliapertura* zone—Early Oligocene; *Globorotalia opima opima* zone—Middle Oligocene; *Catapsydrax dissimilis* zone—late Oligocene to Early Miocene; *Globorotalia praemenardii* zone—Middle Miocene and *Globorotalia ex gr. margaritae* zone—Late Miocene. Relatively broad sample intervals precluded higher stratigraphical resolution.

The Scotian Shelf and southern Grand Banks record in the 6 wells under discussion takes advantage of detailed sampling, including many sidewall cores (SWC) in the Triumph P-50 and Mohican I-100 wells and of better processing. Triumph P-50 and Brant P-87 were most fossiliferous. Over 70 planktonic taxa were determined in the 6 wells and their exits or partial ranges assembled in more than 15 stratigraphically-successive (with some overlap) clusters of Paleocene through Quaternary age. Figure 11 shows the complete record and its footage in each well. The scant Paleocene well coverage has been complemented by Ascoli's (1977) Paleocene determinations in adjacent wells.

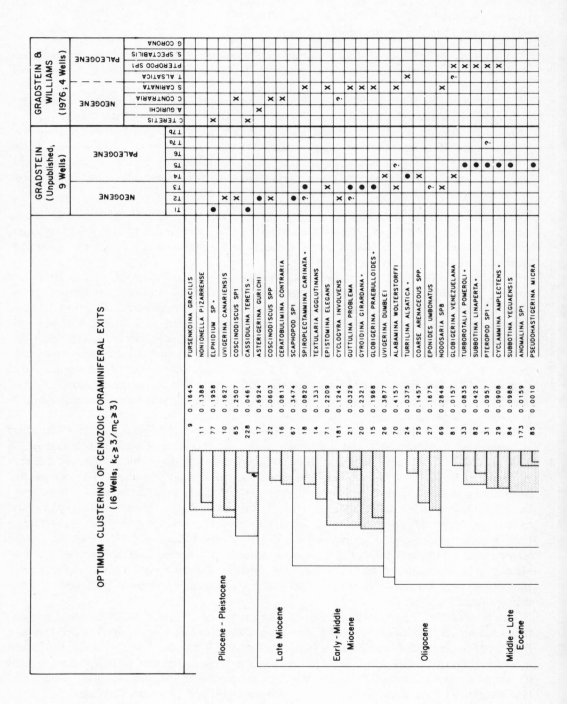

OPTIMUM CLUSTERING OF CENOZOIC FORAMINIFERAL EXITS
(16 Wells; $k_c \geq 3 / m_c \geq 3$)

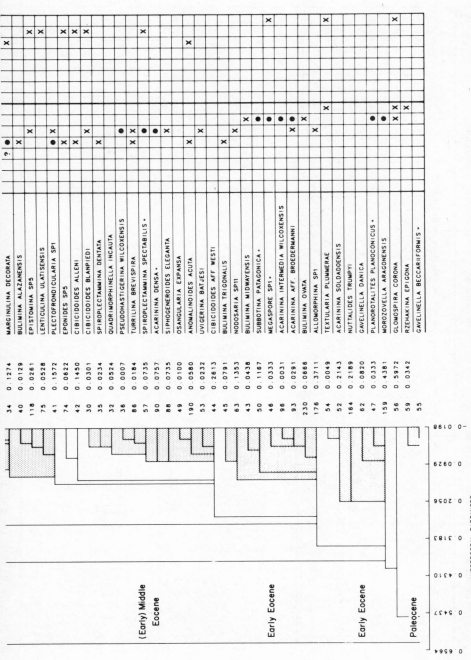

Fig. 10 Comparison of $k_c \geq 3/m_c \geq 3$ optimum clustering based on 16 wells of Group 1 on Labrador Shelf-northern Grand Banks with conventional zonations of Gradstein and Williams (1976) based on four of wells, and of Gradstein (unpublished) based on 9 of same wells. Both four and 9 wells database produced 8-fold biostratigraphic scheme. Optimum clustering retains 68 events (exits of taxa) grouped in 9 distinct and progressively-younger clusters, resembling assemblage zones. Shading pattern was used to enhance stratigraphically useful part of individual clusters. Sequence of species (marked with x or 0) common to optimum clustering and conventional zonations shows stratigraphically-comparable sequential arrangement, with little deviating scores. This is particularly obvious from sequence of dotted key markers in T zonation. For further explanation, see text

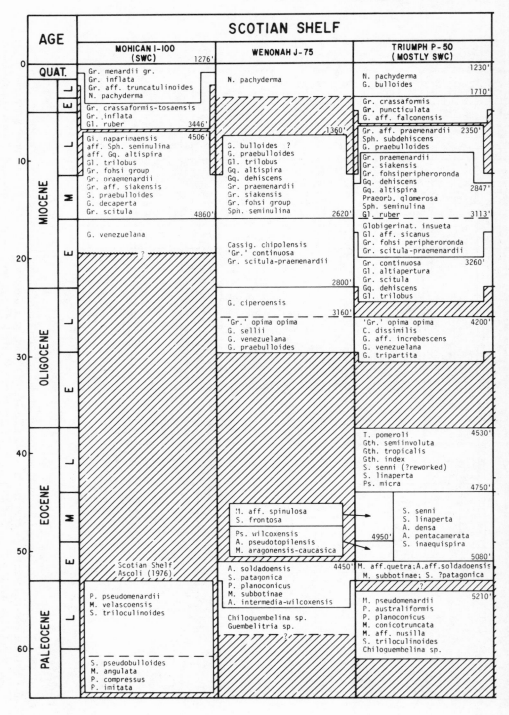

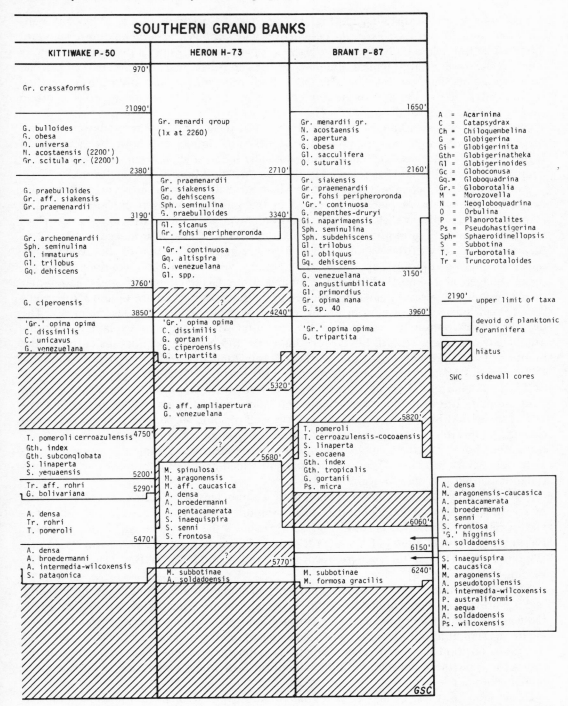

SOUTHERN GRAND BANKS

KITTIWAKE P-50	HERON H-73	BRANT P-87

Fig. 11. Stratigraphic distribution and correlation of Cenozoic planktonic foraminifera in 6 wells of Group 2 on Scotian Shelf-southern Grand Banks

KITTIWAKE P-50

970'

Gr. crassaformis

?1090'

G. bulloides
G. obesa
O. universa
N. acostaensis (2200')
Gr. scitula gr. (2200')

2380'

G. praebulloides
Gr. aff. siakensis
Gr. praemenardii

3190'

Gr. archeomenardii
Sph. seminulina
Gl. immaturus
Gl. trilobus
Gq. dehiscens

3760'

G. ciperoensis

3850'

'Gr.' opima opima
C. dissimilis
C. unicavus
G. venezuelana

T. pomeroli cerroazulensis 4750'
Gth. index
Gth. subconglobata
S. linaperta
S. yeguaensis

5200'

Tr. aff. rohri
G. bolivariana

5290'

A. densa
Tr. rohri
T. pomeroli

5470'

A. densa
A. broedermanni
A. intermedia-wilcoxensis
S. patagonica

HERON H-73

Gr. menardi group
(1x at 2260)

2710'

Gr. praemenardii
Gr. siakensis
Gq. dehiscens
Sph. seminulina
G. praebulloides

3340'

Gl. sicanus
Gr. fohsi peripheroronda

'Gr.' continuosa
Gq. altispira
G. venezuelana
Gl. spp.

3760'

?

4240'

'Gr.' opima opima
C. dissimilis
G. gortanii
G. ciperoensis
G. tripartita

?

5320'

G. aff. ampliapertura
G. venezuelana

?

5680'

M. spinulosa
M. aragonensis
M. aff. caucasica
A. densa
A. broedermanni
A. pentacamerata
S. inaequispira
S. senni
S. frontosa

?

5770'

M. subbotinae
A. soldadoensis

BRANT P-87

1650'

Gr. menardii gr.
N. acostaensis
S. apertura
G. obesa
Gl. sacculifera
O. suturalis

2160'

Gr. siakensis
Gr. praemenardii
Gr. fohsi peripheroronda
'Gr.' continuosa
G. nepenthes-druryi
Gi. naparimaensis
Sph. seminulina
Sph. subdehiscens
Gl. trilobus
Gl. obliquus
Gq. dehiscens

G. venezuelana   3150'
G. angustiumbilicata
Gl. primordius
Gr. opima nana
G. sp. 40   3960'

'Gr.' opima opima
G. tripartita

5820'

T. pomeroli
T. cerroazulensis-cocoaensis
S. linaperta
S. eocaena
Gth. index
Gth. tropicalis
G. gortanii
Ps. micra

6060'

6150'

M. subbotinae   6240'
M. formosa gracilis

A = Acarinina
C = Catapsydrax
Ch = Chiloquembelina
G = Globigerina
Gi = Globigerinita
Gth= Globigerinatheka
Gl = Globigerinoides
Gc = Globoconusa
Gq.= Globoquadrina
Gr.= Globorotalia
M = Morozovella
N = Neogloboquadrina
O = Orbulina
P = Planorotalites
Ps = Pseudohastigerina
Sph= Sphaeroidinellopsis
S = Subbotina
T. = Turborotalia
Tr = Truncorotaloides

2190' ⎯⎯ upper limit of taxa

☐ devoid of planktonic foraminifera

▨ hiatus

SWC   sidewall cores

A. densa
M. aragonensis-caucasica
A. pentacamerata
A. broedermanni
A. senni
S. frontosa
'G.' higginsi
A. soldadoensis

S. inaequispira
M. caucasica
M. aragonensis
A. pseudotopilensis
A. intermedia-wilcoxensis
P. australiformis
M. aequa
A. soldadoensis
Ps. wilcoxensis

GSC

The only major consistent gap (Fig. 11) appears to occur in Lower Oligocene beds corresponding to the *Cassigerinella chipolensis* and *Globigerina ampliapertura* standard Zones. The immediately overlying *Globigerina opima opima* Zone is present consistently whereas the lower stratigraphical boundary of the gap is irregular. Locally, part of Eocene beds is missing. The gap may be a function both of sediment starvation during relative coastal onlap of the sea and of subsequent erosion during the Oligocene coastal offlap (major eustatic sealevel drop).

The abrupt impoverishing or absence (Fig. 11) of the Late Miocene planktonic assemblage—*Globorotalia acostaensis* standard Zone—was influenced both by accelerated higher-latitude water-mass cooling and eustatically-induced shallowing. This explanation complements Bartlett and Hamdan's (1972) post-Middle Miocene cooling trend.

An abrupt post-Eocene and post-Middle Miocene shallowing and impoverished, provincial, or suddenly absent planktonic record in the Labrador wells is explained similarly (Gradstein and Srivastava, 1980). The authors employed the Cenozoic planktonics distribution pattern in the Labrador and northern Grand Banks wells, reproduced in Fig. 12 to furnish an overview of the regional paleobiogeography as a function of the evolutionary development of planktonics, opening of the Labrador Sea and Labrador continental margin facies changes, and of north Atlantic Ocean and Arctic Island climatic trends. The number of stratigraphically-useful planktonic species in the Labrador–northern Grand Banks wells was recognized to diminish from about 75 on the Scotian Shelf, to less than 30. The deepest-water conditions occurred in Eocene time which helps to explain that stratigraphically-consistent record. Shallower Oligocene and Neogene conditions curtailed the number of taxa present. Also, the northern limit along the Western Atlantic continental margin of keeled forms essentially occurs on the northern Grand Banks (±48°N). North of this area Paleocene–Eocene *Morozovella* and Neogene *Globorotalia* s. str. disappear and planorotaliid, turborotaliid, accarininid, and subbotinid/globigerinid taxa remain. We attribute the odd Eocene record in the Freydis well of some specimens of *Hantkenina* and *Globigerinatheka* to a northward 'sweep of the Gulf Stream'. Maximum low-latitude planktonic incursions along the Canadian margin occurred in Early to Middle Eocene and less so in Middle Miocene times. The Labrador current which engulfs the present day Canadian Atlantic margin is a post-Miocene feature.

Figure 12 shows the correlation of planktonic assemblages in the Cenozoic sediments in 8 of 16 wells on the northern Grand Banks–Labrador Shelf. The record in the other 8 is almost nil. For correlative purpose Fig. 12 also shows the coeval record in the southern Labrador Sea DSDP Sites 111A and 112, modified after Berggren (1972b) and Poore and Berggren (1974), and the meagre record on land around Baffin Bay (Hansen, 1969; Feyling-Hanssen, 1976).

The interrelation of the southern and northern planktonic assemblages in Figs. 11 and 12 affords an opportunity to outline a zonal scheme which is shown in Fig. 13. It reduces the abundance of names to a biostratigraphically-simple scenario. Instead of striving for maximum resolution based on all occurrences of markers, however rare, a 'middle of the road' position has been taken. Emphasis is on taxa which occur frequently in the southern or northern continental margins and on taxa which emphasize an extratropical faunal character.

In the interval (Figs. 11 and 12) with *Subbotina pseudobulloides*, *S. triloculinoides*, and *Planorotalites compressus*, the latter taxon is rare; the assemblage correlates with the standard zones of *Subbotina pseudobulloides* through *Morozovella uncinata* or *M. angulata*–Early Paleocene. *Globoconusa daubjergensis* was discovered in the United States Coastal Plain (Olsson, 1969) and on Nugssuaq, West Greenland (Hansen, 1969), but not in any Canadian well examined. Scrutiny of the correct size fraction was to no avail; it may indicate that our zone is essentially post-Danian in age.

An Upper Paleocene interval containing *Planorotalites pseudomenardii* was encountered in Triumph P-50 and in Cumberland B-55. Associated taxa are *P. australiformis, P. planoconicus, P. chapmani,* and rare morozovellids resembling *M. pusilla, M. conicotruncata,* and *M. velascoensis.* In the northern wells, Indian Harbour and Karlsefni, the interval with rare subbotinids (*?S. patagonica* or *S. triloculinoides*) and rare *Acarinina soldadoensis* and no *Pseudohastigerina* may be correlative to Berggren's (1972b) record with *Subbotina velascoensis* and *S. triangularis* and to the standard zone of *Morozovella velascoensis* p.p., below the micropaleontologically-traditional Paleocene–Eocene boundary.

The stratigraphically-immediately-overlying interval is widespread and fossiliferous in the southern wells. Taxa occurring include *Acarinina intermedia-wilcoxensis, A. soldadoensis* (mainly *A. soldadoensis soldadoensis*), *Planorotalites planoconicus, Subbotina patagonica,* and uni- to bi-apertural *Pseudohastigerina wilcoxensis* (the latter two taxa are abundant locally, as in Bonavista C-99, 11 510 ft.). Their highest cooccurrence defines a zone correlative to the mid- to high-latitude interval with *Subbotina patagonica* (? = *Globigerina linaperta* of Norwegian Sea DSDP sites, Schrader and others, 1976) and standard low-latitude Zones of *Morozovella subbotina* to *Acarinina pentacamerata,* Early Eocene.

In the southern wells (Scotian Shelf, southern Grand Banks), the zone can be subdivided into a lower and upper part. In the lower interval, rare *Morozovella subbotinae* and *M. formosa* occur together. As far as determined, *M. aragonensis* is absent. This lower interval correlates to the low-latitude standard Zone of *M. subbotinae* (? and *M. formosa formosa*), early Early Eocene. The upper interval, also recognized in northern Grand Banks wells, includes the tops of the *Planorotalites planoconicus-australiformis* group, *Morozovella* aff. *aequa,* and *Acarinina pseudotopilensis,* together with the taxa listed previously for the zone as a whole. *Acarinina densa* (embracing *A. spinuloinflata* Bandy and *A. bulbrooki* Bolli), which is frequent to abundant as far north as the northern Grand Banks, and *Morozovella aragonensis* range above the zone. The same is true for specimens intergrading between *Morozovella caucasica* and *M. aragonensis.* Such specimens may occupy a stratigraphic position from the upper part of the *Subbotina patagonica–Planorotalites planoconicus–Acarinina intermedia/wilcoxensis* interval (for example, Triumph P-50, 4950 ft.), to the lower part of the *A. densa* interval (for example, Heron H-73, 5680 ft., and Bonavista C-99, 11 010 ft.).

The next higher interval with the upper range of *Acarinina densa* as zone marker is based on the cluster of highest occurrences which also includes *Subbotina inaequispira* (rare), *S. frontosa, Acarinina senni, A. pentacamerata* (rare), *A. aragonensis, A. broedermanni* (rare), *Morozovella* aff. *spinulosa* (rare), and *Truncorotaloides rohri* (rare). The assemblage correlates with the low-latitude standard Zones of *Hantkenina aragonensis* to *Globigerinatheka subconglobata,* early Middle Eocene but may range as high as the *Truncorotaloides rohri* Zone, Middle Eocene.

In the northern region, the *Acarinina densa* interval includes the zone marker and rare *A. senni, A.* aff. *pentacamerata, Subbotina frontosa,* and the odd record of *Hantkenina* sp. and *Globigerinatheka kugleri* in Freydis B-87, 4180 ft. We equate the assemblage with the (late Early to) early Middle Eocene mid-latitude *Acarinina densa* interval of Berggren (1972b).

The top of the *A. densa* assemblage in the southern region coincides with a marked lithological change from carbonates to clastics; a boundary hiatus cannot be excluded. Immediately above occurs an assemblage with *Turborotalia pomeroli-cerroazulensis* group (rare), *Globigerinatheka* spp., *Globigerina linaperta, Pseudohastigerina micra, Globigerina*

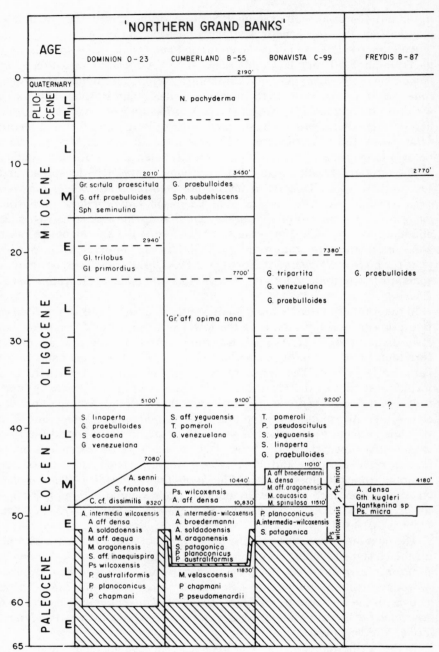

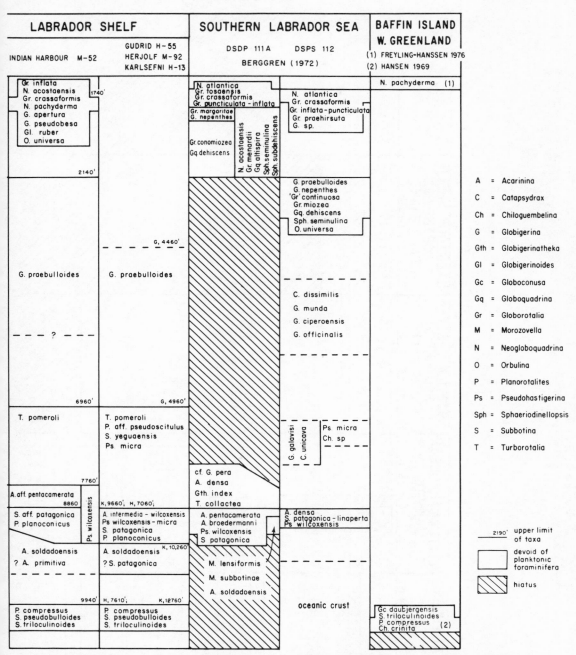

Fig. 12. Stratigraphic distribution and correlation of Cenozoic planktonic foraminifera in 8 wells of Group 1 on Labrador Shelf-northern Grand Banks in DSDP sites 111A and 112, southern Labrador Sea, and in isolated outcrops in West Greenland and Baffin Island

Biostratigraphic correlation chart

AGE	LOW LATITUDES STAINFORTH ET AL 1975 (SLIGHTLY MODIFIED IN LATE NEOGENE)	N. ATLANTIC OCEAN (SLIGHTLY) MODIFIED AFTER BERGGREN (1972); POORE (1979)	SCOTIAN SHELF – SOUTHERN GRAND BANKS (THIS PAPER AND ASCOLI 1976)	LABRADOR SHELF – NORTHERN GRAND BANKS (THIS PAPER)	NORWEGIAN SEA (SCHRADER ET AL 1976)
QUAT. L	Gr. truncatulinoides	N. pachyderma / Gr. inflata	N. pachyderma / Gr. inflata	N. pachyderma / Gr. crassaformis; Gr. inflata	N. pachyderma
QUAT. E	P. obliquiloculata	Sin. / Gr. puncticulata	Gr. crassaformis		N. atlantica S
PLIOC.	Gr. margaritae	N. atlantica / dextr.			N. atlantica D
MIOCENE L	N. acostaensis	N. acostaensis	N. acostaensis		N. acostaensis
MIOCENE M	Gr. menardii / Gr. siakensis	G. druryi / Gr. praemenardii / Gr. praescitula	Gr. praemenardii-Gr. siakensis / Gr. fohsi group-Sph. seminulina	G. praebulloides	"Gr. mayeri – siakensis"
MIOCENE E	Gr. fohsi lobata / fohsi / periph. / Pr. glomerosa - G. insueta / C. stainforthi / C. dissimilis	Orbulina / Praeorbulina / Globoquadrina spp. (baroemoenensis; praedehiscens)	Globoquadrina spp. / 'Gr.' continuosa – Globigerinoides spp.		
OLIGOCENE L	'Gr.' kugleri / G. ciperoensis	Catapsydrax spp. / Chiloguembelina	G. ciperoensis		
OLIGOCENE E	'Gr.' opima opima / G. ampliapertura / Cass. chipolensis / Ps. micra	G. munda / G. ampliapertura	'Gr.' opima opima / G. tripartita		

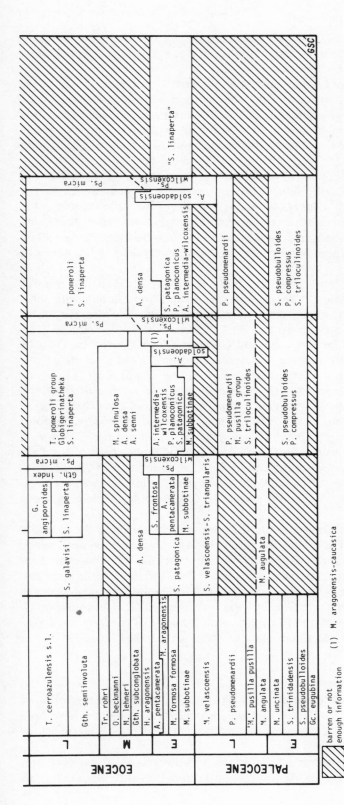

Fig. 13. Latitudinally differentiated zonal scheme based on planktonic foraminiferal assemblages of southern Grand Banks–Scotian Shelf (Fig. 11) and northern Grand Banks–Labrador Shelf (Fig. 12). Instead of striving for maximum resolution based on all occurrences of markers, however rare, 'middle of the road' position has been taken. Emphasis is on those taxa which occur frequently enough to be useful for correlation and which emphasize extra-tropical faunal character. Scheme is related to low-latitude standard zonation (Stainforth and others, 1975) and to informal mid-latitude schemes established for North Atlantic Ocean (Berggren, 1972b; Poore, 1979). Also shown is planktonic record in Norwegian Sea (after Schrader and others, 1976)

yeguaensis, and *G. venezuelana*. The first two genera listed did not reach the northern region. The assemblage correlates to the standard Zones of *Globigerinatheka semiinvoluta* to *Turborotalia cerroazulensis* s.l., Late Eocene.

The standard low-latitude Zone of *Globorotalia opima opima*, Late Oligocene (Hardenbol and Berggren, 1978), with cooccurrence of the zone marker and *Globigerina sellii, G. gortanii, G. venezuelana, G. praebulloides, G. tripartita, G. ciperoensis, Catapsydrax dissimilis*, and *C. unicavus* is present in 5 out of 6 southern wells but not discernible northward. The same is true for the latest Oligocene standard Zone of *Globigerina ciperoensis* in Wenonah F-75, 2800 ft., and Kittiwake P-50, 3760 ft.

Above the Oligocene *Globorotalia opima opima* Zone and below the next events, firmly of Middle Miocene age, is an interval with frequent *Globoquadrina* spp. (including *dehiscens* and *altispira*) and higher up with *Globigerinoides* (including rare *primordius*); large *Globigerina venezuelana* like specimens exit in this assemblage as does a *Globorotalia* listed as *G. continuosa* Blow but probably closer to *G. acrostoma* (Wezel) and scattered small *G. ?archeomenardii* type specimens. There is no persistent coeval equivalent in the northern wells.

In the southern wells Triumph P-50 (3113 ft.) and Heron H-73 (3340 ft.), isolated occurrences of *Globigerinoides sicanus, Globigerinatella insueta*, and *Globorotalia fohsi peripheroronda* correlate to the *Globigerinatella insueta* to *Globorotalis fohsi peripheroronda* Zones just below and above the Early/Middle Miocene boundary. This reflects a warming event listed as *Praeorbulina* and *Globigerinoides sicanus* intervals by Berggren (1972b) and Poore (1979). A more widespread and prominent assemblage correlative with the Middle Miocene standard zones of *Globorotalia fohsi* s.l. to *G. praemenardii* and to the mid-latitude zones of *G. praemenardii* to *Globigerina druryi* (Berggren, 1972b; Poore, 1979) occurs in all 6 southern wells. It consists of the exits clustered with *Globorotalia praemenardii* (frequent), *Globorotalia siakensis, Globorotalia fohsi peripheroronda, Globoquadrina* spp., *Sphaeroidinellopsis seminulina*, and *Globigerina praebuloides*. The latter three taxa were also reported in the Middle Miocene of the Labrador Sea Site 112 and occur on the northern Grand Banks in Dominion O-23, 2010 ft., and Cumberland B-55, 3450 ft. *Globigerina praebulloides* is the only taxon reaching consistently farther northward along the continental margin in (presumably) Middle Miocene time.

Late Miocene and Pliocene, Quaternary standard Zones are recognized in the southern approaches to the Labrador Sea (Poore and Berggren, 1974), but cannot be determined consistently in the continental margin wells. There is evidence of the Late Miocene *Neogloboquadrina acostaensis* Zone in the southern wells Brant P-87 (1630–2160 ft.) and Kittiwake P-11 (1900–2380 ft.) with *N. acostaensis* and without exclusively post-Miocene forms and of a Pliocene interval with rare *Globorotalia crassaformis, G. inflata,* and no *G. truncatulinoides* (Indian Harbour M-52, 1740 ft.; Triumph P-50, 1710 ft.; Mohican I-100, 3446 ft.). The latter form occurs in the Quaternary of Mohican I-100 (1276–3395 ft.).

Conclusions

A combination of several strategies is generally the best approach to obtain information or to solve a problem. In this case two models of Cenozoic foraminiferal zonations have been outlined which provide insight in the Cenozoic biostratigraphy and provincialism of the Northwest Atlantic Continental Margin.

One model is a conventionally-constructed planktonic biostratigraphic framework employing exits and higher parts of ranges. It produces a latitudinally-differentiated 7- to

12-fold scheme (Figs. 11, 12, 13), which works best for the Scotian Shelf–southern Grand Banks region.

The other model is based on optimum sequence and optimum clustering methods which provide stratigraphic solutions employing all of the exits of benthonic and planktonic taxa that are potentially useful in correlation. Consider solutions for the 16 wells of Group 1; the combination of the more rigorous, $k_c \geq 5/m_c \geq 3$, solution (Fig. 9) with the 'lenient' and larger $k_c \geq 3/m_c \geq 3$ (Fig. 10) yields a flexible and pragmatic 9-fold assemblage zona-

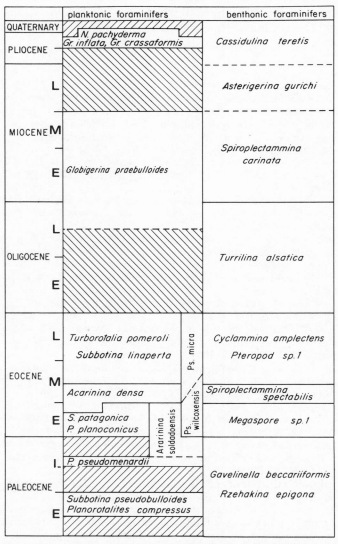

Fig. 14. Zonation for Labrador Shelf and northern Grand Banks. Column to right is based on optimum clustering solutions shown in Figs. 9 and 10. Left column summarizes conventional planktonic scheme of Figs. 11 and 12. Stratigraphic resolution is relatively conservative. For further explanations see Section 8 and 9

tion and an optimum sequence of exits for the Labrador Shelf–northern Grand Banks. Index fossils in the individual successive assemblages provide chronostratigraphic tiepoints and transform the continuous relative time scale as assumed in the method to a stratigraphically-better calibrated one. Studies of cored sections to obtain stratigraphic ranges of benthonic and planktonic taxa are needed to refine substantially calibrations.

Figure 14 summarizes zonations for the Labrador Shelf–northern Grand Banks as based on optimum clustering and on the conventional planktonic biostratigraphic scheme. Stratigraphic resolution is relatively conservative.

We have demonstrated that the statistical techniques produce assemblage zonations, based on as much information as possible, for large scale or detailed correlation; optimum sequences of events serve as a standard to which (new) individual stratigraphic sequences can be compared. The technique produces average range charts. The methods are well suited for large data sets.

References

Ascoli, P., 1977, Foraminiferal and ostracod biostratigraphy of the Mesozoic–Cenozoic, Scotian Shelf, Atlantic Canada, in Proceedings of the 1st International Symposium on Benthonic Foraminifera, Halifax 1975. *Maritime Sediments Spec. Publ.* no. 1 (B), 653–771.

Aubert, J., and Berggren, W. A., 1976, Paleocene benthic foraminiferal biostratigraphy and paleoecology of Tunisia, *Bull. Centre Rech Pau–SNPA*, **10**(2), 379–469.

Barker, R. W., 1960, Taxonomic notes on the species figured by H. B. Brady in his report on the foraminifera dredged by H.M.S. 'Challenger' during the years 1873–1876, *Soc. Econ. Paleont. Miner., Spec. Publ.* no. 9.

Bartlett, G. A., 1973, The Canadian Atlantic Continental Margin: paleogeograhy, paleoclimatology and seafloor spreading, in *Earth Science Symposium on Offshore Eastern Canada, Ottawa 1971*, Geol. Survey Canada Paper 71–23, pp. 43–73.

Bartlett, G. A., and Hamdan, A. R. A., 1972, The Canadian Atlantic Continental margin— biostratigraphy, paleocology and paleooceanography from Cretaceous to Recent, *24th Int. Geol. Congress, Montreal 1972*, Section 8, pp. 3–15.

Batjes, D. A. J., 1958, Foraminifera of the Oligocene of Belgium, *Verh. Kon. Belg. Inst. Natuur Wetensch.*, **143**, 1–168.

Berggren, W. A., 1972a, A Cenozoic time-scale, some implications for regional geology and paleogeography, *Lethaia*, **5**, 195–215.

Berggren, W. A., 1972b, Cenozoic biostratigraphy and paleobiogeography of the North Atlantic, in Laughton, A. S., and others, *Reports Deep Sea Drilling Project 12*. U.S. Government Printing Office, Washington, D.C., pp. 965–1001.

Berggren W. A., 1977, Atlas of Palaeogene planktonic foraminifera—some species of the genera, *Subbotina, Planorotalites, Morozovella, Acarinina* and *Truncorotaloides*, in Ramsay, A. T. S. (ed.), *Ocean Micropaleontology*. Academic Press, N.Y., pp. 205–265.

Berggren, W. A., and Aubert, J., 1975, Paleocene benthonic foraminiferal biostratigraphy and paleoecology of Atlantic–Tethyan regions: Midway-type fauna; *Paleogeogr., Paleoclimatol., Paleoecol.*, **18**, 73–192.

Berggren, W. A., and van Couvering, J. A., 1978, Biochronology, in Cohee, G. V., Glaessner, M. F., and Hedberg, H. (eds.), Contributions to the Geological Time Scale, *Am. Assoc. Petroleum Geol., Studies in Geology*, No. 6, 39–57.

Bettenstaedt, F., Fahrion, H., Hiltermann, H., and Hinsch, W., 1962, Tertiar Norddeutschlands, in *Leitfossilien der Mikropalaontologie*. Borntraeger Berlin, pp. 339–378.

Blow, W. H., 1969, Late middle Eocene to Recent planktonic foraminiferal biostratigraphy, in *Proceedings of the 1st International Conference on Planktonic Microfossils, Geneva 1967*. Bronnimann, P., and Renz, J. H. H. (eds.), E. J. Brill, Leiden, pp. 199–422.

Brady, H. B., 1884, Reports of the scientific results of the voyage of H.M.S. 'Challenger', *Zoology*, **9**, 1–814.

Brower, J. C., and Burroughs, W. A., 1982, A simple method for quantitative biostratigraphy. (This volume.)

Cubitt, J. M., 1978, Quantitative Stratigraphic Correlation, *Computers & Geosciences*, **3**, 215–318.

Cushman, J. A., 1918–(1931), Foraminifera of the Atlantic Ocean, *U.S. Natural History Museum Bull.*, **104**.

Cushman, J. A., 1951, Paleocene foraminifera of the Gulf Coastal Region of the United States and adjacent regions, *U.S. Geol. Survey Prof. Paper*, no. 232, 1–75.

Cushman, J. A., and Renz, H. H., 1946, The foraminiferal fauna of the Lizard Springs Formation of Trinidad, British West Indies, *Contr. Cushman Foundation Foraminifora Res.*, *Spec. Publ.* no. 18, 1–48.

David, H. A., 1963, *The Method of Paired Comparisons*. Griffin, London, England, 124 pp.

Davidson, R. R., 1970, On extending the Bradley–Terry model to accommodate ties in paired comparison experiments, *Jour. Am. Stat. Assoc.*, **65**, 81–95.

Davis, J. C., 1973, *Statistics and Data Analysis in Geology*. Wiley, New York, N.Y., 550 pp.

Drooger, C. W., 1974, The boundaries and limits of stratigraphy, *Proc. K. Ned. Akad. Wet. Ser. 11 B*, **17**, 159–176.

Edwards, L., and Beaver, R., 1978, The use of a paired comparison model in ordering stratigraphic events, *Jour. Math. Geol.*, **10**, 261–272.

Ellis, B. F., Messina, A. R., Charmatz, R., and Ronai, L. E., 1969, Catalogue of Index Smaller Foraminifera (3), Mesozoic–Tertiary Benthonic Foraminifera, *Spec. Publ. Am. Museum Natural History*, New York.

Feyling-Hanssen, R. W., 1976, The Clyde Foreland Formation: A micropaleontological study of Quaternary stratigraphy, in Proceedings of the International Symposium on Benthonic Foraminifera Halifax, 1975, B, 315–377.

Feyling-Hanssen, R. W., Jorgensen, J. A., Knudson, K. L., and Anderson, A. L. L., 1971, Late Quaternary from Vendsyssel, Denmark and Sandnes, Norway, *Bull. Geol. Soc. Denmark*, **21** (Parts 2–3), 1–317.

Gradstein, F. M., Williams, G. L., Jenkins, W. A. M., and Ascoli, P., 1975, Mesozoic and Cenozoic stratigraphy of the Atlantic continental margin, eastern Canada, Yorath, C. T., and others, eds., Canada's Continental Margin and Offshore Petroleum Exploration, *Can. Soc. Petroleum Geol. Mem.*, **4**, 103–121.

Gradstein, F. M, and Williams, G. L., 1976, Biostratigraphy of the Labrador Shelf. *I: Can. Geol. Survey Rept.* 349, 1–40.

Gradstein, F. M., and Berggren, W. A., 1981, Flysch-type agglutinated Foraminifera and the Maestrichtian to Paleogene history of the Labrador and North Seas, *Marine Micropaleont.*, **6**, 211–268.

Gradstein, F. M., and Srivastava, S. P., 1980, Aspects of Cenozoic stratigraphy and paleogeography of the Labrador Sea and Baffin Bay, *Palaeogeogr., Palaeoclimatol., Palaeoecol.*, **30**, 261–295.

Hansen, H. J., 1969, Danian Foraminifera from Nugssuaq, Greenland, *Meddt. Gronland*, **193** (2), 1–132.

Hansen, H. J., 1972, Two species of Foraminifera of the genus *Turrilina* with different wall structure, *Lethaia*, **5**, 39–45.

Hardenbol, J., and Berggren, W. A., 1978, A new Paleogene numerical time scale, *Am. Assoc. Petroleum Geol., Studies in Geology*, no. 6, 213–234.

Hay, W. W., 1972, Probabilistic stratigraphy, *Eclog. geol. Helv.*, **65**(2), 255–266.

Hay, W. W., and Southam, J. R., 1978, Quantifying biostratigraphic correlation, *Ann. Rev. Earth Sci.*, **6**, 353–75.

Hazel, J. E., 1970, Binary coefficients and clustering in biostratigraphy, *Bull. Geol. Soc. Am.*, **81**, 3237–3252.

Hillebrand, A. Von., 1962, Das Paleozan und seine Foraminiferen fauna im Becken von Reichenhall und Salzburg, *Abh. Bayer. Akad. Wiss. NF 108*, 1–182.

Hilterman, H., 1972, Zur Morphology der Benthos-Foraminifere *Spiroplectammina spectabilis* (Grzybowski), *Geol Jb.* **A4**, 43–61.

Howe, H. V., 1939, Louisiana Cook Mountain Eocene Foraminifera, *Louisiana Geol. Survey, Geol. Bull.* **14**, 1–118.

Indans, J., 1962, Foraminiferen-Faunen aus dem Miozän des Niederrheingebietes: *Fortschr. Geol. Rheinld. Westfo.*, **6**, 19–82.

International Subcommission on Stratigraphic Classification, 1976, in Hedberg, H. D. (ed.), *International Stratigraphic Guide*. John Wiley and Sons, New York, 200 pp.

Kaasschieter, J. P. H., 1961, Foraminifera of the Eocene of Belgium, *Verh. Kon. Belg. Inst. Natuurwetensch.*, **147**, 1–271.

Kendall, M., 1975, *Rank Correlation Methods*. Griffin, London, England, 202 pp.

Lamb, J., 1964, The stratigraphic occurrences and relationships of some mid-Tertiary *Uvigerina* and *Siphogenerina*, *Micropaleont.*, **10** (4), 457–476.

Marks, P., 1951, A revision of the smaller Foraminifera from the Miocene of the Vienna Basin, *Contr. Cushman Foundation Foraminifera Res.*, **2**(2), 33–73.

Miller, F. X., 1977, The graphic correlation method in biostratigraphy, in Kauffman, E., and Hazel, J. (eds), *Concepts and Methods of Biostratigraphy*. Dowden, Hutchinson and Ross, Stroudsburg, Pa., pp. 165–186.

Nogan, D. S., 1964, Foraminifera, stratigraphy and paleoecology of the Aquia Formation of Maryland and Virginia, *Cushman Foundation Foraminifera Res., Spec. Publ.* no. 7, 1–50.

Olsson, R. K., 1969, Early Tertiary planktonic foraminiferal zonation of New Jersey, in Bronnimann, P., and Renz, H. H., (eds.), *Proc. 1st Int. Conf. Planktonic Microfossils, Geneva 1967*. E. J. Brill, Leiden, pp. 493–504.

Poore, R. Z., 1979, Oligocene through Quaternary planktonic foraminiferal biostratigraphy of the North Atlantic: DSDP Leg 49, in Luyendyk, B. P., Cann, J. R., and others, *Initial Reports Deep Sea Drilling Project* No. 49, U.S. Government Printing Office, Washington, pp. 447–517.

Poore, R. Z., and Berggren, W. A., 1974, Pliocene biostratigraphy of the Labrador Sea: Calcareous Plankton, *Jour. Foram. Res.*, **4**(3), 91–108.

Postuma, J. A., 1971, *Manual of Planktonic Foraminifera*. Elsevier Publ. Co., Amsterdam, 420 pp.

Schrader, H., Bjorklund, K., Manum, S., Martini, E., and Van Hinte, J., 1976, Cenozoic biostratigraphy, physical stratigraphy and paleooceanography in the Norwegian–Greenland Sea, DSDP Leg 38, Paleontological Synthesis, in Talwani, M., Udintsev, G., and others, *Initial Reports Deep Sea Drilling Project* no. 38, U.S. Government Printing Office, Washington, pp. 1197–1211.

Scott, G. H., 1974, Biometry of the Foraminiferal shell, in Hedley, R. H., and Adams, C. G. (eds.), *Foraminifera 1*. Academic Press, 55–151.

Sen Gupta, B. K., 1971, The benthonic Foraminifera of the tail of the Grand Banks, *Micropaleont.*, **1**, 69–98.

Shaw, A. B., 1964, *Time in Stratigraphy*. McGraw-Hill Book Co., New York, 365 pp.

Shaw, A. B., 1969, Adam and Eve, Paleontology, and the nonobjective arts, *Jour. Paleont.*, **5**, 1085–1098.

Stainforth, R. M., Lamb, J. L., Luterbacher, H., Beard, J. H., and Jeffords, R. M., 1975, Cenozoic planktonic foraminiferal zonation and characteristics of index forms, *Univ. Kansas Paleont. Contr.*, no. 62, 1–162.

Steineck, P. L., and Fleisher, R. L., 1978, Towards the classicial evolutionary reclassification of Cenozoic Globigerinacea (Foraminiferida), *Jour. Paleont.*, **52**(2), 618–635.

Subottina, N. N., 1953, *Fossil Foraminifera of the USSR—Globigerinidae, Hantkeninidae and Globorotaliidae*. Collet's Publ. Ltd., London, 320 pp. (Engl. Trans. 1971).

Tjalsma, R. C., 1976, Cenozoic Foraminifera from the South Atlantic, DSDP Leg 36, in Barker, P. F., Dalziel, I. W. D., and others, *Report Deep Sea Deep Project*. No. 36, U.S. Government Printing Office, Washington, pp. 493–517.

Todd, R., and Kniker, H. T., 1952, An Eocene Foraminiferal Fauna from the Aqua Fresca shale of Magallanes Province, southernmost Chile, *Cushman Foundation Foraminifera Res., Spec. Publ.* no. 1, 1–28.

Torgerson, W. S., 1958, *Theory and Methods of Scaling*. John Wiley, New York, N.Y., 460 pp.

Vetrova, S. V., 1975, New genus *Quadrimorphinella* and its representatives from the Eocene of Azerbaidszhan (in Russian), *Akad. Nauk. Azerb. SSR, Ser. Nauk. Zemle, Baku, USSR*, **2**, 27.

Worsley, T. R., and Jorgens, M. L., 1977, Automated biostratigraphy, in Ramsay, A. T. S. (ed.), *Oceanic Micropalaeontology*, vol. 2. Academic Press, London, pp. 1201–1229.

Zachariasse, W. J., 1975, Planktonic foraminiferal biostratigraphy of the Late Neogene of Crete (Greece), *Utrecht Micropaleont. Bull.*, no. 11, 1–171.

Zachariasse, W. J., Riedel, W. R., Sanfilippo, A., Schmidt, R. R., Brolsma, M. J., Schrader, H. J., Gersonde, R., Drooger, M. M., and Broekman, J. A. 1978, Micropaleontological counting methods and techniques—an exercise on an eight metres section of the Lower Pliocene of Capo Rossello, Sicily, *Utrecht Micropaleont. Bull.* no. 17, 1–265.

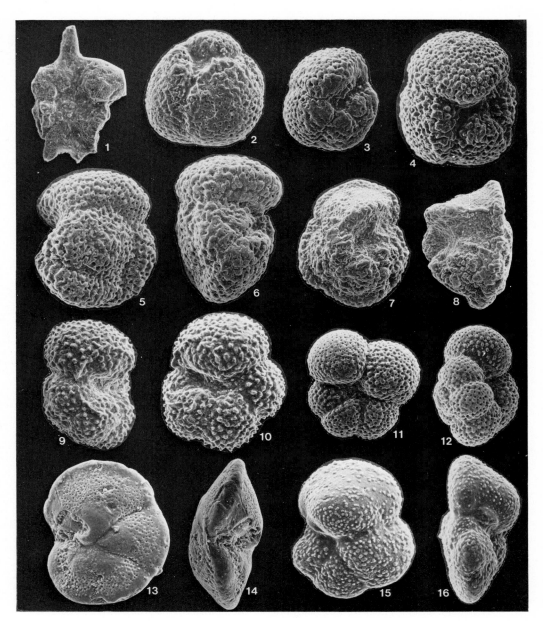

Plate 1. Paleogene Planktonic Foraminifera

1—*Hantkenina* sp. Freydis B-87, 4270', Labrador Shelf. 2—*Acarinina senni* (Beckmann).
Triumph P-50, 4750', Scotian Shelf. 3, 4, 5, 6—*Acarinina densa* (Cushman). Three specimens
from Kittiwake P-11, 5500', Grand Banks. 7, 8—*Morozovella aragonensis* (Nuttall), *M.
caucasica* (Glaessner). Triumph P-50, 4950', Scotian Shelf. 9, 10—*Acarinina intermedia*
(Subbotina), *A. wilcoxensis* (Cushman and Ponton) Bonavista C-99, 11 510', northern Grand
Banks. 11, 12—*Acarinina pentacamerata* (Subbotina) Triumph P-50, S.W.C. 4825', Scotian
Shelf. 13, 14—*Planorotalites pseudomenardii* (Bolli) Triumph P-50, 5460', Scotian Shelf. 15,
16—*Planorotalites planoconicus* (Subbotina) Bonavista C-99, 11 510', northern Grand
Banks.

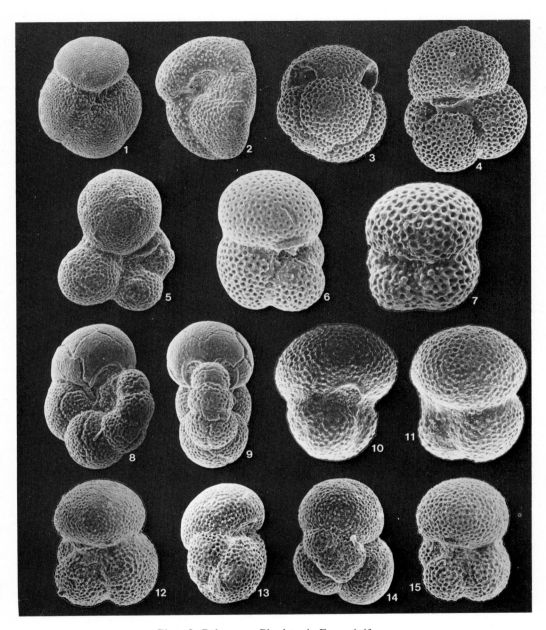

Plate 2. Paleogene Planktonic Foraminifera

1, 2—*Turborotalia pomeroli* (Toumarkine and Bolli) Tern A-68, 8540′, Grand Banks. 3—*Globigerinatheka index* (Finlay) Kittiwake P-11, 4840′, Grand Banks. 4—*Subbotina yeguaensis* (Weinzierl and Applin) Kittiwake P-11, 4930′, Grand Banks. 5—*Subbotina inaequispira* (Subbotina) Triumph P-50, 4950′, Scotian Shelf. 6, 7—*Subbotina linaperta* (Finlay). Two specimens from Kittiwake P-11, S.W.C. 5230′, Grand Banks. 8, 9—*Pseudohastigerina wilcoxensis* (Cushman and Ponton), biumbilicate variety. Dominion O-23, 10 020′, northern Grand Banks. 10, 11—*Subbotina frontosa* (Subbotina) Triumph P-50, S.W.C. 4825′, Scotian Shelf. 12, 13, 14, 15—*Subbotina patagonica* (Todd and Kniker). Four specimens from Bonavista C-99, 11 610-810′, northern Grand Banks.

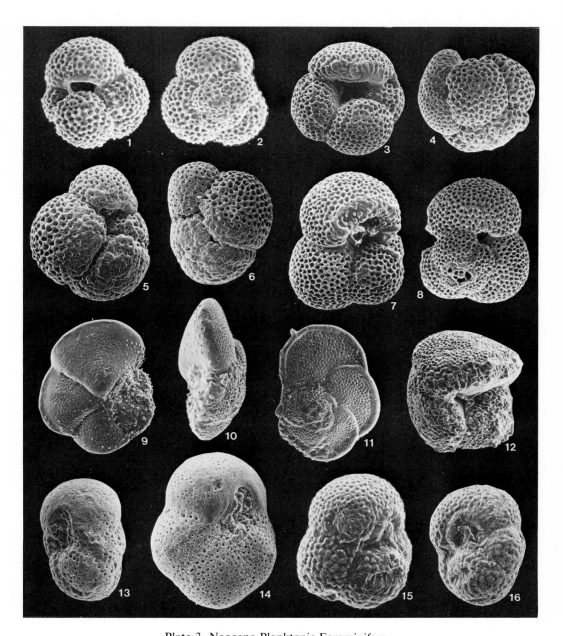

Plate 3. Neogene Planktonic Foraminifera
1, 2—*Globigerina praebulloides* Blow Bjarni H-81, 3960', Labrador Shelf. 3, 4—*Globoquadrina altispira* Bolli Triumph P-50, S.W.C. 3113', Scotian Shelf. 5, 6—*Sphaeroidinellopsis seminulina* (Schwager) Triumph P-50, S.W.C. 2800', Scotian Shelf. 7, 8—*Globigerinoides primordius* Blow and Banner Triumph P-50, 2650', Scotian Shelf. 9, 10, 11—*Globoratalia praemenardii* Cushman and Stainforth. Three specimens from Heron H-73, 2980', Grand Banks. 12—*Globoquadrina dehiscens* Chapman, Parr and Collins Triumph P-50, S.W.C. 3113', Scotian Shelf. 13, 14, 15, 16—*Globorotalia acrostoma* Wezel—Globorotalia mayeri Cushman and Ellisor (recorded as '*Globorotalia' continuosa* Blow). Four specimens from Heron H-73, 3340', Grand Banks.

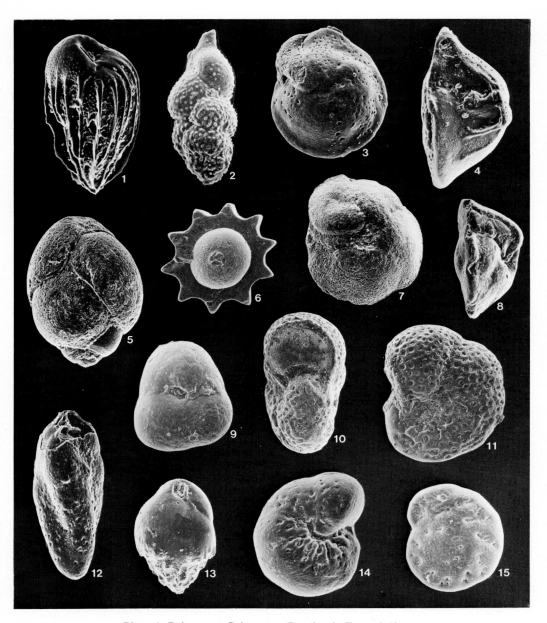

Plate 4. Paleogene Calcareous Benthonic Foraminifera

1—*Bulimina alazanensis* Cushman, Kittiwake P-11, 5340′, Grand Banks. 2—*Uvigerina batjesi* Kaasschieter, Karlsefni H-13, 8960′, Labrador Shelf. 3,4 —*Cibicidoides blanpiedi* (Toulmin). Two specimens from Gudrid H-55, 6800′, Labrador Shelf. 5—*Bulimina ovata* d'Orbigny. Karlsefni H-13, 6800′, Labrador Shelf. 6—*Nodosaria* sp 11 (only observed as fragments) Karlsefni H-13, 9160′, Labrador Shelf. 7, 8—*Cibicidoides westi* Howe (note characteristic umbilical depression and pointed apex). Karlsefni H-13, 9460′, Labrador Shelf. 9,—*Allomorphina* sp 1. Snorri J-90, 9520′, Labrador Shelf. 10, 11—*Gavelinella danica* (Brotzen). Two specimens from Indian Harbour M-52, 10 240′, Labrador Shelf. 12— *Bulimina trigonalis* Ten Dam Freydis B-87, 3910′, Labrador Shelf. 13—*Bulimina midwayensis* Cushman and Parker, Dominion O-23, 9770′, Grand Banks. 14—*Gavelinella beccariiformis* (White) Indian Harbour M-52, 10 540′, Labrador Shelf. 15—*Gavelinella beccariiformis* (White) Herjolf M-92, 7250′, Labrador Shelf.

Plate 5. Paleogene Calcareous Benthonic Foraminifera

1—*Alabamina wolterstorffi* (Franke) Bjarni H-81, 3960', Labrador Shelf. 2—*Turrilina alsatica* Andreae. Leif M-48, 3886', Labrador Shelf. 3—*Turrilina alsatica* Andreae. Murre G67, 1020', Grand Banks. 4—*Nodosaria* sp 8, Leif M-48, 3790', Labrador Shelf. 5—*Marginulina* ex. gr. *decorata* Reuss Leif M-48 4090', Labrador Shelf. 6—*Plectofrondicularia* sp 1, Bjarni H-81, 5360', Labrador Shelf. 7—*Cibicidoides alleni* (Plummer), Leif M-48, 4720–5070', Labrador Shelf. 8—*Cibicidoides alleni* (Plummer). Freydis B-87, 3820', Labrador Shelf. 9—*Lenticulina ulatisensis* (Boyd) Bjarni H-81, 5660', Labrador Shelf. 10—*Anomalina* sp. 1 Freydis B-87, 3820', Labrador Shelf. 11—*Cibicidoides granulosus* (Bykova). Kittiwake P-11, 5110', Grand Banks. 12, 13—*Epistomina* sp 5 (probably, *E. eocenica* Cushman and Hanna). Two specimens from Leif M-48, 5250', Labrador Shelf. 14,15—*Eponides* sp 5 (probably *E. plummerae* Cushman). Two specimens from Leif M-48, 5250', Labrador Shelf.

Plate 6. Calcareous Benthonic Foraminifera
1—*Cassidulina teretis* Norvang. Leif M-48, 1300′, Labrador Shelf. 2—*Cassidulina teretis* Norvang. Leif E-38, 1500′, Labrador Shelf. 3, 4—*Nonionella pizarrensis* (Berry). Heron H-73, 1150′, Grand Banks. 5—*Bulimina* (*?Caucasina*) *elongata* (d'Orbigny) [recorded as *Fursenkoina gracilis* (Collins)]. Kittiwake P-11, 970′, Grand Banks. 6, 7—*Uvigerina canariensis* d'Orbigny. Two specimens from Indian Harbour M-52, 2040′, Labrador Shelf. 8—*Ceratobulimina contraria* (Reuss). Egret K-36, 1500′, Grand Banks. 9—*Ceratobulimina contraria* (Reuss). Leif M-48, 2670′, Labrador Shelf. 10, 11—*Asterigerina* ex gr. *gurichi* (Franke) Reworked at 1210′ in Leif M-48, Labrador Shelf. 12—*Asterigerina* ex gr. *gurichi* (Franke). Osprey H-84, 1280′, Grand Banks. 13—*Asterigerina* ex gr. *gurichi* (Franke). Egret K-36, 950′, Grand Banks. 14—*Asterigerina* ex gr. *gurichi* (Franke). Note chamberlets; Egret K-36, 860′, Grand Banks. 15, 16—*Uvigerina dumblei* Cushman and Applin. Two specimens from Brant H-83, 3360′, Grand Banks. 17, 18—*Gyroidina girardana* (Reuss). Leif E-38, 3430′, Labrador Shelf.

Plate 7. Agglutinated Benthonic Foraminifera and miscellaneous shelly microfossils
1, 2—*Spiroplectammina carinata* d'Orbigny. Leif M-48, 3380', Labrador Shelf. 3—*Spiroplectammina carinata* (d'Orbigny). Leif M-48, 3380', Labrador Shelf. 4, 5—*Cyclammina amplectans* Grzybowski. Bjarni H-81, 5560', Labrador Shelf. 6, 7—*Spiroplectammina navarroana* Cushman [recorded as *Textularia plummerae* (Lalicker)]. Dominion O-23, 8550', Northern Grand Banks. Fig. 7 shows spiral initial part of test. 8, 9—*Spiroplectammina navarroana* Cushman [recorded as *Textularia plummerae* (Lalicker)]. Freydis B-87, 5050', Labrador Shelf. Fig. 9 shows spiral beginning of test. 10, 11—*Ammosphaeroidina* sp 1 (recorded as *Quadrimorphinella incauta* Vetrova). Dominion O-23, 7320', northern Grand Banks. 12, 13—*Spiroplectammina spectabilis* (Grzybowski), two specimens from Bjarni H-81, 6560', Labrador Shelf. 14—*Tritaxia* sp 3. Triumph P-50, 4950', Scotian Shelf. 15, 16—*Glomospira corona* Cushman and Jarvis. Two specimens from Leif M-48, 5820', Labrador Shelf. 17, 18—*Gastropod* sp 1 *(recorded as 'Pteropod' sp 1)*. Two specimens from Petrel A-62, 2580', Grand Banks. 19—*Scaphopod* sp 1. *Leif E-38, 3030'*, Labrador Shelf. 20—*Megaspore* sp 1. Bjarni H-81, 6260', Labrador Shelf. 21—'*Coscinodiscus*' sp 1. Bjarni H-s, 2860', Labrador Shelf.

Appendix 1

Numerical dictionary of Foraminifera and miscellaneous shelly microfossil taxa

1 NEOGLOBOQUADRINA PACHYDERMA	63 NODOSARIA SP11
2 GLOBIGERINA APERTURA	64 CASSIDULINA ISLANDICA
3 GLOBIGERINA PSEUDOBESA	65 COSCINODISCUS SP1
4 GLOBOROTALIA INFLATA	67 SCAPHOPOD SP1
5 GLOBOROTALIA CRASSAFORMIS	69 NODOSARIA SP8
6 NEOGLOBOQUADRINA ACOSTAENSIS	70 ALABAMINA WOLTERSTORFFI
7 GLOBIGERINOIDES RUBER	71 EPISTOMINA ELEGANS
8 ORBULINA UNIVERSA	73 EPONIDES SP3
9 FURSENKOINA GRACILIS	74 EPONIDES SP5
10 UVIGERINA CANARIENSIS	75 LENTICULINA ULATISENSIS
11 NONIONELLA PIZARRENZIS	76 CASSIDULINA SP
12 EHRENBERGINA SERRATA	77 ELPHIDIUM SP
13 HANZAWAIA CONCENTRICA	78 UVIGERINA PEREGRINA
14 TEXTULARIA AGGLUTINANS	79 GLOBIGERINA TRIPARTITA
15 GLOBIGERINA PRAEBULLOIDES	80 CYCLAMMINA CANCELLATA
16 CERATOBULIMINA CONTRARIA	81 GLOBIGERINA VENEZUELANA
17 ASTERIGERINA GURICHI	82 SUBBOTINA LINAPERTA
18 SPIROPLECTAMMINA CARINATA	83 PLANOROTALITES PSEUDOSCITULUS
19 GLOBIGERINOIDES SP	84 SUBBOTINA YEGUAENSIS
20 GYROIDINA GIRARDANA	85 PSEUDOHASTIGERINA MICRA
21 GUTTULINA PROBLEMA	86 TURRILINA BREVISPIRA
22 COSCINODISCUS SPP	87 BULIMINA AFF. JACKSONENSIS
24 TURRILINA ALSATICA	88 SIPHOGENEROIDES ELEGANTA
25 COARSE ARENACEOUS SPP.	89 MOROZOVELLA SPINULOSA
26 UVIGERINA DUMBLEI	90 ACARININA DENSA
27 EPONIDES UMBONATUS	92 MOROZOVELLA CAUCASICA
28 CIBICIDOIDES SP5	93 ACARININA AFF. BROEDERMANNI
29 CYCLAMMINA AMPLECTENS	94 GLOBIGERINATHEKA KUGLERI
30 CIBICIDOIDES BLANPIEDI	96 ACARININA INTERMEDIA WILCOXENSIS
31 PTEROPOD SP1	109 CASSIDULINA CURVATA
32 QUADRIMORPHINELLA INCAUTA	110 GLOBIGERINA BULLOIDES
33 TURBOROTALIA POMEROLI	111 PARAROTALIA SP1
34 MARGINULINA DECORATA	112 MARGINULINA BACHEI
35 SPIROPLECTAMMINA DENTATA	113 GLOBOROTALIA MENARDII GROUP
36 PSEUDOHASTIGERINA WILCOXENSIS	114 GLOBIGERINOIDES SACCULIFER
37 ACARININA AFF PENTACAMERATA	115 GLOBOROTALIA OBESA
38 LENTICULINA SUBPAPILLOSA	116 ORBULINA SUTURALIS
39 ALABAMINA WILCOXENSIS	117 SPHAEROIDINA BULLOIDES
40 BULIMINA ALAZANENSIS	118 EPISTOMINA SP5
41 PLECTOFRONDICULARIA SP1	119 SPHAEROIDINELLOPSIS SUBDEHISCENS
42 CIBICIDOIDES ALLENI	120 GLOBOROTALIA SIAKENSIS
43 BULIMINA MIDWAYENSIS	121 GLOBIGERINA NEPENTHES
44 CIBICIDOIDES AFF WESTI	122 SPHAEROIDINELLOPSIS SEMINULINA
45 BULIMINA TRIGONALIS	123 GLOBIGERINOIDES TRILOBUS
46 MEGASPORE SP1	124 GLOBOQUADRINA DEHISCENS
47 PLANOROTALITES PLANOCONICUS	125 "GLOBOROTALIA" CONTINUOSA
48 ANOMALINA SP5	126 GLOBIGERINOIDES OBLIQUUS
49 OSANGULARIA EXPANSA	127 GLOBIGERINITA NAPARIMAENSIS
50 SUBBOTINA PATAGONICA	128 GLOBOROTALIA PRAEMENARDII
51 ACARININA PRIMITIVA	130 SIPHONINA ADVENA
52 ACARININA SOLDADOENSIS	131 CIBICIDOIDES TENELLUS
53 UVIGERINA BATJESI	132 "GLOBORTALIA" OPIMA NANA
54 TEXTULARIA PLUMMERAE	133 LENTICULINA SP3
55 GAVELINELLA BECCARIIFORMIS	134 LENTICULINA SP4
56 GLOMOSPIRA CORONA	135 GLOBIGERINA SP40
57 SPIROPLECTAMMINA SPECTABILIS	136 MELONIS BARLEANUM
58 EPONIDES SP8	137 GLOBIGERINOIDES PRIMORDIUS
59 RZEHAKINA EPIGONA	138 GLOBIGERINA ANGUSTIUMBILICATA
60 PLANOROTALITES COMPRESSUS	139 "GLOBOROTALIA" OPIMA OPIMA
61 SUBBOTINA PSEUDOBULLOIDES	140 ROTALIATINA BULIMOIDES
62 GAVELINELLA DANICA	141 PLANULINA RENZI

142 GYROIDINA SOLDANII MAMILLIGERA
143 UVIGERINA GALLOWAY
144 TURBOROTALIA CERROAZULENSIS
145 ANOMALINOIDES ALLENI
146 SUBBOTINA EOCAENA
147 CATAPSYDRAX AFF. DISSIMILIS
148 GLOBIGERINATHEKA INDEX
149 GLOBIGERINATHEKA TROPICALIS
150 GLOBIGERINA GORTANII
151 BULIMINA BRADBURYI
153 BULIMINA COOPERENSIS
154 ANOMALINOIDES MIDWAYENIS
155 ANOMALINOIDES GROSSERUGOSA
156 SUBBOTINA FRONTOSA
157 TRITAXIA SP3
158 SUBBOTINA INAEQUISPIRA
159 MOROZOVELLA ARAGONENSIS
160 ACARININA PSEUDOTOPILENSIS
161 PLANOROTALITES AUSTRALIFORMIS
162 MOROZOVELLA AEQUA
164 NUTTALIDES TRUMPYI
166 MOROZOVELLA SUBBOTINAE
167 MOROZOVELLA FORMOSA GRACILIS
172 PSEUDOHASTIGERINA SP
173 ANOMALINA SP1
175 ALLOGROMIA SP
176 ALLOMORPHINA SP1
177 BOLIVINA DILATATA
179 GLOBOROTALIA SCITULA PRAESCITULA
180 GYROIDINA SP4
181 GYCLOGYRA INVOLVENS
182 PLECTOFRONDICULARIA SP3
184 GYROIDINA OCTOCAMERATA
187 CIBICIDOIDES GRANULOSA
188 PLEUROSTOMELLA SP1
190 ANOMALINOIDES ACUTA
191 "GLOBIGERINA" AFF. HIGGINSI
194 PLANOROTALITES CHAPMANI
196 OSANGULARIA SP4
206 EPONIDES POLYGONUS
210 LOXOSTOMOIDES APPLINAE
211 HANTKENINA SP

213 ARENOBULIMINA SP2
216 GLOBIGERINOIDES SICANUS
217 GLOBOROTALIA SCITULA
218 MARGINULINA AMERICANA
219 MARTINOTIELLA COMMUNIS
220 CIBICIDOIDES WUELLERSTORFFI
221 GLOBIGERINOIDES SUBQUADRATUS
222 GLOBOQUADRINA ALTISPIRA
223 GLOBIGERINA CIPEROENSIS
224 UVIGERINA MEXICANA
225 GLOBIGERINA AFF. AMPLIAPERTURA
226 ACARININA SENNI
227 CIBICIDOIDES AFF. TUXPAMENSIS
228 CASSIDULINA TERETIS
230 BULIMINA OVATA
231 UVIGERINA RUSTICA
232 GLOBIGERINOIDES IMMATURUS
233 CATAPSYDRAX UNICAVUS
234 TRUNCAROTALOIDES AFF. ROHRI
235 SUBBOTINA BOLIVARIANA
236 EPONIDES SP4
237 LENTICULINA SP8
238 CIBICIDOIDES SP7
239 NONIONELLA LABRADORICA
240 ELPHIDIUM CLAVATUM
241 GLOBOROTALIA TRUNCATULINOIDES
242 GLOBOROTALIA FOHSI GROUP
243 GLOBIGERINA DECAPERTA
244 GAUDRYINA SP10
245 PRAEORBULINA GLOMEROSA
246 GLOBIGERINATELLA INSUETA
247 GLOBIGERINOIDES ALTIAPERTURA
248 "GLOBOROTALIA" AFF. INCREBESCENS
249 GLOBIGERINATHEKA SEMIINVOLUTA
250 VULVULINA JARVISI
251 ANOMALINA SP4
252 MOROZOVELLA AFF. QUETRA
253 SUBBOTINA TRILOCULINOIDES
254 PLANOROTALITES PSEUDOMENARDII
255 MOROZOVELLA CONICOTRUNCATA
256 "MOROZOVELLA" AFF. PUSILLA
257 CHILOGUEMBELINA SP

Appendix 2

Alphabetical dictionary of Foraminifera and miscellaneous microfossil taxa

37 *Acarinina* aff. *pentacamerata* (Subbotina)
51 *Acarinina primitiva* (Finlay)
52 *Acarinina soldadoensis soldadoensis* (Bronnimann)
90 *Acarinina densa* (Cushman)
93 *Acarinina* aff. *broedermanni* (Cushman and Bermudes)
96 *Acarinina intermedia* Subbotina–*A. wilcoxenis* (Cushman and Ponton)
160 *Acarinina pseudotopilensis* (Subbotina)
226 *Acarinina senni* (Beckmann)
39 *Alabamina wilcoxensis* Toulmin
70 *Alabamina wolterstorffi* (Franke)
175 *Allogromia* sp.
176 *Allomorphina* sp. 1
251 *Anomalina* sp. 4
173 *Anomalina* sp. 1
48 *Anomalina* sp. 5
154 *Anomalinoides midwayensis* (Plummer)
155 *Anomalinoides grosserugosa* (Gümbel)
190 *Anomalinoides acuta* (Plummer)
213 *Arenobulimina* sp. 2
17 *Asterigerina gurichi* group (Franke)

177 *Bolivina dilatata* Reuss
40 *Bulimina alazanensis* Cushman
43 *Bulimina midwayensis* Cushman and Parker
45 *Bulimina trigonalis* Ten Dam
87 *Bulimina* aff. *jacksonensis* Cushman
151 *Bulimina bradburyi* Martin
153 *Bulimina cooperensis* Cushman
230 *Bulimina ovata* d'Orbigny

228 *Cassidulina tereis* Tappan
109 *Cassidulina curvata* Phleger and Parker
76 *Cassidulina* sp.
64 *Cassidulina islandica* Norvang
233 *Catapsydrax unicavus* Bolli, Loeblich and Tappan

147 *Catapsydrax* aff. *dissimilis* (Cushman and Bermudez)
16 *Ceratobulimina contraria* (Reuss)
238 *Cibicidoides* sp. 7
227 *Cibicidoides* aff. *tuxpamensis* Cole
220 *Cibicidoides wuellerstorfi* (Schwager)
187 *Cibicidoides granulosus* (Bykova)
131 *Cibicidoides tenellus* (Reuss)
44 *Cibicidoides* aff. *westi* (Howe)
42 *Cibicidoides alleni* (Plummer)
30 *Cibicidoides blanpiedi* (Toulmin)
28 *Cibicidoides* sp. 5
257 *Chiloguembelina* sp.
25 Coarse arenaceous spp.
22 *Coscinodiscus* spp.
65 *Coscinodiscus* sp. 1 Bettenstaedt
29 *Cyclammina amplectens* Grzybowski
80 *Cyclammina cancellata* Brady
181 *Cycloygra involvens* (Reuss)

12 *Ehrenbergina serrata* Reuss
77 *Elphidium* sp.
240 *Elphidium clavatum* Cushman
118 *Epistomina* sp. 5
71 *Epistomina elegans* (Orbigny)
27 *Eponides umbonatus* (Reuss)
58 *Eponides* sp. 8
73 *Eponides* sp. 3
74 *Eponides* sp. 5
206 *Eponides polygonus* Le Calvez
236 *Eponides* sp. 4

9 *Fursenkoina gracilis* (Collins)

244 *Gaudryina* sp. 10
55 *Gavelinella beccariiformis* (White)

Appendix 3

Taxonomic emendations

74. *Eponides* sp. 5. This species illustrated in Plate 5, items 14, 15 resembles *Eponides plummerae* Cushman.

118. *Epistomina* sp. 5. This species, illustrated in Plate 5, items 12, 13 resembles *Epistomina eocenica* Cushman and Hanna.

9. *Fursenkoina gracilis* (Collins). Specimens recorded in the wells as *Fursenkoina gracilis* (Plate 6, item 5) belong in *Bulimina* or *Caucasina* and closely resemble illustrations of *Bulimina elongata* (d'Orbigny) in Marks (1951) or Batjes (1958).
   The initial part of the test is either trochoid or multiserial, followed by an often elongate, twisted triserial to twisted biserial part; the test is smooth, the aperture is loop-shaped. Common in Miocene or Pliocene deposits.

125. *'Globorotalia' continuosa'* Blow. Specimens referred to this form (Plate 3, item 13–16) more likely belong in the plexus of *Globorotalia mayeri* Cushman and Ellisor—*G. acrostoma* Wezel.

32. *Quadrimorphinella incauta* (Vetrova). This strongly-involute form is illustrated in Plate 7. The test appears to be insoluble in 10% HCl. Since the genus *Quadrimorphinella* as stated in the original description is thin-walled, calcareous (although frequently pyritized) our specimens more appropriately belong in *Ammosphaeroidina*, and are being classified as A. sp. 1

24. *Turrilina alsatica* Andreae and 86. *T. brevispira* Ten Dam [= *T. robertsi* (Howe and Ellis)].
   Turrilina alsatica Andreae (Plate 5, items 2, 3) and *T. brevispira* Ten Dam are distinguished on difference in wall structure; also *T. brevispira* is not much longer than wide, whereas *T. alsatica* shows considerable variation in length versus width, with length exceeding width up to 2–2.5 times.
   T. brevispira has a random orientation of calcite crystals in the wall or an orientation where the c-axis of the crystals deviates from normal to the wall, resulting in a granular extinction pattern of (crushed) tests under the light microscope between crossed nichols. *T. alsatica* has the calcite crystals oriented with the longest (c-) axis normal to the wall resulting in a black cross extinction pattern (radial pattern). Hansen (1972) discussed this wall structure and the taxonomic and systematic implications.
   T. robertsi (Howe and Ellis) (originally *Bulimina robertsi* Howe and Ellis) from the Eocene of Louisiana probably is a senior synonym of *T. brevispira* Ten Dam. Specimens from *T. (Bulimina) robertsi* as described by Howe (1939) and by Nogan (1964) closely resemble *T. brevispira* and *T. robertsi* as described by Todd and Kniker (1952) from (Eocene) Agua Fresca shale of S. Chile shows a granular extinction pattern.
   T. alsatica is a useful Oligocene subsurface marker for the Labrador Shelf and Grand Banks; *T. robertsi* occurs in the E–M Eocene of several wells studied.

54. *Textularia plummerae* (Lalicker). A detailed study of (complete) specimens recorded under this name shows a spiral beginning of the test (Plate 7, items 6–9); the specimens are referred to *Spiroplectammina navarroana* Cushman. For a discussion of the differentiation of *Spiroplectammina* and *Textularia* the reader is referred to Gradstein and Berggren (1981).

31. Pteropod sp. 1. Specimens recorded under this name (Plate 7, items 17, 18) more likely belong to the Gastropoda and are now referred to as gastropod sp. 1. The original designation was inspired on the widespread occurrence of such specimens in an open marine environment, reminiscent of a pelagic mode of distribution.

Appendix 4

File of the original sequences of exits of microfossils as recorded in each of the 16 wells of Group 1, northern Grand Banks and Labrador Shelf, and for the Group 2 of 6 wells, southern Grand Banks and Scotian Shelf. The sequence in each well reads stratigraphically downward, proceeding from left to right. Dictionary numbers are as in Appendix 1. A minus sign in front of a taxon number indicates that the species exit occurs in the same sample as that of the taxon immediately to the left.

ORIGINAL SEQUENCE DATA

```
Heron
  11  -109    -9   112    14    78    -8  -116    -2  -113  -117    18  -120  -111   217   -67   -71    15    65  -128
-124  -122   125  -132  -242  -216  -218   219  -220   -81   221    21  -123  -222    26  -147   -70   -20   142  -150
 -16   223    79  -224   225   159   -92   -90   -93   -37   -89  -226  -156  -158  -227   -57  -154  -190  -155  -151
 -43   -35  -157    49  -166

Kittiwake
   5   -11    -9   112    14  -115    78     8    -6  -217    15    18  -219  -113  -120   -25    70   128  -117
 231  -232   -20  -122  -124  -220   223  -130   233   -81   -21  -139   -29    33   -31  -148   -94   -82   -84
 -69   187  -234  -235  -142  -140  -190    90    40    50   -96    93

Mohican
 228   -64   -77    -1   113    -4  -239   240   241     5    -7   115     9   219   242   243   -15  -127    20  -128
 -16  -120  -122    81  -231

Triumph
 228   -78   -77     1     5     9   -67    17    15  -119  -128    18   -11  -245  -115   219   117   222  -120  -137
 242   124  -122   -16  -123   -71   -26  -231   -21  -220   -20    -7  -127   217  -246   125  -216   247   143    25
  70  -147  -139   -81   -79  -248    33  -149  -249  -148  -226   -82  -250    85   -29  -187   -34    90   -37  -158
-153   -54    93   -89   -43  -190    36   -92  -160  -164  -154   -59  -210  -230  -157  -251  -227   159    57  -252
 166   -96   -52    30  -244   -56    53   253   -60  -161   -47  -254  -255  -256  -257

Brant
 109    11   -77    -9   111   -14  -112   -65     2    -6  -114  -115  -116  -117   -71   -78    18  -119  -120
-121  -122  -123   -67  -124  -125  -126   127   242    -8    21   -15    12  -130  -131   -27  -132  -133  -134
  20   -16  -135   -26  -136  -137  -138    25    79   139   140   -29  -141   142   143    44   -34    33  -144  -145
 -82  -146  -147  -148  -149  -150   -31   -85    90  -159   -37   -93  -153  -154  -155  -156   -52   -49  -157
 158  -160   -96  -161  -162   -92   -36  -164   166  -167   -62

Wenonah
  11    14   128  -219  -123   -78    -9   -18   -20  -124   -17   -10  -122   120  -222    15  -242  -112   220
-231  -217   -16  -125   -80    21   -40  -131  -223  -139   143   -81   147    52   -50   -56  -244    57   -47  -166
 -96   -54
```

ORIGINAL SEQUENCE DATA

Bjarni H81
```
64   65  -16   22   67   20  -21   18  -69  -70  -71   15   24   25  -29   34   41   42  -73  -74
75  -30   57   46   56
```

Cartier
```
65   16   18   15   21  -70   67   69   24   24   25   31   34   36 -118   29  -46  -42  -75
35  -50   41  -57   54   56  175 -176  -59 -172                   -118 -173
```

Freydis
```
16  181  -67  -21  -18   20   69  -27   15  -70   31   29  -25  -34 -206  -42  -74 -173   45
90  -81  -41  -75 -176 -210  211  -85  -94  -50   57  -88  -86  -46  -35   56   54  213  -55
59
```

Gudrid
```
10  -17   20  -21  -18  -16   69   70  -15  -25   33   31   41   40  -29  -34   84  -85  -83
35  -42  -45  -74   57  -88  -30   43 -176  -46  -50   56  -59   55
```

In. Harb.
```
 1   -2   -3   -4   -5   -6   -7   -8   -9  -10  -11  -12  -13   15  -16  -17  -18  -19   20
-21  -22   24  -25   26  -27  -28   29  -30  -31   32   33   34   36  -37  -38  -39  -40  -41
-42   43  -44   45  -46  -47  -48   49  -50   51  -52  -53  -54  -56  -57  -58  -59  -60  -61
-62  -63
```

Karlsefni
```
228   22   67   25   41 -118   69   29   57  -39   53 -206 -173  -30  -63  -31  -34  230  -44
-42  -46   96  -36  164  -50   52   45  -54   55  -43  -62  -56 -253
```

Leif M48
```
228  -77   22  181   16  -67   15   20  -21  -18   70   69   85   34   73 -236   25 -237 -238
 42   29   57  -74 -118  -50   30  -41   46  -56
```

Leif E38
```
228  -64  -77   65   67  -16   18  -21   20
```

Sample																				
Snorri	77	228	65	16	67	15	−21	18	25	57	−34	−29	−53	−41	−30	−173	27	−46	118	230
	86	−63	31	−42	43	45	−176	56	59	−54										
Herjolf	65	67	18	−15	−20	−16	78	70	25	85	−145	−71	−40	45	−35	41	−53	−36	−30	−29
	86	57	54	46	32	−190	31	47	−157	−56	−140	55	60	59						
Bonavista	76	−77	10	−17	−16	21	25	−20	18	79	24	−26	81	15	82	31	83	−70	40	33
	−84	−27	29	85	−86	−69	−87	41	−34	57	−88	−42	−30	−89	−90	−159	−92	−93	−94	56
	−50	47	−96	−36	46															
Dominion	177	−109	11	−9	17	10	−117	−78	112	18	179	−16	−15	−71	−122	180	26	−123	−137	14
	−136	27	20	−21	−181	24	−131	25	−31	−29	34	−182	38	142	−81	−184	−82	−30	−146	69
	−32	57	187	−49	−188	−54	−147	−190	−140	56	−40	191	−156	151	250	−226	36	−44	194	−90
	50	−47	−158	−161	−52	−46	37	−159	−162	196	45	−230	53	164						
Egret K-36	17	26	16	20	−21	−18	−71	−15	24	27	−42	31	69	82						
Osprey	17	18	−20	15	−16	26	−181	81	31	82	84	−147	−69	−148	90	−89	−33	−62	−234	−34
	−244	52	−51	−162	−159	−166	−50	−93												
Cumberland B-55	22	−76	228	−1	10	−11	−9	−109	−71	17	−16	−20	18	15	−119	25	−80	−117	219	26
	24	132	42	41	−182	84	29	43	144	31	49	32	−57	−36	−54	90	52	161	−93	−96
	−151	−164	−157	46	−50	−159	55	−56	−254	−194										
Egret N-46	11	−16	−18	14	−27	−71	−22	23	−26	−20	15	−24	31	−172						

Quantitative Stratigraphic Correlation
Edited by J. M. Cubitt and R. A. Reyment
© 1982 John Wiley & Sons Ltd.

Analysis of Paleontologic Time Series and its Application in Stratigraphic Correlation — A Case Study Based on Orbulina Data from DSDP Samples

BENOY K. GHOSE

Indian Institute of Technology, Kharagpur-721302, India

Abstract

Temporal distribution of the abundance of a species varies in response to climatic fluctuations through time. The time-dependant oscillatory movement of the abundance can be fitted to one of four theoretical models presented herein, depending on specific responses. A paleontologic time series, based on an ordered set of observations of the abundance of species obtained from regularly-spaced samples, yields information on frequency and length of the oscillatory cycles. As climatic factors are global in their extent, the distribution cycles are synchronous, providing a base for stratigraphic correlation.

Analysis of the Recent *Orbulina universa* distribution, a holoplanktonic foraminifer chosen for the case study, shows that it thrives when winter and summer temperatures are between 19.5–21.5 °C and 24.5–26.5 °C, and salinity is within $36.3^0/_{00}$ and $37.3^0/_{00}$, and thus fits a model appropriate for the temperate species.

An univariate time series analysis of the temporal frequency distribution data of this species from Late Pleistocene samples from DSDP core 216, 217, and 219, and of published data from Caribbean core, shows that the 216 and 217 series are similar in respect of the length of dominant periodic components; the other two series are different. Bivariate analysis indicated the need for stretching to achieve acceptable correlability. Analysis of the stretched series improved the degree of intercorrelation and allowed the construction of a line of stratigraphic equivalence on the depth plots of the series. Autospectral analysis helped determine the stretch factors. Use of spline function yielded stretched series with minimum distortion.

An analytical study on the sampling interval showed that optimum SI is related directly to cycle length of interest and the length of the series should be preferably 50 or more.

Introduction

Paleontologic time series may be defined as an ordered set of observations on fossils ideally obtained from samples taken at equal intervals of time. As this ideal is rarely realizable, data yielded by regularly-spaced samples from a sedimentary sequence may also be used to construct a series. Here, depth may be regarded as a time metameter simply and linearly related to time. The observational data may include morphometric attributes of a species, its relative or absolute abundance, ratios of abundances of two groups of species, species diversity, or any other data derived from the above.

In recent years, a number of papers have appeared in which paleontologic time series based on planktonic foraminiferal data have been used to estimate paleoclimatic trends (Hunkins and others, 1971; Lynts, 1971; Wollin, Ericson, and Ewing, 1971; Hecht, 1973; Malmgren and Kennett, 1978; and Malmgren and Healy-Williams, 1978). In these papers mean size, abundance, and other paleontological data have been taken to form the series.

The purpose of the present paper is to investigate the applicability of the time series analytic method in stratigraphic correlation. The work is based on temporal distribution data of *Orbulina universa*, a holoplanktonic foraminifer common in Recent tropical and subtropical seas. In the first section, Recent distribution of the species has been analyzed to determine the optimum condition under which it thrives and to understand its environmental response. Broad periodicities in the distributional pattern, which are assumed to be related to paleoclimatic cycles, are then determined from univariate time series analysis. This also helped form an idea about the nature of the series and the appropriate model for it. To establish a relation between the two series, bivariate analytic method was applied. This not only showed the degree of correlation or lack of it, but also indicated if stretching was needed to improve the correlatebility. The correlated time series formed the basis of stratigraphic correlation. In the final section, an extended analytical discussion is given on the optimum sampling interval for analysis of paleontologic time series.

Theory

It is universally accepted that the abundance of a species is a function of environmental variables. This leads to the spatial variation which is responsible for the specific nature of geographic distribution of Recent species. In the marine realm, important physical environmental factors are salinity, summer temperature, winter temperature, depth, turbidity, nature of the substrate, and turbulence. Of these factors, the last four are purely local, conditioned by spatial distribution of lands and seas. The first three are primarily global in their extent, although partly, and in some cases considerably, affected by local conditions.

It is also well known from studies based on fossils, oxygen isotope ratios, and other considerations that every location on the earth has witnessed changes in environment through time. In this temporal change, local factors appear to be transitory in nature, whereas effects of global factors persist.

The environmental changes discussed previously must have affected the species of that locality and induced changes in their relative and absolute abundance. Organisms thrive under conditions which are optimum for their growth. Consequently, the abundance of a species will increase when the environment changes towards this optimum and decrease when conditions deteriorate. The temporal variation in abundance of different species will be in or out of step depending on whether they have similar or unlike optimum growth conditions.

It is logical to assume that changes in the abundance of a species will be broadly oscillatory in nature with well defined peaks and troughs coinciding with optimum and adverse conditions. This broad oscillatory movement of abundance values will be variously modified due to the multiplicity of environmental factors. Generally speaking, global factors may be supposed to be responsible for broad movements and local factors, for minor oscillations.

It is obvious that, temperature is the most important of global factors, which is again related to paleoclimatic cycles. Complications arise as there are cycles within cycles. Consequently, abundance values of a particular species would show a complicated movement in a temporal sequence with occasional peaks and troughs. Although the resulting time series is a complicated one, it is amenable to statistical analysis yielding useful information about the past history of environmental changes at a locality.

If we can identify a species which responds more to changes in global rather than local factors, and define a locality which was little influenced by local factors during the period under consideration, temporal changes in global factors can be followed with considerable precision from a study of change in abundance of that species with time. As, by definition, these factors uniformly affect all parts of the earth, the changes induced by them in abundance of a species or group of species inhabiting different localities would be in the same direction at a particular point of time. This synchroniety of events provides us with an important tool for stratigraphic correlation with greater precision than possible by other methods. However, application of this method depends solely on our ability to establish the influences of global and local factors on the abundance of a species. If we regard the former as the signal and the latter as noise, we may then use the parlance of the spectral analyst and say that our ability to separate signal from noise determines the usefulness of the method. If the time series constructed from the temporal distribution data is such that the noise is as strong as the signal, the analysis would prove to be futile. It must be remembered that the 'noise' mentioned previously may also be created by factors unrelated to the growth condition of the species. For example, accumulation of fossil skeletons, which are supposed to represent the organism in question, is, to some extent, influenced by rate of sedimentation, influx of sediments, and other controlling depositional conditions.

In this context I would like to point out that quantitative correlation based on time series analysis presupposes the existence of a datum which can be fixed with considerable accuracy in all sections being correlated. From that datum we may go forward or backward in time, or both. The reason for this is obvious. Similar general patterns in the distribution of a species is likely to repeat more than once. Hence, unless we have a fixed frame of reference, comparison of patterns, which forms the basis of correlation, would be meaningless.

Before proceeding further, it is necessary to consider the nature of the broad temporal distributional patterns that may arise owing to changes in global factors, that is summer temperature, winter temperature, and salinity. As salinity, to a great extent, depends on prevailing temperature, it is logical to assume that the distributional pattern is mainly controlled by paleotemperature trends.

Theoretically, four patterns of distribution may be recognized.

Pattern 1

Optimum growth conditions occurring at high temperature (tropical species). In this situation, the abundance maxima will coincide with temperature maxima and the temporal distribution pattern would faithfully represent paleoclimatic cycles (Fig. 1a).

Pattern 2

Optimum condition occurring at low temperature (subpolar species). This situation is similar to the previous pattern. The only difference is that abundance maxima coincide with temperature minima (Fig. 1b).

Pattern 3

Optimum condition occurring at intermediate temperatures (temperate species). Here, for each paleoclimatic cycle, there would be two abundance cycles (Fig. 1c). Probably the

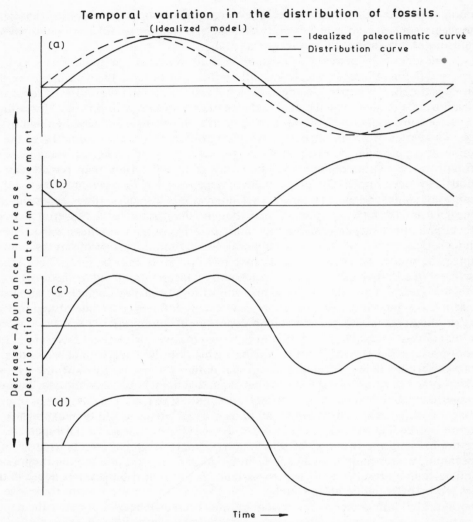

Fig. 1. Models for temporal abundance distribution cycles: (a) for tropical species,
(b) for subpolar species, and (c) and (d) for temperate species

trough corresponding to higher temperature would be at a higher level than that corresponding to the lower one.

Pattern 4

Optimum condition spreads over a wide range of temperature (commonly intermediate to high). In this situation, the abundance values would rapidly reach the maximum and stay there for a long period; hence, the temporal abundance curve would have flat tops (Fig. 1d).

In actual examples, the general models presented above will be considerably modified by paleoclimatic cycles of higher frequency and other factors already referred to. The generalized models may appear to be too simplistic, but it helps us understand the basic nature of temporal distribution of an organism.

It may be noted that, in the figures, a slight lag is shown between paleoclimatic and distributional cycles. This has been calculated on the assumption that environmental response of a species is not instantaneous.

Choosing the Species for Time Series Analysis

There is no organism which is totally insensitive to environmental changes. Nevertheless, some species are more sensitive than others and respond vigorously to changes in environmental condition. Such a species should be selected for the type of analysis we are envisaging.

Terrestrial forms are unsuitable as one can rarely obtain a continuous record of occurrence through time. Moreover, preservation is usually poor and abundance low.

Fresh water organisms are suitable provided they leave a continuous record of their existence, preservable parts are abundant, and distribution uniform. Spores and pollens answer to these requirements well. In fact, palynologists frequently have used temporal frequency-distribution diagrams of spores and pollens, in a semiquantitative manner, for stratigraphic and coal-seam correlation.

It may be emphasized here that the occurrence of a large number of fossil specimens in a small sample is an essential precondition. Because unless the abundance is high, data cannot be collected for the time series analysis. For this reason, marine microorganisms with preservable hard parts such as foraminifera, ostracodes, and radiolarians may be selected for the study.

Benthic nearshore forms, particularly those living near bay mouths, are not suitable as the temporal distribution data for these species are likely to be strongly affected by 'noise' to the point of masking the 'signal'. Outer neritic benthic forms and planktonic organisms are usually free from 'noise' created by the proximity of land.

In this context, it may be mentioned that the distribution of planktonic organisms with calcareous tests tends to be patchy. There may be various reasons for the patchiness, but the major one is depth. For example, Be, Harrison, and Lott, (1973), in their study of the distribution of *Orbulina universa* have shown that there are large areas in the Indian ocean where this species is absent in the bottom sediments; presumably, the water depth in these areas is greater than the depth of compensation. Hence, if there are reasons to believe that the depth of the ocean in the area being investigated remained near to or below the depth of compensation during the period in question, microfossils with siliceous skeletons, such as radiolarians, may be chosen for the study.

In the present investigation, the species selected is *Orbulina universa* and the time interval considered is Late Pleistocene to Recent. Samples have been obtained from DSDP core from holes 216, 217, and 219. Of the three sites, the first two are located on the Ninetyeast ridge and the last one on Laccadive–Chagos ridge (Fig. 2). There are reasons to believe that, during the time interval under question, the depth of the ocean at these sites did not exceed the compensation depth. *Orbulina* data from Caribbean core given in Imbrie and Kipp (1971) were also analyzed. In none of the data is there any difficulty in fixing the datum as core top is the obvious choice.

Sampling Interval

It is difficult to take a decision on the optimum sampling interval. In many situations, the choice does not rest on the analyst. Ideally, the interval should be such that the samples are free from the effect of each other. By the term effect, I do not refer to the correlation

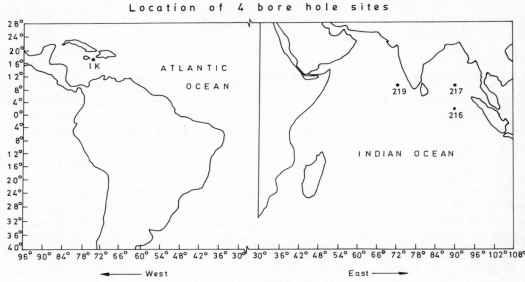

Fig. 2. Sample sites for paleontologic time series data

between neighbouring values of a series that is inherent in the data structure. Here, effect due to bioturbation and other related phenomenon is only inferred.

However, the interval should not be so wide as to cause the loss of valuable information. Obviously, sampling requirement would depend on the nature of the paleoenvironment of deposition of sediments. It would also depend on the purpose of the investigator. In the present situation the interval is 50 cm; it is perhaps too wide. This and other aspects of sampling will be discussed later.

The Species

Orbulina universa is a widely-distributed holoplanktonic foraminifer occurring frequently in tropical and subtropical seas. This species can be readily identified and separated from other foraminifera by a distinctive spherical test. Be, Harrison, and Lott, (1973) have shown from a detailed study of whole and dissected tests that it is a distinct species and its spherical test should not be regarded as a protective chamber developed by diverse forms during reproduction.

Studies of the distribution of Recent *O. universa* show that it thrives in subtropical waters (Hecht, 1973), with a depth range of 25–94 meters, and a salinity between $36.14–36.55\%_{00}$ (Lynts, 1971). To understand the relation between environmental factors and abundance of the species in greater detail, and to fit the temporal distribution pattern to one of the four theoretical models given earlier, the data given in Imbrie and Kipp (1971) have been analyzed in the following section.

The Data

As mentioned earlier, the data for the temporal distribution of *O. universa* was obtained from DSDP core samples taken at an interval of 50 cm. Samples at closer interval could not be obtained owing to certain restrictions placed by DSDP authorities. In holes 216 and 217, only the upper 9 meters of Pleistocene cores were recovered. Hence, the number of samples was only 19. For hole 219, 26 samples from 13.5 meters of Pleistocene core were available for study. In this situation, the core top sample and another sample from the middle were missing.

Although the interval was nearly uniform, in some situations the exact sampling point was 10 to 20 cm off line and needing adjustment. The data were adjusted using a step function as suggested by Davis (1973). For situations needing greater adjustment and for the two missing data values, interpolated values were obtained using a spline function (cubic splines).

In order to make a comparative study, *Orbulina* data from the Caribbean core given in Imbrie and Kipp (1971) were also taken. The authors have tabulated relative abundance values for this and other species of planktonic foraminifera separated from 110 samples taken at an interval of 10 cm. For the sake of uniformity, every fifth value of the series, including the first, were taken. Thus, 11 meters of Caribbean core yielded 23 data points. Analysis was also performed in the full 110 data points in order to understand the underlying structure and fit a suitable model. The data given by the same authors for the distribution of Recent *O. universa* in the bottom sediments were also taken to analyze the relation between 'global' factors (summer and winter temperatures and salinity) and abundance of the species.

In this context, it may be noted that relative abundance data is corrupted by the abundance distribution of other species. If we take the extreme case where two co-existing species are growing under environmental conditions favorable to both, the species whose reproductive efficiency is higher would grow at a faster rate. As a result, the percentage of the other species would show a fall in spite of the fact that it was growing in absolute terms. In actual examples however, there are a much larger number of coexisting species whose

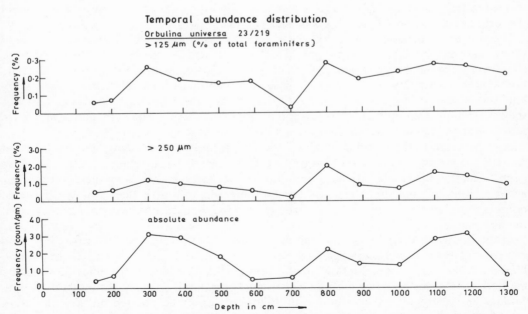

Fig. 3. Frequency polygons showing relation between relative and absolute abundance distribution of *Orbulina universa*

optimum conditions for growth do not coincide. The resulting relation between growth of the community and the individual species cannot be simply stated.

To understand the relationship between absolute and relative abundance, values have been plotted for one core (site 219). Fig. 3 shows that the distributional pattern in the three graphs is broadly similar though peaks and troughs of the relative abundance are subdued. Hence, we may conclude that a series based on relative abundance data can also be used for TS analysis.

Evidently, the series that have been analyzed are all short. Books on time series and spectral analysis mention that the number of observations in a series (that is, the series length) should be more than 50. However, Granger and Hughes (1968) have shown from a simulation study that the spectral analysis of a series as short as 30 observations does give considerable information, if the power at the frequency of interest is large.

It will also be evident from what is presented in this paper that analysis of short series is not a futile exercise, though determination of spectral peaks may occasionally be imprecise.

Preparation of the Sample

About 1 gm of air-dried sample was taken and disintegrated by the usual method. The dispersed material was then wet sieved into three size fractions, which are, $+250\ \mu$m, 125–$250\ \mu$m, and $-125\ \mu$m. All the tests of *O. universa*, which are mostly concentrated in $+250\ \mu$m fraction, were separated and counted. The resulting data were then recalculated to give abundance values in count/gm. These formed the basic raw data, which were adjusted to obtain the data points for the series.

Analytical Method

The series were analyzed both in the time and frequency domains. Filtering was not performed before the analysis as choice of a suitable filter depends on prior information about the structure of the series, which we do not possess.

It has been assumed in this work that the series are stationary. This is based on the *a priori* reasoning that, on average, abundance of a species does not show any systematic change in either mean or variance. Further, it may also be assumed that the multivariate probability density function is normal. These are conditions for complete stationarity (Jenkins and Watts, 1968). Limited interval of time being considered is an additional basis for the assumption. Generally, however, if the paleontologic time series is long, one may have to consider the possible influence of evolutionary trend and take steps to remove it.

As a first step to time-domain analysis, the data were smoothed using Spencer's 5-point formula. This gave a preliminary idea of the trend. Next, serial correlation coefficients have been calculated up to a lag equal to one-third of the series length, using the exact formula given in Kendall and Stuart (1966). The correlograms indicated the nature of the most dominant periodic component of the series even if individual values were not significant at the acceptable level.

In the frequency domain, both harmonic and spectral analysis were performed because a comparison of smoothed spectrogram and spectrum helped fix the value of peak frequency. Herein, I have used 'spectrogram' in the sense of Anderson (1971), and followed the method given in his book for harmonic analysis. For spectral analysis, the method given in Jenkins and Watts (1968) was uniformly followed. It may be pointed out that the auto-covariance estimator used by the previous authors, though biased, has less mean square error than that used by other authors including Granger and Hatanaka (1964).

For smoothing the sample spectral estimate the Tukey–Hanning window was used. In all situations, the technique of window closing has been employed to investigate the consistency of estimated frequency at the neighbourhood of the peak. However, due to shortness of the series, only a two-fold decrease in the band width was possible in some situations. In the present investigation, it has been observed that spectrum estimated with a wide window (band width -0.444 corresponding to lag 3) smooths out minor peaks but not major ones.

Spectra were estimated usually with three different frequency spacings. Of them, one is related to the series length; the other two are 0.05 and 0.025. The first one was calculated so that a comparison between spectrum and spectrogram becomes easy. It may be pointed out that, although the frequency spacing need not bear any relation to the truncation point L (lag), no purpose is served by decreasing the spacing further than $1/N$.

For bivariate analysis in the time domain, cross correlation coefficients were calculated using the exact formula up to lag $= \frac{1}{3} N_2$, N_2 being the length of the shorter series. For significance estimation, the terms correlated were used to calculate degrees of freedom.

As suggested by Jenkins and Watts (1968), coherency estimates were plotted in the 'Z' scale. This was performed to show confidence limits by a single vertical line. The same procedure was followed in plotting spectral estimates, which were drawn on a log scale. Inspite of the criticism levelled against such a procedure (Anderson, 1971; Chatfield, 1975), the advantage of inserting a single confidence scale is obvious. For 'Z' transformation, statistical table (Pearson and Hartley, 1966) rather than transformation formula has been used.

For multivariate analysis, I have followed Finn (1974). In the program that I have written for this purpose, matrix inversion was performed by Cholesky factorization. The advantage of this method is that the Cholesky factor matrix is obtained as an intermediate step, which has uses in step-down analysis. However, this subroutine for inversion is not particularly accurate when there are large differences in sums of squares, as one would obtain in calculating second or higher degree response surfaces (see Bock, 1975, for a comparative account of different inversion methods).

Programs for TS analysis, except that for harmonic analysis, were developed by the author. The latter has been taken from Davis (1973). The programs are in FORTRAN IV.

Analysis of the Distribution of Recent *O. Universa*

Imbrie and Kipp (1971, Table 2) have tabulated relative abundance (%) value of *O. universa* in 61-core top samples. Figure 1 of this paper shows that the sample localities are confined mostly to the Atlantic Ocean and are more or less evenly-distributed over the northern and southern hemispheres. The environmental variables recorded are summer temperature (ST), winter temperature (WT), and salinity (SAL). In the analysis presented, depth of sampling sites has not been taken into consideration. This factor is likely to affect the preservation of all planktonic foraminifera equally and, therefore, would not change the percentage of *O. universa* materially. Of course, I am not ruling out the probability of differential destruction entirely.

It was suspected that the difference between ST and WT also influences abundance of a species. So, for analyzing the relation between abundance of *O. universa* and environmental factors, an additional variable was created by taking the difference of ST and WT and multiplying it by the ratio of WT and ST. The multiplication was performed to avoid the sum of products (SP) matrix being singular creating complication during factorization and inversion. Thus, a 5-element observation vector and 61×5 data matrix were obtained.

The variables are:

Vector elements	Variables
x_{i1}	Summer temperature (°C)
x_{i2}	Winter temperature (°C)
x_{i3}	Salinity (‰)
x_{i4}	Interaction $\dfrac{(ST - WT) \times ST}{WT}$
x_{i5}	Abundance of *O. universa* (%)

Of these, the first four are predictor variables and the last, the dependent variable.

A summary of the multivariate data is presented in Table 1. The correlation matrix brings out the relationships between the predictor variables, some of which are well known. These relationships must be understood in order to discover how the environmental factors influence abundance of a species. As expected, the correlation between ST and WT is high and the relationship is a functional one. On the other hand, the relations between SAL, and ST and WT, though extremely significant, are not as strong as that between the latter two. This can be explained by the nature of the horizontal distribution of salinity. Although, at one spot, salinity is linearly dependant on temperature, that is, T–S curve is approximately straight, the horizontal distribution of surface salinity has a subtropical maximum at about 20° to 30° NS latitude and a temperature of 23 °C. Obviously, spatial T–S relationship would be best represented by a curvilinear regression (see Defant, 1961, for details). The

Table 1.   Table showing zero-order regression and correlation coefficients and other parameters

	x_1	x_2	x_3	x_4	x_5
Mean	29.3918	16.9213	35.6885	2.8686	0.8365
Standard deviation	7.0781	8.0684	0.9965	1.5236	0.8374

$$N = 61$$

Correlation matrix:

x_2	0.9653			
x_3	0.7629	0.7012		
x_4	0.4217	0.2390	0.5964	
x_5	0.4286	0.4342	0.4405	0.2055
	x_1	x_2	x_3	x_4

Significance level matrix (d.f. = 59):

x_2	0.001			
x_3	0.001	0.001		
x_4	0.001	0.05	0.001	
x_5	0.001	0.001	0.001	0.1
	x_1	x_2	x_3	x_4

Regression coefficients and equations:

(1) $x_{51} = 0.0507 \pm 0.0274$; $x_5 = 0.0507x_1 - 0.2482$
(2) $x_{52} = 0.0451 \pm 0.0239$; $x_5 = 0.0451x_2 + 0.0739$
(3) $x_{53} = 0.3702 \pm 0.1932$; $x_5 = 0.3702x_3 - 12.3740$
(4) $x_{54} = 0.1130 \pm 0.1377$; $x_5 = 0.1130x_4 + 0.5124$

interaction term (SWT) is significantly (at ~5 percent level) related to SAL and ST. This relationship needs further investigation.

Before turning to the joint influence of all predictor variables, it is necessary to examine the zero-order correlation and regression between each of the predictor and dependent variables. Table 1 shows that abundance (A) is significantly correlated with ST, WT, and SAL. Moreover, the coefficients are equal approximately. It may be construed that all three predictor variables equally affect the abundance.

The above conclusion is not, however borne out by partial correlation analysis. The coefficients are,

$$r_{14.32} = 0.066546 \ (n.s.)$$
$$r_{24.12} = 0.132746 \ (n.s.) \quad (1 = ST, 2 = WT, 3 = SAL \text{ and } 4 = A)$$
$$r_{34.12} = 0.337349 \ (\text{significant at 1 percent level})$$

(in the above *SWT* has not been considered).

The previous analysis clearly indicates that as a predictor variable salinity is the most important. The result is surprising and needs further investigation; the result, perhaps partly reflects inadequacy of the model in which only three environmental factors have been considered. It may be noted that, although the relation between A and SAL is significant, there is an amount of residual variation which has not been explained. Moreover, it should be remembered that T and SAL are highly correlated. Hence, at this stage, it would be unwise to exclude, ST and WT as predictor variables. The only permissible conclusion is that, of the factors considered, salinity has the greatest influence on abundance of the species and that, in multivariate analysis, the variables should be taken in the order, SAL–WT–ST–SWT. Although the zero-order correlation between A and SWT is not significant, this predictor variable is retained for reasons to be explained later.

Zero-order regression coefficients and equations are given in Table 1. It may be seen from the 95 percent confidence limits of β that, except b_{54} all are significant. Figure 4 shows the relationship graphically.

Table 2 gives the statistical parameters and other relevant data for multiple regression analysis. The multiple correlation coefficient, though significant at the 1 percent level, cannot be regarded as high. Only 23 percent of the variance could be explained by this linear regression model. It is interesting to note that the value of this coefficient is margin-

---

Table 2. Table showing multiple regression and correlation coefficients and other related data

Equation: $x_5 = -7.8875 - 0.0438x_1 + 0.0601x_2 + 0.2399x_3 + 0.0292x_4$

Standard error: $b_1 = 0.0808; b_2 = 0.0676; b_3 = 0.1761; b_4 = 0.1140$

Error sum of products (S_E) = 32.4290 d.f. = 56 mean = 0.5791 (1)

Regression sum of products (S_R) = 9.6462 d.f. = 4 mean = 2.4115 (2)

Test of significance: (2)/(1) = $F_{4,56}$ = 4.164; Significance level: 0.5%

Standardized regression coefficients:

$$b_1 = -0.3702; b_2 = 0.5791; b_3 = 0.2855; b_4 = 0.0537$$

Conditional standard deviation = 0.7610

Multiple correlation coefficients: $R^2 = 0.2293; R = 0.4788$

ally higher than the zero-order coefficients of correlation between dependant (A) and the first three regressor variables (ST, WT, and SAL). It may indicate that the increase of correlation is not appreciable when all the factors are considered together.

From the standard errors of β it appears that, individually, none of the coefficients are significant. But, it would be wrong to conclude that the preditor variables do not exert any influence. We have already observed that the zero-order and multiple correlation coefficients are statistically significant. Finn (1974) has pointed out that, the multivariate relationship is complex and, in such situations, signs are more important than the actual coefficients. In the present case, b_1 has a negative sign indicating that the abundance of *O. universa* decreases with increasing summer temperature. The result may appear surprising but not unexpected. It perhaps, reflects the well-known fact that this species has a maximum abundance in the subtropical, and not tropical zone (Be, Harrison, and Lott, 1973).

Analysis of the residuals was not followed up in detail, but certain salient features may be pointed out. Table 3 shows the location and other environmental parameters. It can be seen that situations of overprediction are fewer. The table clearly indicates that the depth of core tops has nothing to do with high or low abundance values. For example, a high positive residual is associated with a site whose core top depth is highest. Hence, the exclusion of core top depth as a preditor variable is justified. It is not suggested that depth does not affect preservation of tests, only that, in the present suit of samples, depth is not an important factor.

Other points of interest are, two of the sites associated with high negative residuals are located near the equator, and are characterized by high WT and ST and comparitively low salinity. On the other hand, four of the 6 sites associated with high positive residuals are characterized by moderate summer and winter temperatures and comparatively high salinity. Moreover, they fall within middle latitudes. A little consideration will convince one that, although the residuals are high, the distribution conforms to the general pattern, that is the abundance of *O. universa* is highest in the subtropical zones and high temperature is not conducive to the growth of the species. The abnormally-high value associated with site 12 has, perhaps, something to do with the North Atlantic Drift.

High positive residual for locality 58 situated on the northwest of Trinidad is difficult to explain. Here, the environmental conditions (low salinity and high WT and ST) are not

Table 3. Table showing location of high residuals
and environmental parameters

Location	Lat.	Long.	Depth (m)	Salinity ⁰/oo	W.T. °C	S.T. °C	Abundance %	Residual
A. High negative residuals (overprediction)								
44	23°02′N	43°48′W	3523	37.2	22.5	26.0	0.329	−1.01
50	0°46′S	26°02′W	3512	35.7	25.0	27.0	0.0	−1.05
57	1°57′S	39°02′W	3294	36.5	26.0	27.5	0.164	−1.10
B. High positive residuals (underprediction)								
12	54°15′N	16°50′W	2393	35.5	9.0	14.5	2.353	1.72
30	28°50′S	41°02′W	3781	36.5	19.0	24.0	2.490	1.42
37	29°01′N	41°24′W	3197	37.0	20.5	26.0	2.450	1.24
40	27°55′N	43°38′W	2582	37.2	21.0	26.2	2.793	1.52
41	15°20′S	19°43′W	4360	37.0	22.5	25.0	2.433	1.12
58	11°33′N	60°31′W	1018	35.0	26.0	27.5	3.061	2.15

W.T. Winter temperature   S.T. Summer temperature.

Table 4. Table showing stepwise analysis

Multiple correlation coefficients:
  (1) Including ST, WT, SAL, and SWT: 0.4788; $F_{4,57} = 4.16$ $(1\%)^*$
  (2) Eliminating SWT: 0.4779; $F_{3,57} = 5.62$ $(0.5\%)^*$
  (3) Eliminating SWT + SAL: 0.4357; $F_{2,57} = 6.80$ $(0.5\%)^*$
  (4) Eliminating SWT + ST: 0.4742; $F_{2,57} = 8.42$ $(0.1\%)^*$
*Significance level

Semipartial regression coefficients (order of predictor variables: SAL, WT, ST, and SWT):

Coeff.	Proportion of variance
2.857	0.194
1.140	0.031
−0.380	0.003
0.195	0.001

conducive to the growth of *O. universa*. The depth is also shallowest and the site is close to the mainland. Imbrie and Kipp (1971), while discussing the probable source of error, mention sampling and loss of surface layer. It is likely that the sample represents sediments deposited at a time when environmental parameters assumed values different from those prevailing now, and conducive to the growth of the species. It is equally difficult to explain the high negative residual (abnormally low concentration) at site 44.

To gain a further understanding about the relative importance of different environmental factors, step-wise analysis was performed which involves the gradual elimination of predictor variables and recalculation of multiple regression and correlation coefficients. The relevant data are presented in Table 4.

A comparative study shows that elimination of x_4 (SWT) decreases the value of the multiple correlation coefficient by only one unit in the third decimal place. In fact, the level of significance is increased to 0.5 percent. Of the regression coefficients, b_1 remains negative but has a smaller value; the value of b_2 is smaller, while that of b_3 is larger, indicating that the importance of salinity as a predictor has increased. Standard errors are large. Residuals having high positive and negative values do not show any appreciable change.

Elimination of salinity (b_3) as the predictor variable does bring in noticeable change. The value of R decreases though it is still significant at the 0.5 percent level; b_1 changes sign from negative to positive. Predictability decreases noticeably as shown by the values of the high positive and negative residuals.

Changing the order of the variables with SAL coming first, followed by WT, ST, and SWT (see Table 5 for the parameters and relevant data), brings out the relationship between the predictor and dependant variables further. It may be observed that, elimination of the last two variables does not change materially the value of R. On the other hand, the level of significance is higher (at 0.1 percent level). From semipartial regression coefficients, we see that the proportion of variation attributable to SAL and WT are 0.194 and 0.031; little (0.004) remains additionally to be attributable to ST and SWT.

From the linear regression model, the following conclusions can be drawn:

(a) Of the predictor variables considered, salinity exerts maximum influence on the abundance of *O. universa*. Higher salinity is conducive to the proliferation of the species.

(b) Winter temperature is next in importance and is positively correlated with abundance.

Table 5.   Table showing statistical parameters and other relevant data for log values

Correlation matrix ($n = 60$)

x_2	0.9000			
x_3	0.7342	0.6005		
x_4	0.6816	0.8696	0.4752	
x_5	0.6141	0.5161	0.5912	0.3182
	x_1	x_2	x_3	x_4

(r corresponding to
   2% and 0.1% significance points for d.f. 58 are
   0.30 and 0.415)

	b_1	b_2	b_3	b_4
Regression Coeff:	2.0274	0.9205	49.8462	-1.0118
Standard error:	2.6000	0.9002	22.6824	0.6448

$$a = -188.2057$$

R (including ST, WT, SAL, and SWT): 0.6682 (0.1%)[*]
R (eliminating SWT):   0.6494
R (eliminating SWT and SAL):   0.6198
R (eliminating SWT and ST):   0.6246

[*]Level of significance

(c) These two predictor variables can, together, account for a considerable proportion of the variation in the distribution of the species.

(d) Importance of summer temperature is marginal. It exerts an adverse influence on the abundance, that is, high summer temperatures decrease the abundance of the species.

(e) The difference between ST and WT is not important at least in the linear model.

(f) The linear model is not adequate to explain the variability of the abundance of the species, as there is a large residual variance which has to be accounted for.

The last conclusion led the author to try a nonlinear regression model of the form:

$$y = ax_1^p x_2^q x_3^r \ldots$$

For this, natural log values of the data were taken, and regression analysis was performed in the manner outlined previously. A summary of the results is given in Table 5.

A comparative study of the correlation matrices of the non-linear and linear models shows that there is weakening of correlation among first three of the predictor variables (ST, WT, SAL). On the other hand, the correlation between SWT, and ST and WT are much stronger.

There is a considerable increase in the strength of correlation between dependent variable A and each of the other predictor variables. In this model, correlation between the inter-action term SWT and A is significant at 2 percent level; it justifies the retention of the interaction term as a predictor variable.

The value of multiple correlation coefficient is higher than that for the linear model; it is significant at less than 0.1 percent level. About 45 percent of the variance is accounted for in this model. Estimate of β coefficients show that the allometry is positive for the first three predictor variables (ST, WT, and SAL) and negative for the fourth (SWT). Unlike that in the linear model, b_3 (for SAL) is significant at the 95 percent level. This again points to the fact that salinity is the most important of the predictor variables. Step-wise analysis reveals that elimination of two of the four predictor variables reduces the value of R from 0.668 to 0.619 only. The conclusion is that, either salinity or summer temperature combined with winter temperature are adequate as predictor variables. This is, perhaps, partly due to the fact that the predictor variables are highly correlated among themselves.

In spite of the improvement in the predictability achieved by the nonlinear regression model, there is a large unexplained residual variance. This may be due to inadequacy of the model. A second degree response surface model was also tried. The multiple correlation (R) is 0.5689. It is clear that the other nonlinear model shows a better fit. The significance level is also lower as the number of variables in this situation are larger. The regression equation is (x_1 = ST, x_2 = WT, x_3 = SAL, and y = A),

$$y = 452.5 + 1.4967x_1 + 1.3358x_2 - 27.1825x_3 - 0.0417x_1^2$$
$$-0.0324x_2^2 + 0.4069x_3^2 + 0.0775x_1x_2 - 0.0295x_1x_3$$
$$-0.0523x_2x_3.$$

The multiple regression coefficients, particularly their signs, pose certain interesting questions, but a discussion on that is beyond the scope of the paper.

It is also likely that much of the residual variance can be accounted for if we include one or two other environmental factors as the predictor variable. I believe that turbidity and availability of food are two such factors which should also be considered. Finally, it must be remembered that the relationship between dependent and predictor variables are structural and not functional; hence, absolute predictability cannot be achieved.

What emerges from the statistical exercise is that, of the three important global climatic factors, ST, WT, and SAL, the last one exerts maximum influence on the abundance of *O. universa*.

Conclusions reached from the analysis is not inconsistent with those reached empirically by several authors from observational data. However, the emphasis laid in this paper is different. For example, Be, Harrison, and Lott (1973) state that *O. universa* is known to thrive in subtropical waters. Obviously, the emphasis is on temperature. The present analysis shows that it is salinity which is responsible for the abundance and not temperature. Oceanographic studies have shown that salinity maximum is at the middle latitude (see Defant, 1961, and discussion given earlier). Incidentally, this is the zone where subtropical conditions prevail. Hence, abundance of the species in this zone is explained.

This shift in emphasis may not appear important when Recent distribution of *O. universa* is only considered, but, a little consideration will show that the paleoecologic implication is considerable.

It is generally known that paleotemperature and paleosalinity cycles are out of phase with one another. High temperature causes deglaciation and melting of the polar ice cap. Low temperature produces the opposite effect. At intermediate temperature, as is prevalent at the present moment, salinity is a function of evaporation, precipitation, surface circulation, and mixing.

Let us now consider the temporal distribution of *O. universa* at one location. If, to start with, it is assumed that WT is low and ST is also low, and glaciation is at its highest, this species will be absent or extremely rare even though salinity is high. This state will continue until WT and ST reach the cut off point, that is 9 °C and 14 °C (see Fig. 4). By then, average surface salinity must have decreased by deglaciation. From that time onwards, paleosalinity would largely control the distribution. Abundance of the species will increase at first slowly and then rapidly, till an optimum condition is reached. With further rise of ST, abundance will decrease though not to a great extent. When conditions are reversed,

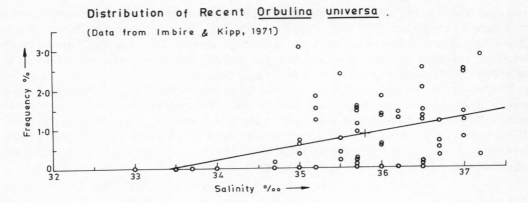

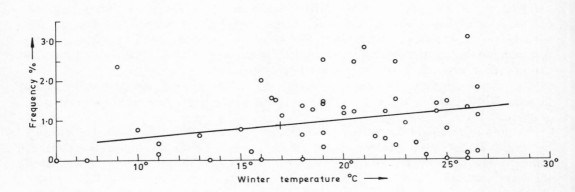

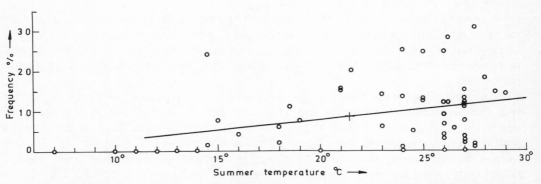

Fig. 4 Bivariate plots of frequency against salinity, winter temperature and summer temperature; regression lines are shown; in all situations, frequency is dependent variable

decrease of ST and WT and increase of SAL will lead to the optimum condition again. Later, further lowering of ST and WT will cause abundance to decrease. It will be logical to assume that the decrease in abundance at high temperature will be of a lower order. Hence, the temporal distribution curve will have nearly flat tops and deep valleys, and will be similar to the theoretical model presented in situation 3 (Fig. 1). It is to be noted that salinity would influence abundance considerably and would determine the point of time at which optimum conditions are reached. The abundance distribution model presented previously will induce logically broad oscillatory movement of the paleontologic time series in step with paleoclimatic cycles. In the next section, it will be shown by spectral analysis that temporal distribution of *O. universa* does show such oscillations.

In the previous discussion, I have referred to the optimum condition. The regression analysis only indicates its presence, but does not help us fix the condition. While rigorous optimization methods developed in recent years have not been followed, a graphical solution is presented. Figures 5 and 6, which are self-explanatory, define the optimum condi-

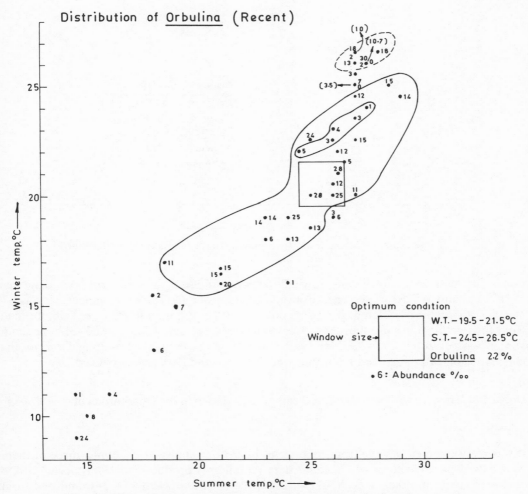

Fig. 5. Optimum growth conditions—optimal summer and winter temperatures

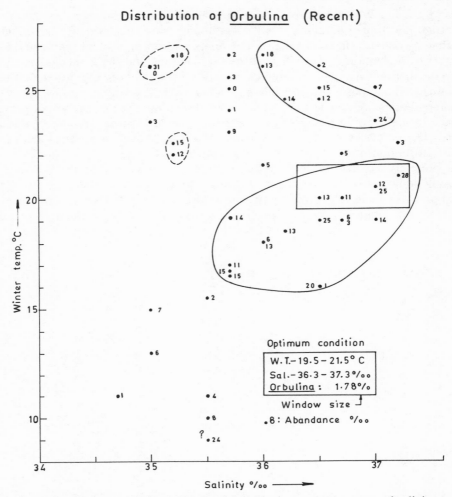

Fig. 6. Optimum growth conditions—optimal winter temperature and salinity

tions. The method involves use of a window to determine peak area. Only two diagrams (WT/SAL amd WT/ST) have been drawn. The third one (ST/SAL) in unnecessary as ST and WT are highly correlated. It may be noted that the optimum limits of WT determined from the two diagrams are the same. The ideal conditions are, salinity: 3.63–3.73 percent, winter temperature: 19.5–21.5 °C, and summer temperature: 24.5–26.5 °C. Had the concentration of points been high, the window size could have been reduced.

Univariate Time Series Analysis of the Temporal Variation in the Distribution of *O. universa*

Introduction

It was mentioned earlier that four time series have been analyzed in details; three of them were based on collections of *Orbulina* data (count/gm) produced for Pleistocene DSDP core samples from holes 216 ($N = 19$), 217 ($N = 19$), and 219 ($N = 28$), and one from published data (percent; $N = 23$) given in Imbrie and Kipp (1971; shortened in the following discussion as IK). For the purpose of present analyses, it has been assumed that the series are second-order stationary, that is there is no systematic trend either in mean or variance. The justifiability of the assumption from a theoretical point of view has been discussed earlier. The sampling interval is 50 cm. Hence, the shortest cycle recognizable in the series, has a length of 100 cm corresponding to the Nyquist frequency of period 2. The problem of aliasing and other related matters will be discussed later.

Analysis in the time domain

Figure 7 shows the plot of four series against depth. A hard look at the graph will convince one that there are broad periodicities of low frequency and well-defined peaks and troughs for three of the four series (216, 219, and IK). The fourth (for hole 217) is somewhat different. It appears that in this series, there is a linear trend. The linear trend line and the detrended series are also shown in the figure. It is interesting to note that the trend-free series retains the general feature of the original series. The reason lies in the low slope of the trend line. In the present work, both original and detrended series have been analyzed.

To gain a preliminary understanding of the series, the data were smoothed using a 5-point moving average. Figure 7 shows that the curves based on smoothed data retain general characteristic features of the series. Even the peak positions are not shifted considerably. In the situation of the 217 series, the smoothed and original curves (not shown in Fig. 7) virtually overlap.

The indication is that analysis based on filtered series would not be much different from those based on original series. So, herin, for further analysis original series was used, as that did not result in loss of data, which is vital in the situation of a short series.

Serial correlation analysis

Figures 8 to 12 are correlograms of the 5 series constructed from values based on exact formula given in Kendall and Stuart (1966). The confidence limits shown are only approximate and are biased downward.

Correlogram for the 216 series (Fig. 8) shows that all coefficients beyond lag zero are well within the confidence limits, indicating that the series is a realization of discrete stochastic process (Jenkins and Watts, 1968). Further, it may also be said that the series was generated by a white noise process. However, it is to be noted that, even though the values could be calculated up to only lag 6, periodic nature of up and down swings is noticeable with a period 4 (cycle length 2 m). Whether this points to actual state of affairs or not can only be determined from spectral analysis which is to follow. It is quite obvious that the shortness of the series renders proper interpretation difficult.

Situation is somewhat better for the 219 series (Fig. 9). Here, the peak and trough values of the correlogram are closer to the upper and lower confidence lines though statistically not significant at the 5 percent level. Further, the correlogram oscillates with a period of 2 metres. As the series is longer, two complete cycles are seen. Hence, it may be logically assumed that there is a well-developed periodic element in the series.

In the IK series, oscillations, even indication of them, are absent (Fig. 10). The persistance of negative correlation beyond lag 2 appears to be significant. It cannot be explained by assuming that the series was generated by purely random process. Because for a series generated by a white noise process, auto-correlation coefficients vary irregularly within confidence limits (see figures given in Jenkin and Watts, 1968 and Davis, 1973).

The 217 series is different from the previous three (Fig. 11); both for the original and detrended data (Fig. 12), r_1 is significant at 2 percent level (Anderson, 1971, Table, 6.3; he

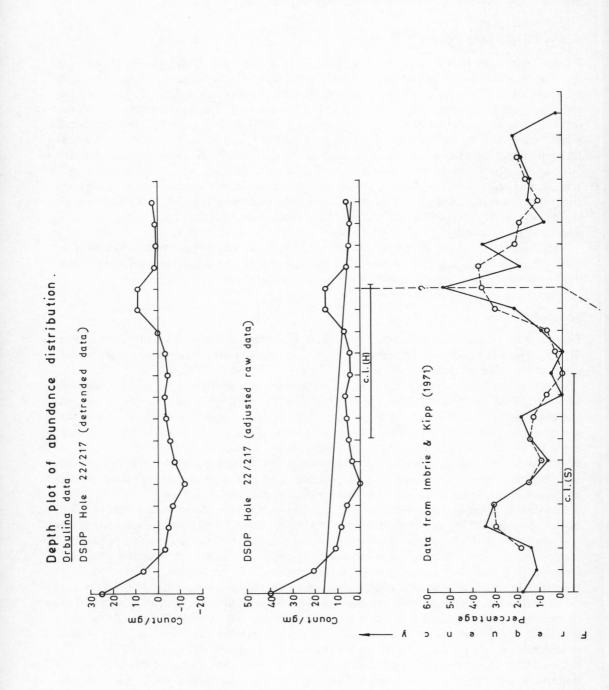

Depth plot of abundance distribution.
Orbulina data
DSDP Hole 22/217 (detrended data)
DSDP Hole 22/217 (adjusted raw data)
Data from Imbrie & Kipp (1971)

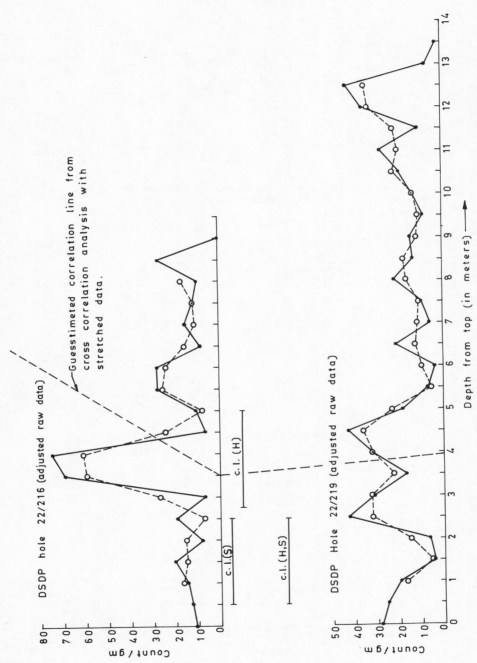

Fig. 7. Depth plots of abundance (adjusted) data for paleontologic times series analyzed; cycle lengths and line of stratigraphic equivalence are shown

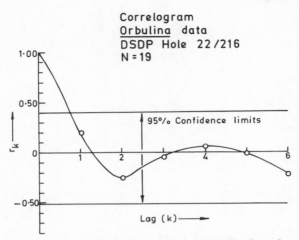

Fig. 8. Correlogram for *Orbulina* distribution data
from DSDP site 216

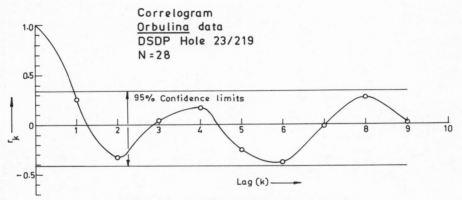

Fig. 9. Correlogram for data from site 219

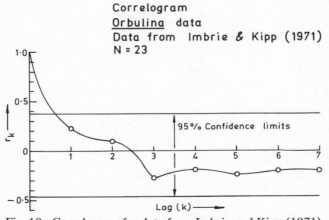

Fig. 10. Correlogram for data from Imbrie and Kipp (1971)

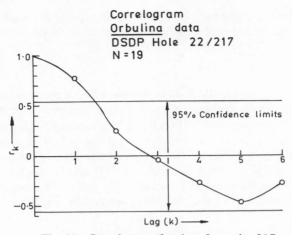

Fig. 11. Correlogram for data from site 217

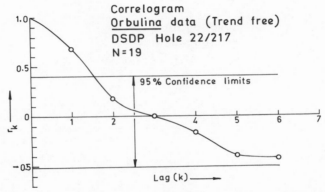

Fig. 12. As in Fig. 11, but correlogram is for series without trend

has tabulated one sided significance points up to 0.001). Using Bartlett's test statistic given in Box and Jenkins (1976), we get the standard error of r_k beyond r_1, which is 0.341; clearly this is high compared to the values of r_k greater than lag 1. The nature and general shape of the correlograms indicate the presence of a trend component in the series. It, perhaps, represents a first-order autoregressive series. The upswing observed after lag 6 in the situation of original series may or may not indicate the presence of a low frequency oscillatory component in the series. The similarity between correlograms of original and detrended series clearly shows that the trend cannot be removed from the series by applying a linear filter.

Analysis in the frequency domain

In the frequency domain, both harmonic and spectral analysis were performed in order to probe the nature of the series. A comparative study of the spectrogram (I have used the term in the sense of Anderson, 1971) and spectrum gives us an idea about the line and continuous spectra for the series and help us in the decision-making process for the next stage of analysis. As the sampling interval (Δ) is 0.5 meter, the Nyquist frequency is 1 cycle per meter. Evidently cycle length corresponding to Nyquist frequency is 1 meter.

Harmonic analysis

Figures 13–17 give the spectrograms. For the convenience of discussion and interpretation, both raw and smoothed curves are given. Generally, it is observed that the maximum power is concentrated at the lowest frequency (that is the fundamental). This appearance is due to the fact that mean is not zero. To remove the effect due to mean and get a true picture, harmonic analysis was also performed with standardized data (shown by dashed line). An inspection of the figures will convince one that, apart from the absence of the mean effect, the general pattern of standardized curves faithfully follow the original curve. Another advantage is that, although the scale of the IK series is different from the other series, the

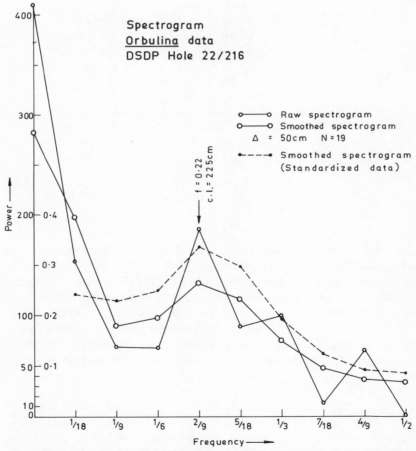

Fig. 13. Spectrogram for time series data from site 216. Raw and smoothed spectrograms for untreated data, and smoothed spectrogram for standardized data are shown. *C.l.* = Cycle lengths, f = frequency, Δ = sampling interval

Table 6.   Table showing parameters for predicted value from Harmonic analysis

DSDP site	N	Series mean	Series s.d.	C.V. mean	C.V. s.d.	r	Sig. level
22/216	19	20.262	19.827	37.592	31.473	0.4562	0.05
22/217	19	9.281	8.917	− 2.024	14.906	0.5193	0.05
22/217 (t.f.)*	19	0.0	8.135	−24.932	17.292	0.4871	0.05
23/219	28	19.158	12.482	24.752	20.011	0.4320	0.02
I and K	23	1.695	1.252	1.650	1.822	0.5044	0.02
I and K (full ser.)	110	1.691	1.157	1.209	3.190	0.2552	0.01

C.V.   = Computed value.
*t.f.   = trend free.
I and K = Caribbean core given in Imbrie and Kipp (1971), no. V12–122; series formed of every fifth data point.

estimated powers at different harmonic frequencies are of the same order, and, hence, are comparable. A study of the figures reveals the following:

(1) For the 216 series both raw and smoothed spectrograms for original and standard-ized data (Fig. 13) show a distinct peak at the fourth harmonic. Later, the smoothed curves gently descend to a low value at the Nyquist frequency, indicating the absence of any marked influence due to aliases. Table 6, which gives other relevant data, shows that there is a significant, though not strong, correlation between original and computed values (based on harmonic coefficients) of the series; hence the fit may be regarded as satisfactory. The analysis also indicates that the random term cannot be disregarded and the series was generated by a process which had both deterministic (periodic) and stochastic elements. This conclusion is important as it will help in model building.

(2) With the 219 series (Fig. 14), there is a strong peak at a frequency of 0.5 cycles/m (cm). The value of the peak frequency is close to that for 216 series; the small difference could be attributed to a difference in the series length. It may be noted that the strength of the peak increases for the standardized data. Here also, the effect of aliasing is not apparent. The smoothed curves show the presence of a distinct but small peak at $f = 0.14$ c/m. This corresponds to a cycle length of 7 m. It hints at the possible existence of a low frequency component. The fit of the computed curve is satisfactory as correlation between the original and computed data is significant at nearly 2 percent level.

(3) The spectrograms for the 217 series (Fig. 15) is different. The raw spectrogram for the original data shows a peak at $f = 0.344$ c/m which fails to show up in the smoothed curve. The smoothed spectrogram for standardized data, on the other hand, clearly points to the existence of a peak at a frequency intermediate between $f = 0.111$ c/m and 0.344 c/m (possibly at $f = 0.278$ c/m corresponding to a cycle length of 3.6 m). Other features of the series are similar to the above two.

In this situation, correlation between original and computed values is significant at 5 percent level. Smoothed spectrogram of the detrended series (Fig. 16) fails to bring out any distinct peak. It was pointed out earlier that the trend, if there is any, is not linear. For further analyses, the detrended series will not be considered.

(4) IK series (Fig. 17) brings out clearly the advantage of working with standardized

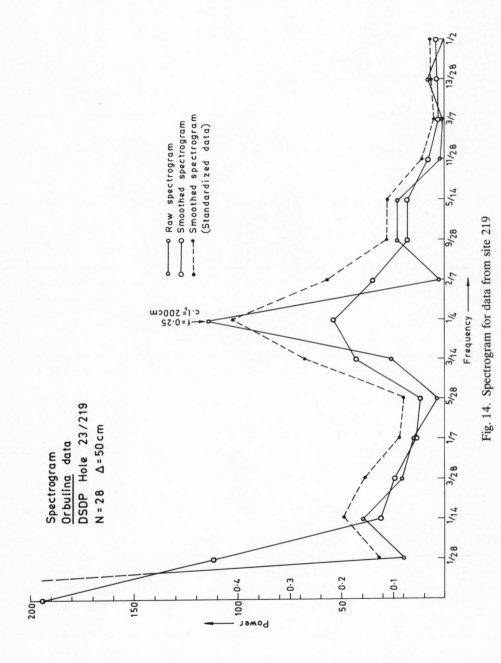

Fig. 14. Spectrogram for data from site 219

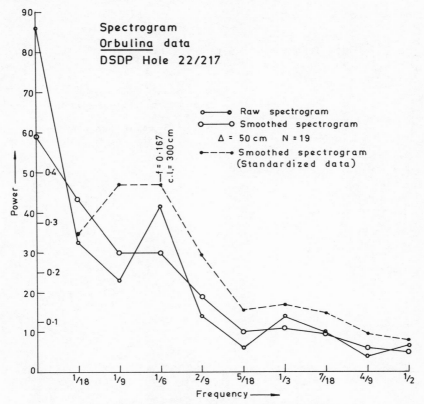

Fig. 15. Spectrogram for data from site 217

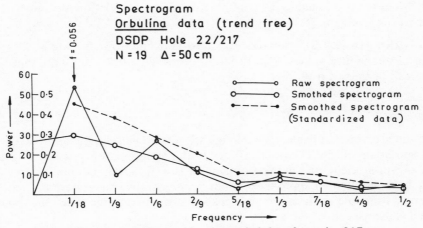

Fig. 16. Spectrogram for detrended data from site 217

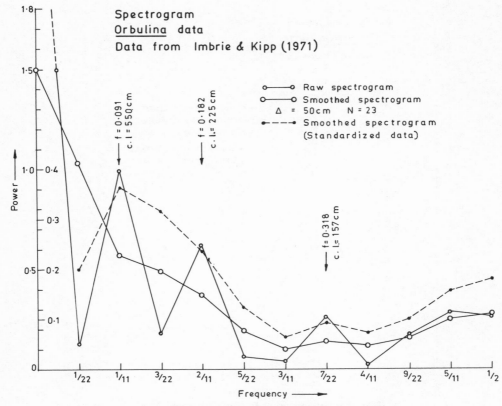

Fig. 17. Spectrogram for data from Imbrie and Kipp

data. Although the smoothed spectrograms based on original and standardized data show similar patterns near the high frequency end, the former fails to reveal the presence of a distinct peak at the second harmonic ($f = 0.182$ c/m; c.l. = 5.5 m). Concentration of considerable power at the Nyquist frequency points to the presence of a higher frequency component and the influence of aliases on the spectrogram. The fit of the computed curve is satisfactory as the correlation is significant at nearly 1 percent level.

For the series 216 and 219, analysis was also performed for the log series, that is by taking log of abundance values. The results (not presented herein) show that no special advantage can be gained from this.

Spectral analysis

Although the spectral and harmonic analyses are related closely, the approach is different (Kendall and Stuart, 1966; Chatfield, 1975). In this situation, it is possible to gain an idea about the power at the nonharmonic frequencies as well. For an infinite or large series the difference between the two approaches is not important, but for a short and finite series it is. The results from one analysis can be checked against that of the other, and both of them may reveal some aspects of the series.

In the following analysis, spectra was estimated with different lag windows and different

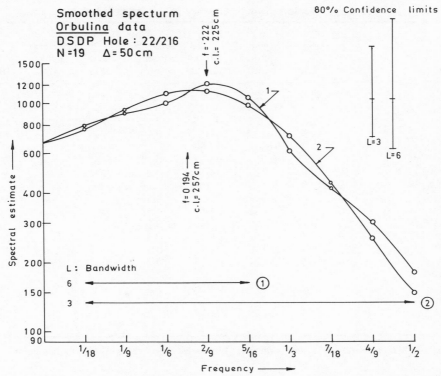

Fig. 18. Smoothed spectra for time-series data from site 216. Two curves are for different lag windows. Estimated and computed peak positions are shown

frequency spacings including that related to the series length. This was performed to compare results of spectrogram and spectral analysis. The narrowest lag window used corresponds to $L = \frac{1}{3} N$. The window was opened up to a maximum of $L = 3$. In all the figures, log of estimates were plotted against frequency and 80 percent confidence limits are shown; hence, the chances of taking a spurious peak as real is 1 in 5. I had to accept the greater risk as the series are short; this will not mislead one as we know what is being calculated.

The smoothed spectrum for the 216 series (Fig. 18) shows the presence of a distinct peak at $f = 0.222$ c/m. This is exactly the same peak frequency obtained from harmonic analysis If the band width is increased, there is a slight shift of the peak towards the left, but the existence of the peak cannot be doubted. The shape of the spectrum suggests that the series might have been generated by a second-order autoregressive process. The slight shift is partly caused by the smoothing effect of the large lag window used (band width = 0.444) and partly due to the assumption that the free hand curve represents actual spectrum. We can also reduce the frequency spacing to get the exact position of the peaks. Table 7 shows that, with a frequency spacing 0.025, there is a strong peak at 0.5 c/m (c.l. $-$ 2 m). Considering the shape of the spectrum with $L = 6$, I have taken this value to be a more acceptable one. With the same frequency spacing, two other peaks appear. They are neither strong nor significant. Longer series is necessary to check the validity of these peaks. The spectrum does not show any effect of aliasing. For the 219 series (Fig. 19), the smoothed spectrum for $L = 9$ shows the presence of two distinct and statistically-significant peaks at about 0.2 c/m and 0.5 c/m. Opening the window gradually smooths the first peak out (see curves for $L = 6$

Table 7.   Table showing frequency peaks in smoothed spectrum

Core site	N	Lag	Band width	Freq. spacing	Peak at cycle/meter			Remarks
					Peak 1 c/m	Peak 2 c/m	Peak 3 c/m	
22/216	19	6	0.222	0.05	0.5	—	—	Strong
22/216	19	3	0.444	0.05	0.4	—	—	Moderate
22/216	19	6	0.222	0.025	0.2*	0.5*	0.8	*Strong
22/217	19	6	0.222	0.05	0.2	—	—	Moderate
22/217	19	3	0.444	0.05		no peak		
22/217	19	6	0.222	0.025	0.2*	0.55	—	*Strong
22/217(t.f.)	19	6	0.222	0.05	0.2	—	—	Weak
22/217(t.f.)	19	3	0.444	0.05		no peak		
23/219	28	6	0.222	0.05	0.2	0.5*	—	*Strong
23/219	28	3	0.444	0.05	0.5	—	—	Strong
V12–122**	23	7	0.190	0.05	0.2	—	—	Strong
V12–122	23	4	0.333	0.05	0.3	—	—	Moderate
V12–122	23	7	0.190	0.025	0.2	—	—	V. strong
V12–122 (full data)	110	18	0.074	0.025	0.5	1.5*	3.25	*Strong
V12–122 (full data)	110	9	0.148	0.025	1.5	3.5	—	Strong
V12–122 (full data)	110	5	0.267	0.025	1	3.75*	—	Moderate

**Data from Imbrie and Kipp (1971) for Caribbean Core; series formed with every fifth data point.

and 3). The reason is that, with the window opening, the width of the peak becomes smaller than the band width (Jenkins and Watts, 1968). It may be concluded that there are two distinct cycles corresponding to cycle lengths of 2 m and 5 m. It may be recalled that there are indications for the existence of a 5 m cycle in the 216 series also. Peaks at the same frequencies show up in spectra estimated with other frequency spacings. Analysis on the basis of standardized data reveals the presence of much stronger peaks at the same frequencies. Upswing of the curves at the high frequency side distinctly shows the effect of aliasing. The presence of more than one intermediate peaks suggest that an autoregressive moving average (ARMA) process could have given rise to the series.

It is interesting to note that the correlograms for both 216 and 219 series show oscillations having a period of 2 m. This coincidence is not fortuitous. It suggests that, even if there is no significant autocorrelation, it should not be concluded that the series lack periodic components.

Spectra of the 217 series (ordinary and detrended data) reveal the presence of a minor weak peak at a frequency between 0.18 to 0.22 c/m corresponding to $L = 6$ (Figs 20 and 21). When the spectral window is made wider ($L = 3$), the spectra are smooth curves gradually descending from the low frequency to high frequency ends. It is evident that the maximum power is concentrated towards the former. Such spectra are characteristic of a first-order autoregressive process with positive α_1. A second-order process (α_1 +ve and α_2 −ve) may also give rise to such a spectrum.

However, if the frequency spacing is increased to get more details of the curve, it is observed that the peak is located at $f = 0.2$ c/m (cycle length = 5 meters, see Table 7). It may be recalled that the spectrogram for standardized data of the series also show a low frequency peak though at a different frequency. It may be concluded that there is a low frequency component in the series. The spectra do not show the effect of aliasing.

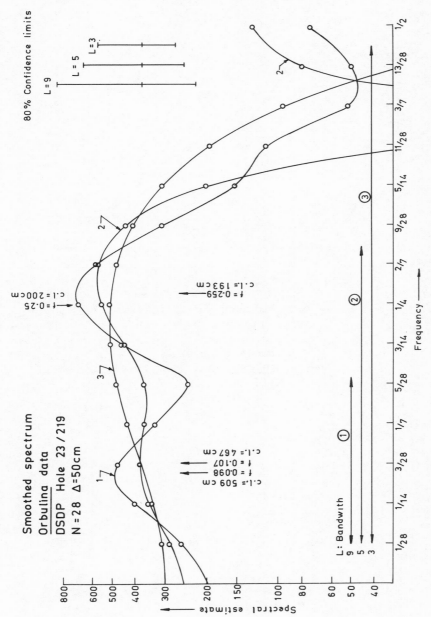

Fig. 19. Smoothed spectra for data from site 219; curves correspond to three different lag windows

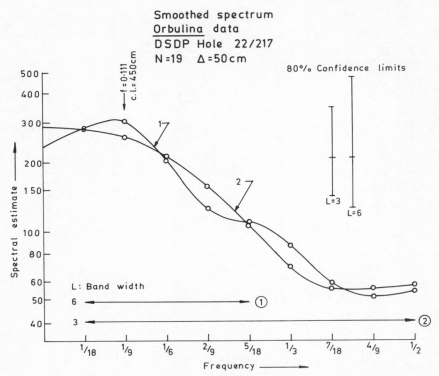

Fig. 20. Smoothed spectra for data from site 217

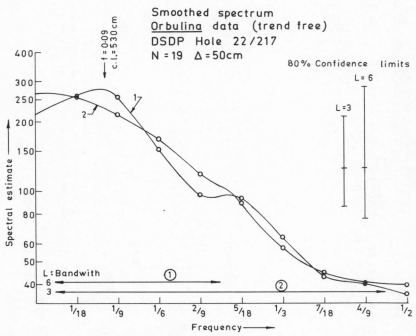

Fig. 21. Smoothed spectra for data with trend removed from site 217

The IK series is different from the above three. Using spectral windows corresponding to $L = 7$ and $L = 4$, spectra are obtained with a strong peak at the low frequency side (Fig. 22), the peak frequency must have a value somewhere between 0.18 c/m and 0.272 c/m. Calculation with a frequency spacing of 0.025 fixes the position of the peak at 0.2 c/m (cycle length 5 m; see Table 7). Further, there is a strong upswing of the curves and consequently concentration of power at the Nyquist frequency; this is due to aliasing and points to the presence of a strong high frequency peak. In this context, it may be mentioned that the spectrogram for standardized data of the series shows the presence of a low frequency peak at $f = 0.182$ c/m (cycle length 5.5 m). Both the curves in Fig. 22 are similar to trough spectrum, which is generated by a second-order *ar* process with $\alpha_1 < |1.0|$ and α_2 + ve.

From the analysis of the time series in the time and frequency domains, the following conclusions may be drawn:

(1) The 216 and 219 series are similar in many respects and appear to be on an equal footing.

(2) Both the series show prominent 2-m cycles and there are enough indications that a second longer cycle of 5 m length also exists. However, the exact length of cycles in one series may be slightly more or less than that in the other series. To estimate this, longer series will be needed. As a logical extension to this, it may be said that the rate of sedimentation in the two were more or less the same during the period under consideration. Incidentally, Malmgren and Kenett (1978) also showed the presence of a 2-m cycle based on the study of mean size variation of *Globigerina bulloides*.

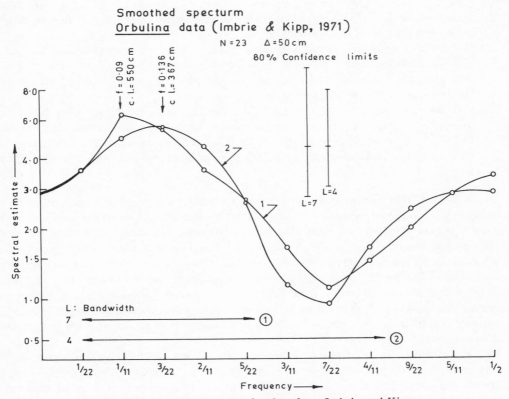

Fig. 22. Smoothed spectra for data from Imbrie and Kipp

(3) The IK series is characterized by a longer cycle of about 5 m and a shorter cycle of undeterminable length. However, this cycle cannot be taken to be equivalent to 5-m cycle of the 216 and 219 series. Because, the 5-m cycle of the IK series is strong, it may be taken to be equivalent to the 2-m cycle of the other two series.

(4) The 217 series is different. No distinct cycles could be recognized. A look at the distribution pattern (Fig. 7) would convince one about the reality of the conclusion. This distribution pattern does not imply that the series represents an extremely improbable realization of the ensemble. We have also studied the distribution patterns of *Globigerinoides sacculifer* and *G. quadrilobatus* from this core; they are broadly similar to the above. The cause for this pattern is being studied, but there are reasons to believe that this has something to do with the spatial position of DSDP site 217, which is located in the northern part of the Ninety-east ridge near the fringe of the Bengal alluvial fan.

Converting cycle lengths mentioned previously to absolute time the following results are obtained:

(1) For IK series, accepting the rate of sedimentation calculated by Broecker and Van Donk for Caribbean core (given in Imbrie and Kipp, 1971), the 5-m cycle length is equivalent to 212766 years or roughly 0.213 my.

(2) For the 219 series no exact data are available. Average rate for Pleistocene sedimentation calculated with the assumption that Pleistocene/Pliocene boundary is at 1.8 m yrs (Berggren, 1978) comes to 1.67 cm/10^3 yrs. Hence the 2-m cycle is equivalent to 0.12 my (see Whitmarsh and others, 1974). But, certainly this rate varied and, perhaps, the Late Pleistocene and Early Pleistocene rates were different. So, the estimated time span is approximate.

(3) For the 216 series, the rate has been calculated with the assumption that the span of *Emiliana huxleyi* zone is $\simeq 0.2$ my (see Borch and others, 1974, and Gartner, 1974). This gives the length of the 2-m cycle, in terms of absolute time, as 0.12 my, which is comparable to the previous estimate.

(4) For the 217 hole the data are unreliable. The rate of sedimentation on the basis of the extent *E. huxleyi* zone comes to 0.5 cm/10^3 yr, low figure (see Borch and others, 1974). So time equivalent for the cycle has not been calculated. I will discuss the validity of the above findings in a later section when dealing with cross-correlation analysis.

Before leaving this section, it is necessary to discuss the nature of the series further. One may argue that, in the analyses presented so far, there is nothing against the assumption that the series were generated by a purely random (white noise) process. The test statistic given in Jenkins and Watts, (1968, p. 234) is conservative and the associated Kolmogoroff–Smirnov test of significance can be applied only when N is large. The sample integrated spectrum was calculated for the 216 series and plotted against harmonic frequencies (k). The points showed a considerable deviation from linearity (Fig. 23), but the K–S test of significance could not be applied as N is small. A little consideration will show that the test would give significant results only when there is a large concentration of power at a single frequency and the sample is large.

In spite of the failure of the test, other features of the spectrogram and spectrum distinctly show that the process responsible for the series was not purely random. Thus, the shape of the spectrum points to a second-order *ar* process; in the situation of a series generated by purely random process, both sample and smoothed spectra fluctuate violently

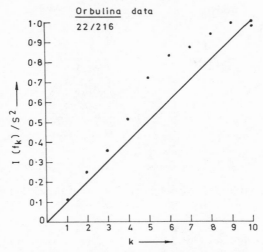

Fig. 23. Test of *Orbulina* data from site 216
for randomness; sample values (shown by
dots) and theoretical line are shown

about the theoretical spectra which is a line parallel to the abscissa (that is, spectral estimate is constant over all frequencies). Further indications are provided by the significant fit of the observed and predicted values obtained from harmonic analysis, and the presence of distinct and statistically-significant peaks in the spectrum.

This question may be viewed from a different standpoint. From *a priori* reasoning, it may be said that the temporal distribution should have stochastic and deterministic components (see discussion on the theoretical model given earlier). Hence, the failure to get a significant result from the tests would indicate only that either the sampling interval was too large, giving appearance of independence of individual observation, or too small, causing high frequency noise to appear. In the above situations, truth lies near the former.

Finally, a close look at Fig. 7, wherein cycle lengths have been shown against the plot of the series, will convince one of the periodicities that have been revealed by the spectral analysis.

Bivariate Analysis

In the previous section I have attempted to determine the similarity between different series on the basis of univariate analysis where possible. Nevertheless, for detailed analysis of the relationship between different series, bivariate methods are also appropriate.

It may be noted that the series are of different lengths. Hence, to determine truncation points, L was taken equal to one-third of the shorter series. For confidence limits, the number of observations (N) of the shorter series only were considered. As all the series are short, cross-correlation coefficients were calculated using the exact formula. For significance testing, tests applicable to ordinary product-moment correlation coefficient were used.

Cross-correlation analysis

In Figs. 24 to 29 cross-correlograms are presented showing the relationship between different series. The figures are self-explanatory. The salient features are presented below. In this

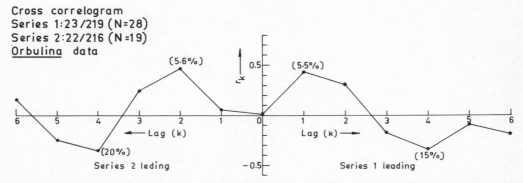

Fig. 24. Cross-correlogram for time-series pairs. Data from sites 219 and 216. Figures within brackets are levels of significance

context, it may be noted that Jenkins and Watts (1968) recommend the use of a filter to convert the two series to white noise before performing cross-correlation analysis because if the two series have a large autocovariance, spurious cross-correlation may be generated between two uncorrelated series. In the present situation, it was observed earlier that the series do not show significant autocorrelation. Hence, the cross-correlations are not spurious. As the series are short, I have uniformly taken 10 percent as the significance level.

Figure 24 shows that the 219 and 216 series are significantly cross-correlated at lags 1 and 2. At lag 0 the correlation coefficient is zero. This distinctly indicates that stratigraphic correlation is not possible on a one to one basis. The reasons may be ascribed to a difference in the rate of sedimentation. As the upper datum is fixed, alignment is not possible and significant correlation at 0 lag may be achieved only by stretching one series relative to the other. It may be noted here that, if two series are identical in all respects, the cross-correlogram will be symmetrical and r_0 will be identically equal to 1.0. Moreover, one half of the cross-correlogram will be a mirror image of the other half. Minor departures will give rise to a value of $r_0 < 1.0$ and a cross-correlogram which is not symmetrical. If the departure is large, the maximum value of r_k may not be at lag 0. Large differences may be caused by (1) large differences between observational values at different times, (2) difference in cycle lengths of major cycles (represented by a significant difference in peak frequencies as shown by spectral analysis), and (3) a difference in the frequency phases (see also Agterberg, 1974, p. 372). The importance and cause of phase difference in paleontologic time series will be discussed later.

It was shown earlier that, the series 219 and 216 have major cycles of similar if not

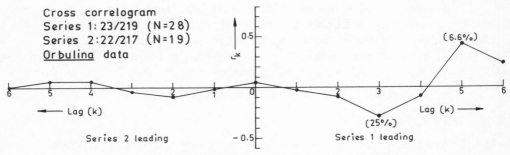

Fig. 25. Cross-correlogram for data from sites 219 and 217

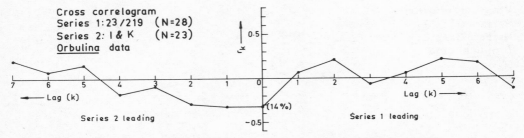

Fig. 26. Cross-correlogram for data from site 219 and Imbrie and Kipp

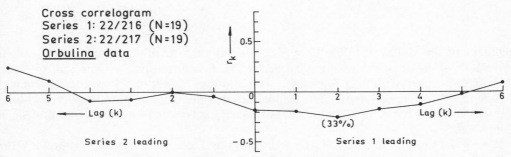

Fig. 27. Cross-correlogram for data from sites 216 and 217

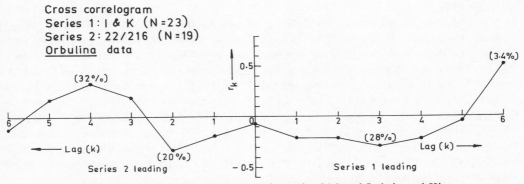

Fig. 28. Cross-correlogram for data from site 216 and Imbrie and Kipp

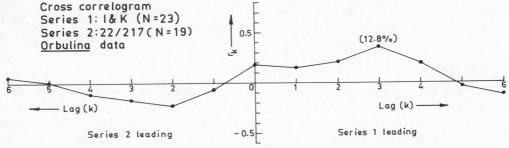

Fig. 29. Cross-correlogram for data from site 217 and Imbrie and Kipp

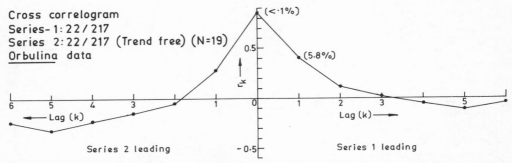

Fig. 30. Cross-correlogram showing relation between original and trend-free time series data from site 217. Large value at r_0 and near-symmetric pattern of curve may be noted

identical length. Hence, the difference may be ascribed to both (1) and (3). The near zero correlation at lag 0 is, most likely, caused by (3). It is interesting to note that the cross-correlogram shows oscillatory movement with a period of four or 5 lags, corresponding to a cycle length between 2 and 2.5 m. It indicates that the cross-amplitude spectrum should show a peak between 0.4 to 0.5 c/m. Cross-spectral analysis will reveal whether this conclusion is valid.

It is logical to suppose that, for accurate stratigraphic correlation, two paleontologic time series for a single species constructed from observational data from two sections being correlated should show highest cross-correlation at lag 0 and should also be similar in respect of frequency and phase. However, stretching (to be described later) may be needed to achieve this.

The cross-correlogram (Fig. 25) showing the relationship between the 219 and 217 series reveals one significant value of lag +5. While one significant value in 12 may arise by chance, we should also consider the possibility of improving the correlation at r_0 by stretching the 219 series. The indications are that, here the stretch coefficient should be high. It may be noted that r_0 is virtually zero. Figure 26 shows that r_k for the 219 and the IK series is not significant for any value of k up to 7. r_0 is negative and has a value which may be considered to be significant in 6 times out of 7. While none of the r_k values are significant for the 216 and 217 series ($k = 1,6$; Fig. 27), the former shows a moderately significant correlation with the IK series at lag 7 (Fig. 28). It indicates that, to obtain a moderately significant value for r_0, the 216 series has to be stretched considerably. The 217 and IK series are uncorrelated (Fig. 29).

Figure 30 shows the relation between original and detrended 217 series. It may be recalled that, removal of the trend did not destroy the essential features of the 217 series. The expected high correlation between the two series at r_0 actually is obtained. A comparative study of this figure and the correlogram for the 217 series given earlier (Fig. 11) supports the discussion of cross-correlograms given at the beginning of this section.

Cross-spectral analysis

Jenkins and Watts (1968) recommend that, for cross-spectral analysis, the two constituent series should, at first, be aligned in such a manner that $r_{12}(\theta)$ has the highest value. This type of alignment is not permissible in paleontologic time series, because of the basic assumption that there is a fixed datum common to both series from which they are supposed to have started. This we cannot change relatively. In geology, the philosophy behind time series analysis is different from that in other subjects. Here, we are not interested only in the structure of the series but also need to know the nature of periodic environmental inputs, and how two or more series behave in relation to one another as time passes. Hence, the absence of alignment is important, and aligning the two series may produce results which are not consistent with the actual state of affairs, particularly, when the sampling interval is not small. In the present situation, alignment was not performed although that has introduced bias in the estimates.

It must be remembered that, as the series are short, the interpretation has considerable limitations. However, the results presented in this section will throw light on the implications of bivariate analysis in the frequency domain.

Coherency. Squared coherency spectrum describes the relationship between the two series at each frequency. Consistently high values at each frequency would indicate that the two series are nearly identical. As we are here comparing two series which are spatially separ-

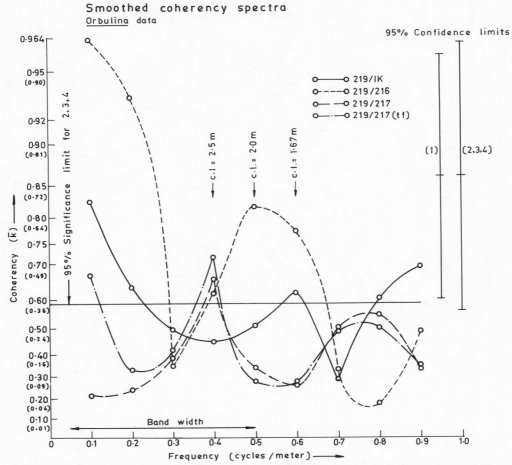

Fig. 31. Smoothed coherency spectra. Composite diagram showing relation between 219 and other series, including significance and confidence limits. Figures within brackets are squared coherency values

ated and, hence, differ slightly in environmental inputs, this ideal is not realizable in practice.

Figures 31 and 32 show the coherency spectra for the four series being considered. In one, the detrended 217 series is also included for comparison. Coherency ($\bar{K}$, *values for* $\bar{K}^2$ are also given) has been plotted in Fisher's 'Z' scale as suggested by Jenkins and Watts (1968). As the analysis was performed with wide window ($b = 0.222$), it is expected that spuriously high coherency will not occur.

The spectrum for the 219/216 series show high coherency values at low (0.1 c/m) and intermediate frequencies (0.5 c/m). This significance of high coherency at extremely low frequency is doubtful. This is partly because, due to shortness of the series, values for extremely low frequency peaks cannot be accurately estimated from the autospectral analysis. However, a high coherency value may indicate that the series being compared have a considerable concentration of power at the low-frequency side.

The significance of the intermediate peak at 0.5 c/m cannot be questioned. The peak is strong, and the value is much above the lower (95 percent) significance limit. It may be recalled that autospectral analysis also showed that both series have a periodic component corresponding to $f \simeq 0.5$ c/m.

For the 219/IK series, significant intermediate peak is absent. The significance of the

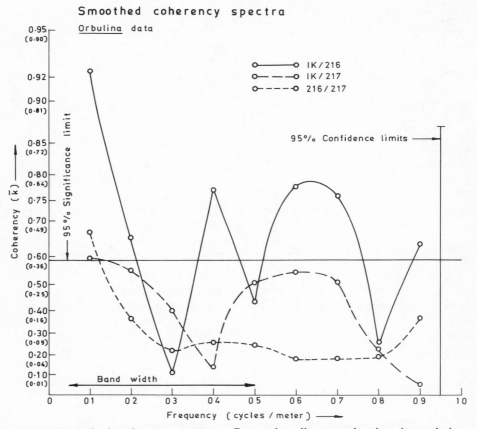

Fig. 32. Smoothed coherency spectra. Composite diagram showing interrelation between 216, 217, and IK series

small peak in the coherency spectrum of 219/217 series at $f = 0.4$ c/m is not clear Autospectral analysis fails to reveal any peak frequency corresponding to this. Perhaps, the high power in the neighbourhood of $f = 0.5$ c/m for the 219 series has resonated with the moderate power of the 217 series to give rise to the peak.

Figure 32 shows that none of the coherency values for the 216/217 and 217/IK spectra are significant. The existence of two distinct and significant peaks at 0.4 c/m and 0.6 c/m for the IK/216 spectra cannot be explained on the basis of autospectral analysis of the two series. Further work is needed to gain an understanding of the relation between auto-spectra, cross-spectra and coherency, in the present situations.

Phase spectra and cross-amplitude spectra

Jenkins and Watts (1968) state that cross-amplitude spectra are not as important as coherency. In my opinion, both spectra should be considered together to understand the implication of cross-spectral analysis. This approach will put a check on spurious or biased coherency estimates that may arise due to various reasons (see above and Jenkins and Watts, 1968). In the following discussion an attempt will be made to understand the significance of the high coherency values obtained at different frequencies.

(1) *The 219/216 series.* Cross-amplitude spectra ($L = 6, b = 0.222$) shows two peaks (Fig. 33) at 0.2 c/m and 0.5 c/m, the latter being strong. Corresponding coherency values are high and the coherency spectrum shows a peak at the latter. At the lower frequency, phase angle (Fig. 34) is not different significantly from $0°$ ($-8° \pm 8°$; values read off from Fig. 9.3 of Jenkins and Watts (1968) showing 95 percent confidence limits of phase spectrum). Phase angle for $f = 0.5$ c/m is $145° \pm 22°$. Perhaps, this large phase difference, for the most important periodic component common to both series, is responsible for the low value of $r_{12}(0)$. In this context, it may be noted that, with a wider band width (0.444), the cross-spectral analysis shows an amplitude peak at the same frequency. Corresponding phase angle is $134°$ and coherency (K^2) = 0.49, which is significant. The peak frequency of the coherency spectrum has however shifted to 0.6 c/m.

The conclusion is that the two series are highly correlated at low frequency (0.2 c/m). For a periodic component corresponding to $f = 0.5$ c/m, the correlation is high but the 219 series leads the 216 series by three-eighths of a cycle length. The reason for this lead is difficult to understand in geologic terms. We could explain the lead or lag if the two sites were latitudinally separated. In that situation, it could be argued that, because of the nature of spatial separation, environmental inputs that are controlled by climatic cycles were different from the start. The temporal distribution pattern at the two sites then would be similar but the dominant periodic element would show a lead or lag. But, here, both the sites are near the equator with a latitudinal separation of only $8°$. Further, the time in geologic terms is so short that we cannot ascribe this to differential plate movement. The only logical explanation is that the cycle lengths of the periodic elements of the two series were different. This difference could have been detected if the series were longer. Within the limited length of the series, the lead or lag of the periodic element appeared as a phase difference.

(2) *The 219/IK series.* In spite of significant coherency values at several frequencies the cross-amplitude spectrum (Fig. 33) fails to show the presence of any significant peak. The shape of the curves only indicates that, for both series, a considerable amount of power is concentrated at the high frequency end, a fact already known from autospectral analysis. The phase spectrum (Fig. 34) is interesting in the sense that low-frequency components of

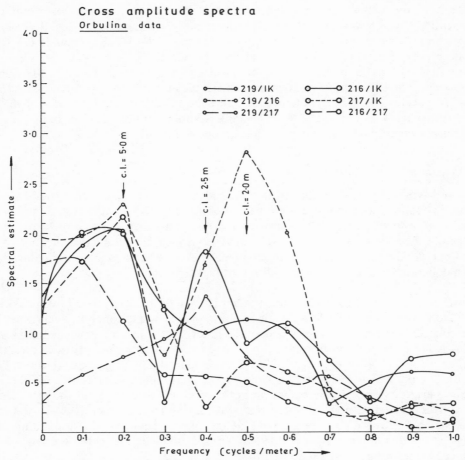

Fig. 33. Smoothed cross-amplitude spectra. Composite diagram for 6 time-series pairs

series 1 lead those of series 2 by about 135° on average, but the high ones tend to lag by more than 90°, and the slope of the spectra is steep between 0.6 and 0.7 cycles/m.

(3) *The 219/217 series*. Figure 33 shows that the spectra has a distinct, sharp peak at 0.4 c/m (cycle length 2.5 m). The phase (Fig. 34) is not significantly different from 0, and the coherency spectrum shows a distinct significant peak at that frequency. The results are unexpected, as autospectral analysis does not show any indication of this feature, although the autospectra of the 217 series (original and trend free) show a tendency to form a peak at this frequency. However, the resonance effect mentioned previously may be responsible .

(4) *The 216/217 series*. Cross-amplitude spectrum for the series does not have any peak (Fig. 33). Its shape is similar to the spectra of the first-order autoregressive process. Coherency is significant only at the lowest frequency (0.1 c/m).

(5) *The 217/IK series*. The cross-amplitude spectrum (Fig. 33) shows two peaks, one of which (at $f = 0.5$ c/m) is significant. But the corresponding coherency estimates are not significant. Hence, there is no indication that the two series are correlated.

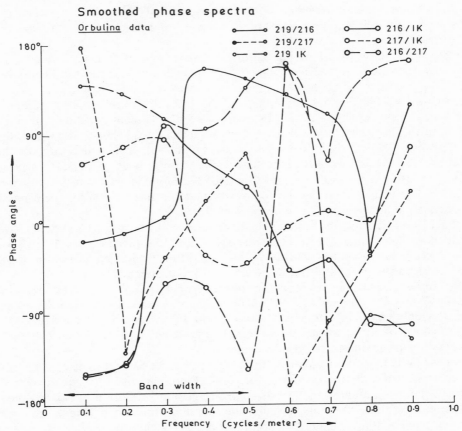

Fig. 34. Smoothed phase spectra, with phase angles in degrees; composite diagram for
6 time series pairs

(6) *The 216/IK series.* In this situation, besides the high value of the lowest freqency, the cross-amplitude spectra show a distinct peak at 0.4 c/m. The corresponding coherency (K^2) estimate is 0.59, which is significant. The phase difference at this frequency is $64° \pm 25°$. The coherency spectra shows that, out of 9 values 6 are significant statistically. Hence, from the bivariate analysis it may be concluded that the two series are correlated.

Synthesis of the Results of Uni-and Bivariate Time Series Analysis and Conclusions Relevant to Stratigraphic Correlation

It may appear from the results and discussion given so far that conclusions arrived at in the different sections are not compatible. But the following discussion will show that study from different angles reveal different aspects of the relationship between two series.

Both autocorrelation and autospectral analysis clearly show that the 219 and 216 series are similar generally. Both show broad periodicities at low and intermediate frequencies. The value of the latter is about 0.5 c/m. Cross-spectral analysis corroborates this. However, the nature of the cross-correlogram and phase difference at the peak frequency indicates that the two series have detectable differences and one series has to be stretched against the other to achieve better correspondence and a significant value at $r_{12}(0)$. The obvious implication is that sedimentation rates at the two sites are different during the time interval under consideration.

The 217 series has a distributional pattern different from the others, as shown by both uni- and bivariate analysis. The significant 0.4 c/m peak has no relevance to the relationship between the 217 and 219 series.

The relationship between the IK series, and the 219 and 216 series needs careful study. Autospectral analyses of the 219 and IK series show a prominent peak at 0.2 c/m, but the spectrum of the IK series lacks the second peak (at 0.5 c/m). This clearly shows that the 0.2 c/m peaks of the two series are not equivalent. Alternatively, we may suppose that the 0.2 c/m peak of the IK series is equivalent to the 0.5 c/m peak of the 219 series. If the previous supposition is correct, we also should obtain an indication that the stretch coefficient is 2.5, that is the 219 series has to be stretched 2.5 times for better correspondence. Cross-correlation analysis broadly corroborates this.

As the 219 and 216 series are similar, it may be said that the 216 and IK series are similarly related. Although autospectral analysis of the 216 series does not clearly show any peak at 0.2 c/m, cross-spectral analysis of the 216/IK series gives an indication of its presence. A cross-correlogram also indicates the need for stretching.

In this context, it may be noted that the coherency spectrum of the 219/IK series shows a frequency peak at 0.6 c/m and that of 216/IK series, at 0.4 c/m and 0.6 c/m. These may be regarded as higher harmonic frequencies for 0.2 c/m. Further study is needed to understand the significance of the previous discussion, but Chatfield (1975) points out that such phenomenon is frequent in autospectral analysis.

As the analysis is in the frequency domain, it enhances the relationship between two series at a particular frequency. However, it does not describe the equivalence of the periodic element represented by the frequency. In geology, we are mostly concerned with periodic elements and their geologic significance, and not frequency. Hence, cross-spectral analysis employed alone is likely to lead to erroneous conclusions.

Cross-Correlation Analysis with Stretched Data

I have previously referred to the need for stretching. Determination of stretch coefficient is difficult, particularly if there are more than two series to be compared. Here, we cannot use the trial and error method or that proposed by Kwon and Rudman (1979) because we have to determine a set of coefficients which will simultaneously improve paired intercorrelations, the number of which increase rapidly with the inclusion of an additional series for comparison. For example, with 5 series there are 10 sets of cross-correlation, with 6, 15, and with 10, 45. Eyeball estimation is possible up to a certain limit. So we must discover an objective criterion. Such a criterion can be recognized in autospectral analysis of the two series. It is also possible to obtain the stretch factor on the basis of information from other sources, for example from prior knowledge about the rate of sedimentation. Usually, however, the rate is calculated on the basis of information which is not complete. Moreover, one of the purpose of time series analysis is to determine the rate accurately.

Table 8 shows the stretch factors used and the basis for them. In this work, cross-correlation analyses on the basis of ordinary and stretched data were programmed separately. It is hoped that, with further work, it will be possible to develop an integrated program in which a decision-making process will be incorporated.

For stretching and interpolation, I have used a spline function with $n = 2$ (for cubic spline) and $m = 3$ (for data points, see Sard and Weintraub, 1971). A higher-order spline or

Table 8.   Table showing stretch factors and the bases

Core site	Spectral peaks	Eyeball from graph	Sedimentation rate
22/216	2.5	1.875	1.426 (1.65 cm/*10^3 years)
22/219	2.5	1.667	1.262 (1.86 cm/*10^3 years)
22/217	1.0	1.0	2.6 (0.90 cm/*10^3 years)
V12–122 (Imbrie and Kipp, 1971)	1.0	1.071	1.0 (2.35 cm/10^3 years)

N.B. Sedimentation rates are shown within brackets.
*Calculated on the basis of data given in Initial reports of DSDP (see Borch, and others, 1974; Gartner, 1979; and Whitmarsh, and others, 1974).

larger number of data points were not used to avoid distortion of the stretched data and because, the objective of stretching is not to generate new data, but to stretch the series thus retaining its essential features. A spline function is eminently suitable as it has minimizing properties. The program was written such that, except at the end of the series, three contiguous points were used for interpolation. For example, let 1, 2, 3, 4, . . . be the contiguous points; for interpolation up to 2.5, the original data for 1, 2, 3 were used; for values >2.5 and <3.5, those for 2, 3, and 4, were used. This approach causes minimal distortion as the method accounts for points of inflection. Neidell's method referred to in Schwarzacher (1975) has not been attempted.

Results

(1) *The 219/216 series.* For the first set, both series were equally stretched by a factor of 2.5 so the relative stretch factor is 1. A look at the cross-correlogram (Fig. 35c, solid line) indicates that stretching by spline function introduces minimal distortion. Its shape is identical to that of the cross-correlogram based on unstretched data. Here, $r_{12}(0)$ is 0.052 against 0.009; large values at $r_{12}(K)$ is at lag -5 ($= -2 \times 2.5$) and lag 3 (1.2×2.5) as against -2 and 1. Evidently, no new conclusions can be drawn.

For the second set, the stretch factors for the two series (219 and 216) are 1.667 and 1.875 and relative stretch factor (219 : 216) is 1.125. For the third set, the values are 1.262, 1.424, and 1.128. It is evident that eyeball estimation and sedimentation rate gives approximately the same relative stretch factor. Spectral analysis fails in this respect. Perhaps this is due to shortness of the series leading to inaccurate estimates of cycle lengths.

The cross-correlogram for the second set (Fig. 35c, small dashes) and third set (Fig. 35c, large dashes) distinctly shows that stretching has improved matters, as $r_{12}(0)$ is now 0.42 which is significant at ~8 percent level even when the degrees of freedom is calculated on the basis of N for the original data of the shorter series (19).

(2) *The 219/IK series.* In this situation absolute and relative stretch factors are as follows (first, second, first/second series).

(a) first set : 2.5, 1.0, and 2.5 (Fig. 35d, solid line);
(b) second set : 1.667, 1.071, and 1.556 (Fig. 35d, small dashes);
(c) third set : 1.262, 1.0, and 1.262 (Fig. 35d, large dashes).

A study of the figure and comparison with the correlogram based on unstretched data indicates that stretching has improved correlation with a relative stretch coefficient of 2.5.

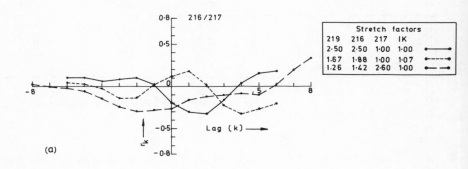

(a)

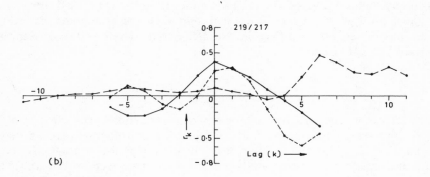

(b)

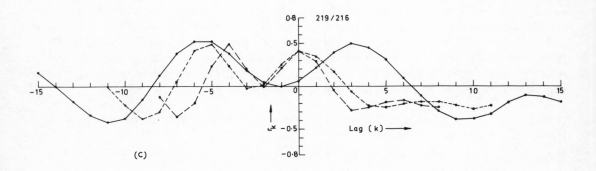

(c)

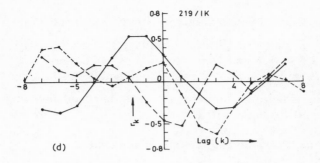

(d)

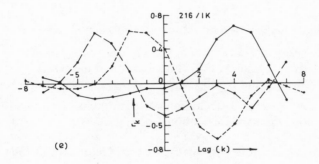

(e)

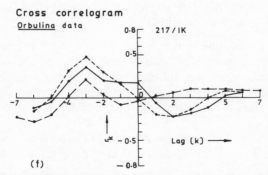

(f)

Fig. 35. Cross-correlogram with stretched data. Three sets of stretch factors used for different series are shown, in table against Fig. 1(a). Series pairs compared are indicated in top center of diagrams

But the final result remains unsatisfactory as $r_{12}(0)$ is not high (0.32); the highest value is at $r_{12}(2)$ and $r_{12}(3)$.

(3) *The 219/217 series.* Absolute and relative stretch factors are:

   (a) first set : 2.5, 1.0, and 2.5 (Fig. 35b, solid line);
   (b) second set : 1.667, 1.0, and 1.667 (Fig. 35b, small dashes);
   (c) third set : 1.262, 2.6, and 0.486 (Fig. 35b, large dashes).

The cross-correlogram clearly shows improvement in correlation with a relative stretch factor of 2.5; here $r_{12}(0)$ is highest (0.4).

(4) *The 216/IK series.* Absolute and relative stretch factors are:

   (a) first set: 2.5, 1.0, and 2.5 (Fig. 35e, solid line);
   (b) second set: 1.875, 1.07, and 1.75 (Fig. 35e, small dashes);
   (c) third set: 1.424, 1.0, and 1.424 (Fig. 35e, large dashes).

Figure 35e distinctly shows that improvement occurs with the second set of coefficients; here $r_{12}(0) = 0.40$.

(5) *The 216/217 series.* Absolute and relative stretch factors are:

   (a) first set: 2.5, 1.0, and 2.5 (Fig. 35a, solid line);
   (b) second set: 1.875, 1.0, and 1.875 (Fig. 35a, small dashes);
   (c) third set: 1.424, 2.6, and 0.548 (Fig. 35a, large dashes).

A comparative study of cross-correlograms with stretched and unstretched data shows that stretching has not improved correlation (see Fig. 35a).

(6) *The 217/IK series.* Absolute and relative stretch factors are as follows:

   (a) first set: 1.0, 1.0, 1.0 (Fig. 35f, solid line);
   (b) second set: 1.0, 1.071, 0.933 (Fig. 35f, small dashes);
   (c) third set: 2.6, 1.0, 2.6 (Fig. 35f, large dashes).

A study of Fig. 35f shows that with stretching, the value of $r_{12}(0)$ decreases.
From the discussion, it may be concluded that,

   (1) with stretching, correlation between two series can be improved considerably.
   (2) Optimum stretching factor is difficult to determine; spectral analysis has the promise to provide objective criterion. Further work is needed with longer series.
   (3) Eyeball estimation provides a good guide if there are only a few series.
   (4) Rate of sedimentation may give incorrect information as we see here in the 217 series.
   (5) Distribution pattern of *O. universa* for site 217 is atypical. Such a series should call for analysis with a much larger data and closer sampling interval.

From the results of the cross-correlation analysis, the best possible solution for a line of stratigraphic equivalence is shown in Fig. 7. The assumed stretch factors are, for 216–2.0, and for 219–1.786. The IK and 217 series are left unstretched. These values have been guesstimated from the three sets of cross-correlograms. The position of the 217 series *vis a vis* the other series is doubtful for reasons explained earlier. One may argue that from eyeball estimation one may arrive at similar results. This contention is not valid, as one would not know which peak of one series should correspond to a given peak of another. However, I did make use of eyeball estimation to fix stretch factors, but this was guided basically by autospectral analysis which gave an idea of cycle lengths.

If the above SF estimates are accepted, the sedimentation rates of the 216 and 219 sites for Late Pleistocene are 1.175 cm/10^3 yrs. and 1.316 cm/10^3 yrs. On this basis, the time span for the 2m cycle for the 219 series is 152,000 years. The estimate given earlier is 120 000 years. I do not have any data to check the correctness of the estimates.

Attention may be drawn to the fact that the method outlined here has been successful in the correlation of Late Pleistocene sediments of sites 216 and 219. In the case of the Caribbean core, the method is only moderately successful. Shortness of series is not the only reason. We must also consider the latitudinal separation between DSDP and Caribbean core sites and the location of the latter, which is in the subtropical zone. Its consequent influence on the distribution of *O. universa* has already been discussed. This aspect needs careful study. In the 217 series, the method has not been successful for reasons explained earlier.

It must be remembered in this context that, if this method has failed, other paleontological methods would not have been successful either, because paleontologic record in such situations is inadequate.

Finally, I would like to emphasize the point that, if the fossil record is complete, the method presented here would give more accurate correlation than is possible with other methods. Additionally, we can obtain information on relative rates of sedimentation and paleoclimatic cycles.

Sampling Interval and Series Length

In the present work, time series analysis was performed on short series sampled at an interval of 0.5 m. It has imposed certain limitations on interpretation of the results.

It was mentioned earlier that Imbrie and Kipp (1971) have tabulated relative abundance distribution data of *O. universa* in the Caribbean core of 10.9 m length sampled at an interval of 10 cm. Thus, 110 data points are available for analysis. From this series, every fifth value was taken to form the IK series ($\Delta = 0.5$ m, $N = 23$). This was performed for the sake of comparison. In this section, analyses on the basis of all ($N = 110$) data and on data formed of every alternate value ($N = 55$, $\Delta = 20$ cm) are presented in order to show the effect of increasing sampling interval (SI) and consequently decreasing N. The three series will now be referred to as the IKF ($N = 110$), IKH ($N = 55$), and IK ($N = 23$) series.

An inspection of the depth plots for original and smoothed data of the three series will convince one of the presence of periodic elements although the exact lengths of the periods are not apparent (Figs. 36 and 37). It also may be seen that the IKF series has high frequency noise, which is smoothed out by the 5-point moving average. By increasing the SI to 0.2 m, the noise is reduced further and the periodic nature is enhanced in the IKH series. Further increase of the SI($\Delta = 0.5$ m) results in loss of information and cycles having a long wave length remain (see Fig. 7).

Serial correlation analysis reveals this aspect in a more appropriate manner. The correlogram of the IKF series (Fig. 38) shows that, besides r_1 all other values up to lag 18 fail to exceed the confidence limits. Hence, one may conclude that the series was generated by a purely random process. Further, the correlogram does not show any regular oscillation.

It may be recalled that similar conclusion was drawn from a study of the IK series. A comparison of the correlograms of the two series (Figs. 10 and 38) would show that, in spite of difference in N, the pattern of the curves has some similarities.

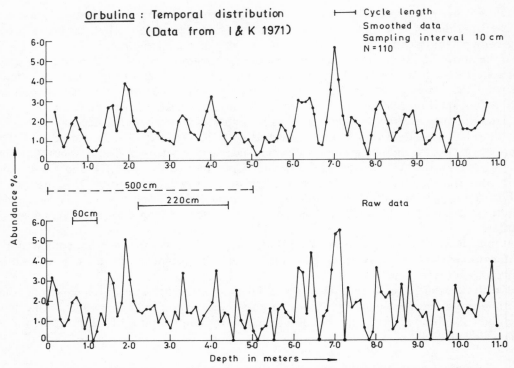

Fig. 36. Depth plots with raw and smoothed data. Full data from Imbrie and Kipp (1971) was used to form series (IKF series); 5-point-smoothing formula has been employed. Cycle lengths are shown by horizontal solid and dashed lines. Presence of low-frequency component is apparent

In contrast to this, the IKH series shows significant auto-correlation (Fig. 39). In this situation, r_2 and r_3 values fall beyond the 95 percent limit, and r_9 is close to it. Moreover, the correlogram shows oscillatory movement indicating the presence of a periodic component of c.l. between 0.6 and 0.7 meters.

This difference in behaviour can be explained if we consider the relative elimination of noise and loss of information when SI is increased. Elimination of noise in the IKH series

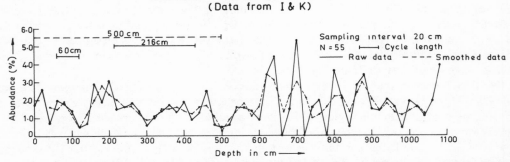

Fig. 37. Depth plots with raw and smoothed data. Series formed by taking alternate value from *Orbulina* data tabulated by Imbrie and Kipp (IKH series)

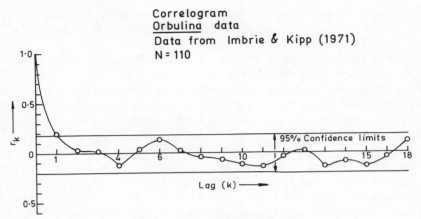

Fig. 38. Correlogram for IKF series

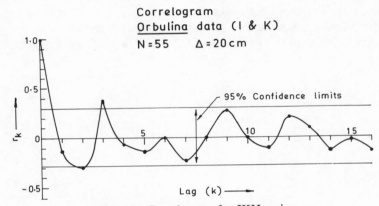

Fig. 39. Correlogram for IKH series

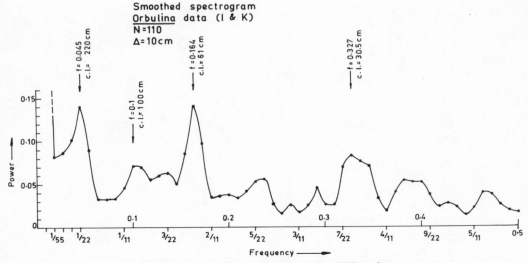

Fig. 40. Smoothed spectrogram for IKF series

improves autocorrelation. Further increase of SI in the IK series results in loss of information leading to destruction of autocorrelation. This conclusion corroborates the earlier contention.

One may argue that, if a series is available with small SI, one could filter the series. This, of course, is the normal procedure, but, to do this, one must have prior knowledge of the model which fits approximately the series. Work is in progress on the aspect. However, the present study indicates that a second-order autoregressive process could have generated a series such as the ones being considered here. It also is possible that autoregressive moving average model may fit the data better. To start with, one may use simple moving average or first difference filter.

Analysis of the three series in the frequency domain further reveals the effect of increasing SI. It has been shown already from harmonic and spectral analyses that the IK series has a low frequency component of 0.2 c/m (c.l. = 5 m). The smoothed spectrogram of the IKF series show two strong peaks at 0.45 c/m and 1.64 c/m (Fig. 40). A third distinct peak occurs at 3.27 c/m. Other peaks are weak. For the IKH series, peaks in the spectrogram are at 0.463 c/m, 1.21 c/m, and 1.66 c/m, the third peak being the strongest (Fig. 41). It is clear that both the strong peaks of IKF spectrogram are repeated here. The third has a frequency higher than the Nyquist frequency, so we cannot expect its presence in the IKH spectrogram. The problem lies with the peak at $f = 1.2$ c/m. The IKF spectrogram does show the

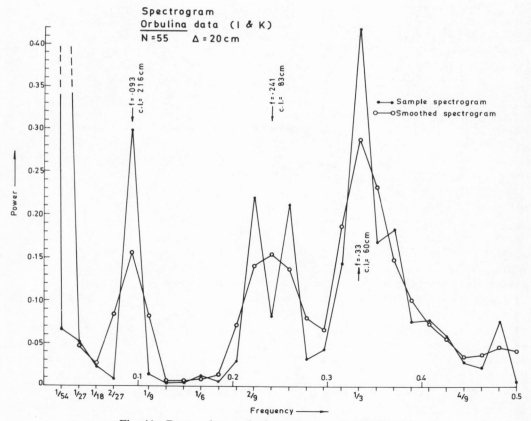

Fig. 41. Raw and smoothed spectrogram for IKH series

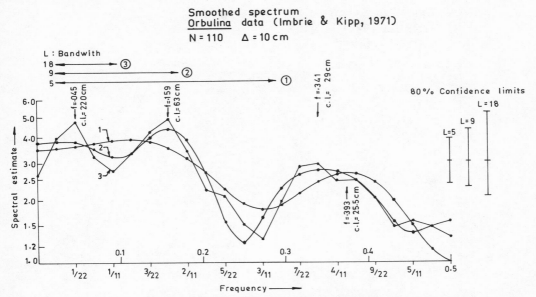

Fig. 42. Smoothed spectra for IKF series using three different lag windows

presence of a small peak at 1 c/m. It is possible that the peak at $f = 1.2$ c/m is spurious and has come into existence due to the increased SI. This point has to be investigated further. In this connection, it may be pointed out that the peak at 3.27 c/m may be regarded as the higher harmonic peak of the strong 1.64 c/m peak. I like to draw attention to Table 6 which shows that the 5-fold increase in SI only marginally affected the series mean, standard deviation, and other parameters.

Figure 42 shows the smoothed spectra for the IKF series estimated for different band widths. It is clear from the figure that window-closing operation stabilizes the peaks towards a central value. As a window corresponding to $L = 18$ ($\simeq \frac{1}{6}N$) cannot be considered to be too narrow, we may accept the corresponding peaks as real. They are also significant. The peaks are at 0.45 c/m, 1.59 c/m, and 3.41 c/m.

Spectra of the IKH series (Fig. 43) estimated with L corresponding to different window sizes show that, with a narrow window, side lobes appear. The reason is, $L \simeq \frac{1}{3}N$; consequently, for such a narrow window, spurious peaks appear. It also shows that window size corresponding to $L = \frac{1}{16}N$ is satisfactory in the present situation and, perhaps, similar other examples. The spectrum with a window corresponding to $L = 8$ reveals the presence of two peaks at 0.38 c/m and 1.63 c/m, the second one being stronger. Here, 1.2 c/m peak is absent. It confirms the suspicion that 1.2 c/m peak is spurious.

A comparative study of the results of harmonic and spectral analysis of the two series shows that frequency of the stronger peak remains virtually constant even if the SI is increased and different analytic methods are followed. Spurious peaks occur inconsistently and tend to disappear. Distinct but weak peaks tends to shift positions. It is likely that with the same SI, a longer series is needed to fix the value. In this situation, the low frequency peak is at 0.45 c/m. Spectrogram of the IKH series also show this. The difference may or may not be considered significant. It may be noted that, if the window is closed corresponding to $L = 16$, the peak frequency shifts to 0.5 c/m; 0.45 c/m is a mean of these two estimates.

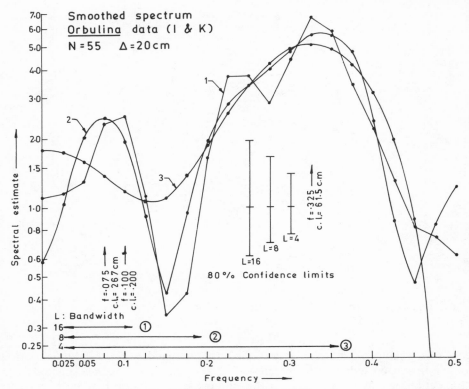

Fig. 43. Smoothed spectra for IKH series

We may now examine what happens when the SI is increased further to 0.5 m. As the Nyquist frequency corresponding to SI = 0.5 m is 1 c/m, one would not expect the spectrogram or spectrum to show a 1.67 c/m peak. However, both spectrogram and spectrum show a prominant peak at 0.182 c/m. Computation with decreased frequency spacing of 0.025 fixes the peak at f = 0.2 c/m. Other peaks are absent. The sample spectrogram, however reveals the presence of another peak at 0.36 c/m, which is comparable to that given by spectral analysis of the IKH series. This frequency may be considered to be a higher harmonic of the 182 c/m. This finding leads to the important question whether the low frequency component of the IKF spectrum may be considered to be the higher harmonic of the lowest frequency, and the real value of this is 0.22 c/m?

One may now argue that the IKF and IKH spectra should have shown the existence of the low frequency peak. A little consideration will show that, unless a much longer series is available, extremely low frequency peaks cannot be properly determined when the SI is small. For example, with 100 data points and SI = 0.1 m, the lowest frequency theoretically determinable from harmonic analysis is 0.1 c/m; but smoothing the sample estimates would remove peaks at this end and the next frequency unless they are strong. High frequency peaks are usually weak, so in practice, peaks of frequencies >0.3 c/m can only be detected. On the other hand, with only 50 data and SI 0.5 m, 0.1 c/m peaks can be detected with ease. The above discussion has an important bearing on the choice of sampling interval. If we are interested in a high frequency component, a smaller SI is needed. On the other hand, it is better to sample at longer interval if we want to determine a geologically more meaningful low frequency component. This, however, requires a greater length of core or a thick section. To estimate the required length, we may apply 'half a dozen' criterion suggested by Granger and Hatanaka (1964). By this, it is meant that the series must contain at least 6 cycles of the desired cycle length. For example, if we need establish the presence of 5-m cycles, the length of the core or the thickness of the section should be $5 \times 6 = 30$ m.

In this context, it may be pointed out that TS analysis based on samples taken from too long a core is hazardous because the rate of sedimentation will not remain uniform for long periods of time, thus changing the relation between time and depth. If a large suit of samples are available to develop a long paleontologic TS, it is appropriate to apply a jackknife and analyze the resulting two or three parts separately. If they are comparable, one may then use the total set for analysis. Further, if the series is long, one may have to remove evolutionary trends.

From the previous discussion, the following conclusions may be drawn:

(1) Too close a sampling interval should be avoided as it may introduce unwanted high frequency noise.
(2) SI should not be so large as to cause loss of information, or destruction of the essential structure of the series.
(3) To establish the presence of low frequency components, larger SI or longer series is necessary.
(4) Length of the series should preferably be above 50, but not less than 30. Smaller SI would require longer series.
(5) If only a short series is available, spectral analysis may be performed with profit if one bases conclusions on stronger peaks; in this situation one should look for spurious peaks. If the 'signal' is strong, it can be detected.
(6) For deep sea samples, a sampling interval of 0.25 m may be considered adequate. This SI would also guard against turbational effect.

Before concluding this section, I would like to draw attention to the points of similarity between results obtained in this paper on the basis of TS analysis of *O. universa* abundance data and that given in Imbrie and Kipp (1971). From a spectral analysis of estimated summer and winter temperatures they arrive at frequency peaks 2.7 c/m, 1.4 c/m, and 0.4 c/m. Corresponding cycle lengths are 0.37 m, 0.71 m, and 2.5 m. In the present paper, the determined cycle lengths are 0.31 m, 0.61 m, and 2.2 m. Corresponding time spans are (in round figures) 13 000 years, 26 000 years, and 94 000 years. The last one is close to the average period of eccentricity (92 000 yrs) and the second one to precession. We have noted already that the first peak has no significance. Perhaps, the coincidence is not fortuitous.

Discussion

The purpose of the present paper was to investigate the usefulness and limitations of the time series analytic method in stratigraphic correlation. On the basis of discussions given in each section and conclusions drawn, it is now possible to outline the steps that should be taken for analysis.

(1) A suitable species which is known to respond to global environmental factors (temperature and salinity) should be chosen.
(2) Recent distribution of the species should be analyzed to understand the relation between global factors and abundance, and to fit the appropriate theoretical model.
(3) Absolute abundance of the species should be determined from equally-spaced strati-

graphic samples; the collection of data thus obtained will form the paleontologic time series.

(4) Sampling interval should not be either too small or too large. Actual value should depend on the nature of the sediments.

(5) The series should be analyzed both in the time and frequency domains.

(6) Information from the analysis should be used to define the most dominant periodic element of the series.

(7) The above procedure is followed for all the stratigraphic sections.

(8) A reliable datum is fixed and the two series should be aligned before attempting to make bivariate TS analysis.

(9) To find the relation between the different series cross-correlation is performed. If the correlation is satisfactory, the two series may be regarded as stratigraphically equivalent.

(10) If no significant result is obtained from cross-correlation analysis, cross-spectral analysis should be performed to identify the periodic element, if present, and investigate the phase difference.

(11) If the series are uncorrelated, one series should be stretched with respect to the other to improve the cross-correlation.

(12) To determine the stretch factor, cycle lengths of dominant periodic components should be taken as a guide. Information from cross-correlation and cross-spectral analysis may also be used.

(13) If there are only two series, iterative methods may be followed.

(14) If the series is too long (in the geological sense), it should be analyzed in sections to avoid complications due to change in rate of sedimentation.

(15) Number of samples should be, preferably, more than 40 and the core length or section thickness should be at least four to 5 times the length of the dominant periodic component.

(16) If the series is shorter than 40, interpretation should be made with due caution.

Acknowledgments. The author is thankful to the Deep Sea Drilling Project authorities of the Scripps Institution of Oceanography for the core samples on which the work is based. Financial support came from the Council of Scientific and Industrial Research, India. The work was carried out in the Micropaleontology Laboratory of the Indian Institute of Technology, Kharagpur.

I am thankful to my students, R. N. Hazra and V. Parichha who helped me in getting data; S. Ghose of Computer Center helped in developing programs.

References

Agterberg, F. P., 1974, *Geomathematics.* Elsevier, Amsterdam, 596 pp.

Anderson, T. W., 1971, *The Statistical Analysis of Time Series.* John Wiley & Sons, New York, 704 pp.

Be, A. W. H., Harrison, S. M., and Lott, L., 1973, *Orbulina universa* d'Orbigny in the Indian Ocean, *Micropaleont.,* **19**, 150–192.

Berggren, W. A., 1978, Recent advances in planktonic foraminiferal biostratigraphy, biochronology, and biogeography, Atlantic Ocean, *Micropaleont.,* **24**, 337–370.

Bock, R. D., 1975, *Multivariate Statistical Methods in Behavioral Research.* McGraw Hill Book Co., New York, 623 pp.

Borch, von der, Sclater, C. C., John, G., and others, 1974, *Initial Reports of the Deep Sea Drilling Project,* vol. 22. U.S. Government Printing Office, Washington, 890 pp.

Box, G. E. P., and Jenkins, G. M., 1976, *Time Series Analysis: Forecasting and Control.* Holden-Day, San Francisco, 575 pp.

Chatfield, C., 1975, *The Analysis of Time Series.* Chapman and Hall, London, 263 pp.

Davis, J. C., 1973, *Statistics and Data Analysis in Geology.* John Wiley & Sons, New York, 550 pp.

Defant, A., 1961, *Physical Oceanography,* vol. 1. Pergamon Press, London, 729 pp.

Finn, J. D., 1974, *A General Model for Multivariate Analysis.* Holt, Rinehart and Winston, New York, 423 pp.

Gartner, S., 1974, Nannofossil biostratigraphy, Leg 22, Deep Sea Drilling Project, in Borch, von der, Schlater, C. C., and others (eds.), *Initial Reports of Deep Sea Drilling Project*, vol. 22. U.S. Government Printing Office, Washington, pp. 557–599.

Granger, C. W. J., and Hatanaka, M., 1964, *Spectral Analysis of Economic Time Series.* Princeton University Press, New Jersey, 299 pp.

Granger, C. W. J., and Hughes, A. O., 1968, Spectral analysis of short series—a simulation study, *Jour. Roy. Stat. Soc., Ser. A*, **131**, 83–99.

Hecht, A. D., 1973, A model for determining Pleistocene paleotemperatures from planktonic foraminiferal assemblages, *Micropaleont.,* **19**, 68–77.

Hunkins, K., Be, A. W. H., Opdyke, N. D., and Mathieu, G., 1971, The Late Cenozoic history of the Arctic Ocean, in Turekian, K. K. (ed.), *The Late Cenozoic Glacial Ages.* Yale Univ. Press, New Haven, pp. 215–237.

Imbrie, J., and Kipp, N. G., 1971, A new micropaleontological method for quantitative paleoclimatology: application to a Late Pleistocene Caribbean core, in Turekian, K. K. (ed.), *Late Cenozoic Glacial Ages.* Yale Univ. Press, New Haven, pp. 71–181.

Jenkins, G. M., and Watts, D. G., 1968, *Spectral Analysis and its Application.* Holden-Day, San Francisco, 525 pp.

Kendall, M. G., and Stuart, A., 1966, *The Advanced Theory of Statistics,* vol. 3. S. Charles Griffins Co., London, 552 pp.

Kwon, B., and Rudman, A. J., 1979, Correlation of Geologic Logs with spectral methods, *Jour. Math., Geology,* **11**, 373–379.

Lynts, G. W., 1971, Analysis of the planktonic foraminiferal fauna of core 6275, Tongue of the Ocean, Bahamas, *Micropaleont.,* **17**, 152–166.

Malmgren, B., and Healy-Williams, N., 1978, Variation in test diameter of *Orbulina universa* in the paleoclimatology of the Late Quaternary of the Gulf of Mexico, *Paleogr. Paleoclimalot. Paleoecol.,* **52**, 235–240.

Malmgren, B. A., and Kennett, J. P., 1978, Late Quaternary paleoclimatic applications of mean size variations in *Globigerina bulloides* D'Orbigny in the southern Indian Ocean, *Jour. Paleont.,* **52**, 1195–1207.

Pearson, E. S., and Hartley, H. O. (eds.), 1966, *Biometrika Tables for Statisticians,* vol. 1. Cambridge Univ. Press, Cambridge, 264 pp.

Sard, A., and Weintraub, S., 1971, *A Book of Splines.* John Wiley & Sons, New York, 817 pp.

Schwarzacher, W., 1975, *Sedimentation Models and Quantitative Stratigraphy.* Elsevier, Amsterdam, 328 pp.

Whitmarsh, R. B., Weser, O. E., Roso, D. A., and others, 1974, *Initial Reports of the Deep Sea Drilling Project,* vol. 23. U.S. Government Printing Office, Washington, 1180 pp.

Wollin, G., Ericson, D. B., and Ewing, M., 1971, Late Pleistocene climates recorded in Atlantic and Pacific deep sea sediments, in Turekian, K. K. (ed.), *The Late Cenozoic Glacial Ages,* Yale Univ. Press, New Haven, pp. 199–214.

Quantitative Stratigraphic Correlation
Edited by J. M. Cubitt and R. A. Reyment
© 1982 John Wiley & Sons Ltd.

Correlating Between Electrical Borehole Logs in Paleoecology

RICHARD REYMENT

Paleontologiska Institutionen, Uppsala Universitet, Box 558, S-75122, Uppsala, Sweden

Abstract

Pirson (1977) has demonstrated the paleoecological usefulness of electrical
borehole logs. Quantitative paleoecological analyses based on electrical logs and
morphological variations of organisms frequently face the problem of proving significant association in patterns of oscillation in variational curves which are obvious
from visual inspection of the plots. The usual method of cross-correlation tends to
provide unsatisfactory results because of trends in the observations, frequently of a
pseudo-cyclic nature. Results for the product-moment correlation coefficient can
be improved by dividing the sequence of observations into subsamples. Nonparametric tests provide a better solution although they lose the paleoecologically-
important information inherent in the stratigraphical ordering of observations.
Tests recognized as useful are a variant of the runs test for paired observations, the
logistic test applied to binary data and the chi-squared test.

Introduction

Electrical logging of boreholes has been a standard procedure in the petroleum industry for
60 years. It has long been known that the resistivity and self-potential logs are useful
indicators of sediment properties. The standard Schlumberger manual states that electrical
logs have two general uses: (1) correlation and stratigraphical studies and (2) the evaluation
of formation fluids and lithology.

This broad subdivision does not take paleobiological aspects into account; however, it
does succeed in indicating that lithology and fluids, in the minds of log-analysts, are inter-
woven conceptually in the interpretation of electrical borehole logs. For paleoecological
work, it is necessary to be aware of this interpretational overlap, as the only useful line of
development lies with the lithological significance of the logs.

Measuring Agreement Between Logs

Ideally, one would like to measure agreements between logs by means of cross-correlation.
Unfortunately, this technique frequently yields misleading results in that the correlations
obtained may be substantially lower or higher than expected from visual inspection of the
logs.

Raup and Gould (1974) have demonstrated by computer-simulation experiments that
randomly-varying series in time can show significant correlations. Even if the effect is due to
nonmultivariate normality of the data and the presence of atypical values, it does not
detract from the importance of the lesson they provide.

One of the reasons for the poor performance of the product-moment correlation coefficient for borehole sequences is the irregularities usually present in such data. This can be ameliorated by calculating correlations for contiguous subsets of the data, but this is not helpful for analyses in which interest is directed towards studying long-term ecological effects.

Clearly, a nonparametric technique would be useful or, as is demonstrated in the paper by Allan Gordon in this volume, special regression methods can be put to advantageous effect.

In the present connexion, results of analyses of borehole logs made by nonparametric methods are presented. These methods are not new, but some do not seem to have been applied extensively, in geology.

Dichotomization

As a paleoecological interpretation of a borehole log will be based on the steps to the right and left shown by the sequence of observations, and magnitudes of the amplitudes will normally be of subordinate interest, little information on reactions of the organism to small fluctuations in the environment will be lost by dichotomizing the data and statistically analyzing the coded changes in sign.

The effect of dichotomizing the data is to codify changes in sign displayed by a series in time. However, dichotomization effectively removes trend in the series, at the expense of a total loss of information of potential ecological significance. Nevertheless, for the purposes of recognizing quantitative expressions of likeness between two logs from the same borehole with respect to patterns of fluctuation, dichotomization of observations is useful.

Paleoecological Significance of Electrical Borehole Logs

The interpretation of electrical logs presented is based on the text by Pirson (1977). Pirson prefers to employ the redox log for analyzing the environmental record preserved in sediments. This log is seldom run in commercial operations, but can be substituted by the SP-log, which is a special case of a redoxomorphic log, for many frequently occurring situations. In general terms, Pirson's method of analysis is based on measuring various levels of oxidation-reduction of the sediments penetrated by a borehole.

The redox log is obtained using an inert platinum electrode and a lead reference electrode. It measures the proportions of oxidized and reduced forms of various minerals and ions in sedimentary formations. This potential is interpreted as being dependent on the environment of deposition of the sediments.

Pirson also advocates the use of the short-normal resistivity, induction, and conductivity curves for paleoenvironmental interpretations.

The SP and short-normal resistivity curve shapes are interpreted as indicators of the regressive and transgressive nature of 'sand deposition' processes.

As an example of a practical application of Pirson's (1977) results, the SP and short-normal resistivity curves for a borehole drilled at Gbekebo, Nigeria (Fig. 1) is submitted. The figures at the foot of the illustration expose Pirson's (1977, p. 47) interpretation of the SP and resistivity logs' behaviour in regressional and transgressional environments. The geological information for transgressions and regressions during the Maastrichtian and Early Paleocene is provided on the left margin of the figure.

The example indicates that it is possible to relate the field evidence to Pirson's interpretation of how electrical curves reflect the transgressional–regressional history of a sequence of sediments (Pirson (1977) provides other examples).

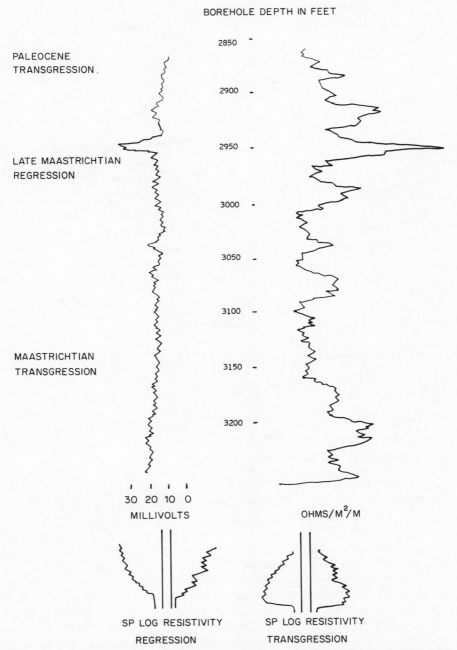

Fig. 1. Example of Pirson's (1977) method of analysis of electrical logs applied to SP-log (left) and short-normal resistivity log (right) for Nigerian borehole Gbekebo I. Relationships between Maastrichtian and Paleocene eustatic events are marked

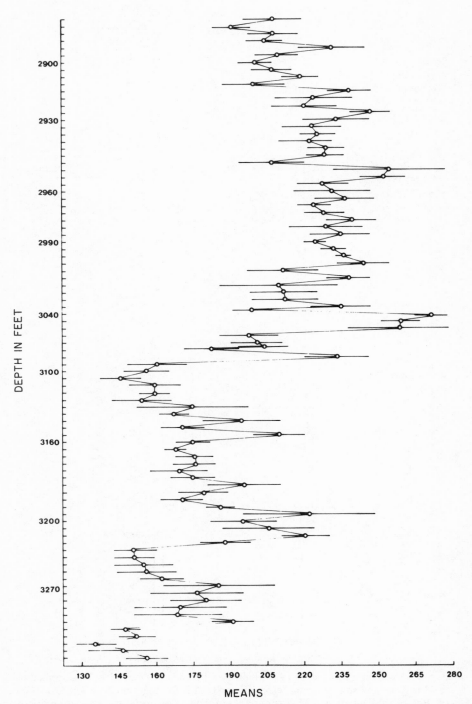

Fig. 2. Basic information for Nigerian example discussed in text. Left-hand curve shows fluctuations in mean diameter of megalospheric proloculi of foraminifer *Afrobolivina afra*,

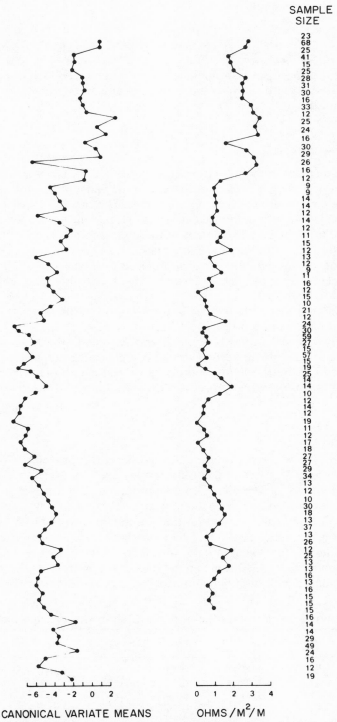

the central curve shows plot of first canonical variate means for 9 variables measured on shell of foraminifer, and right-hand curve is that of short-normal resistivity log. SP-log is not shown. (95% confidence intervals for means drawn)

Statistical Analysis of Example

The basic information for the Nigerian example analyzed here is given in Fig. 2. This illustration shows the variation in the megalospheric proloculus of a species of benthic foraminifers, *Afrobolivina afra*, the first canonical variate means, and the short normal resistivity curve. Even cursory inspection of the plots reveals that the canonical variate means and resistivity log display considerable agreement, whereas the first curve shows divergencies from the other two.

Correlations between curves

Despite close agreement in the patterns of the curves for the canonical variate means and short-normal resistivity log, this is not reflected in the value of the correlation coefficient. This seems to be a result of the trend in both curves. The canonical variate means, referred to in the foregoing section, are the first canonical variate means for 9 variables measured on the species of foraminifers. The details of these measurements as well as the method of statistical analysis are treated in Reyment (1980).

The correlation coefficients for the dichotomized observations for the three curves of Fig. 2 are (proloculus = 1, canonical variate means = 2, resistivity = 3): $r_{12} = 0, r_{13} = 0$, and $r_{23} = 0.8$.

Subdividing the sequences into four subsamples improved the correlations, but not to the extent one might have expected (cf. Table 1).

An analysis for atypical values (Campbell and Reyment, 1980) showed that the first subsample is free of outliers, the second section contains three atypical values, the third has three outliers and the fourth section, 9 outlying values. For this analysis, the total sample size was 81 with the subsamples comprising 20, 20, 20, and 21 observations.

Nonparametric methods

Rank correlation was applied to the data, but without improvement over the standard correlation coefficient. Variants of the 'runs' test are useful for testing for randomness in a series of dichotomized observations. The appropriate null hypothesis is whether the sequence of fluctuations could have arisen by chance, that is are the fluctuations random? (cf. Bradley, 1968). A suitable variant for the present data is a test based on the number of runs up and down, for which Bradley (1968) supplies tables of significant values.

Applying this test to the Nigerian data, with 80 observations ($N - 1$), using changes of sign for the sequence of canonical variate means for morphological variations in the foraminifer, compared with the fluctuations of the short-normal resistivity log, yielded 19

Table 1. Correlation coefficients for material of Fig. 2 partitioned into four contiguous subsamples

Subsample	N	r_{12}	r_{13}	r_{14}	r_{23}	r_{24}	r_{34}
1	20	0.82	0.43	0	0.43	0	0.79
2	20	0.57	−0.17	0.26	0.26	0.23	0.50
3	20	0.80	0.62	0.67	0.64	0.71	0.86
4	21	0.43	0.29	0.22	0	0	0.94

Key: 1 = diameter of proloculus; 2 = first canonical variate means;
3 = short-normal resistivity curve; 4 = SP-curve.

runs. The corresponding value for a comparison between the canonical variate means and the SP-log gave 26 runs. In a series of N observations, of which N_1 denotes the number of agreements a in paired sign comparisons and N_2 denotes the number of lacks of agreement b, the observed a's and b's will occur in random order if each position in the series had equal likelihood of being occupied by an a or b event.

Thus, for the first comparison (means-resistivity), there are 19 runs, $N_1 = 69$ and $N_2 = 11$. The total-number-of-runs test rejects randomness in the sequence (extrapolating from the tables given by Bradley, 1968). The second comparison (SP-means), with 26 runs, $N_1 = 59$ and $N_2 = 21$, also rejects the hypothesis of randomness.

Another test applied to the data is based on the logistic transformation applied to binary data. It was studied by Cox (1970) for a 2×2 contingency table. The test is defined by the relationship

$$Z = [\log(R + \tfrac{1}{2})/(N - R + \tfrac{1}{2})], \tag{1}$$

where R denotes the number of p-ositive signs in a sample of size N. The variance of equation (1) is $[(N + 1)(N + 2)]/[N(R + 1)(N - R + 1)]$.

For two logs being compared for agreement in the patterns of oscillations, one requires the difference $D = Z_1 - Z_2$ with standard error $S_E = \sqrt{(V_1 + V_2)}$, where V_i denotes the variance of Z. The ratio D/S_E is a standardized normal deviate which gives a two-sided test for the hypothesis $D = 0$ (Cox, 1970).

In the present example, it was determined that $R = 0.51$ for the curve formed by the canonical variate means and 0.48 for the short-normal resistivity curve. Performing the necessary computations leads to the conclusion that the difference $Z_1 - Z_2$ is not significant.

The most frequently-used nonparametric test, chi-squared, was also applied to the data. It was found for comparisons of the short-normal resistivity log and SP-log, respectively, with the canonical variate means that the value of chi-squared is very highly significant.

Conclusions

The results of the statistical analyses indicate that there is close correlation in the resistivity log and canonical variate means, and that there is moderate association in the SP-log and canonical variate means. Nonparametric methods provide more realistic results for testing association in patterns of oscillation than the product-moment correlation coefficient, although the usefulness of the latter can be improved if the data are partitioned into contiguous subsamples.

Palaeontological Remarks

It now remains to comment briefly on the significance of the relationships between electrical logs and morphological variations. The overall conclusion is that variations in the electrical logs show close association with variations in morphology of the benthic foraminifer. This suggests that redox conditions of the sediment had an influence on shell morphology but whether this was solely ecophenotypic or whether the changes in the morphology of *Afrobolivina* resulted from selectional response to minor fluctuations in the environment, cannot be determined without further study (genetic tracking).

A third possibility, that of random genetic drift, seems less likely considering the demonstrated association between variation shown by the organism and fluctuations in the electrical logs. The latter seems to reflect the changes in composition of sediment resulting from a cycle of transgressions and regressions at the close of the Cretaceous Period.

References

Bradley, J. V., 1968, *Distribution-free Statistical Tests*. Prentice-Hall, New Jersey, 388 pp.

Campbell, N. A., and Reyment, R. A., 1980, Robust multivariate procedure applied to the interpretation of atypical individuals of a Cretaceous foraminifer, *Cretaceous Research*, 1(1), 207–221.

Cox, D. R., 1970, Analysis of binary data, *Monographs on Applied Probability and Statistics*, Methuen, London, 142 pp.

Pirson, S. J., 1977, *Geological Well Log Analysis*. Gulf Publishing Company, Houston, Texas, 377 pp.

Raup, D. M., and Gould, S. J., 1974, Stochastic simulation and evolution of morphology, *Syst. Zool.*, **23**, 305–322.

Reyment, R. A., 1980, *Morphometrical Methods in Biostratigraphy*. Academic Press, London, 175 pp.

Quantitative Stratigraphic Correlation
Edited by J. M. Cubitt and R. A. Reyment
© 1982 John Wiley & Sons Ltd.

On Measuring and Modelling the Relationship Between Two Stratigraphically-recorded Variables

A. D. GORDON

Department of Statistics, University of St. Andrews, St. Andrews, Fife, Scotland

Abstract

This paper describes methods of measuring and modelling the relationship between two variables which have been measured at specified depths down a single stratigraphical column. The methods take into account the information provided by the depths at which the pairs of readings are recorded.

Introduction

Researchers in the earth sciences are sometimes concerned to investigate the relationship between two quantitative variables, X and Y, which have been measured at n specified depths down a single stratigraphical column: thus, the data are of the form $\{x_i, y_i, d_i, (i = 1, \ldots, n)\}$ where (x_i, y_i) denotes a pair of readings recorded at the ith position (depth d_i) in the core; it will be assumed that the data have been ordered by depth, that is, $d_1 > d_2 > d_3 > \ldots > d_n$.

It is of interest to measure and model the relationship between X and Y; this paper describes briefly methods of analysis which take into account the information provided by the depth variable, D, and gives references which present the work in detail. The assumptions underlying the use of each method are emphasized, in order to assist an investigator in the selection of appropriate methods for data analysis. The methods are illustrated by application to data provided by R. A. Reyment comprising 81 pairs of measurements of resistivity (X) and 'biolog' (Y) in a stratigraphical column.

Partial Correlation

The correlation ρ_{XY} between two variables X and Y is a parameter which indicates the strength of linear relationship between the variables. An estimate of ρ_{XY} can be obtained from the sample correlation coefficient,

$$r_{XY} \equiv S_{xy}/(S_{xx}S_{yy})^{1/2},$$

where

$$S_{xx} \equiv \sum_{i=1}^{n} (x_i - \bar{x})^2; \qquad S_{yy} \equiv \sum_{i=1}^{n} (y_i - \bar{y})^2; \qquad S_{xy} \equiv \sum_{i=1}^{n}(x_i - \bar{x})(y_i - \bar{y}); \qquad (1)$$

$$\bar{x} \equiv \sum_{i=1}^{n} x_i/n; \qquad \bar{y} \equiv \sum_{i=1}^{n} y_i/n.$$

This is a biased estimator, and small correction factors have been proposed (Kendall and Stuart, 1973, Chapter 26). More importantly, it is possible that any correlation between variables X and Y arises, in part, from the relationship of each of them to some other variable; conversely, it might be that the effect of this third variable is to mask, or reduce, a stronger correlation between X and Y. An example of such a third variable is 'time'. In stratigraphical studies, usually 'time' is not measurable directly; however, depth, D, can be employed frequently as a surrogate, an approach used in the following example.

As the main aim is to compare X and Y, one is led to examine their correlation conditional upon other variables taking certain fixed values: the partial correlation of X and Y with D fixed, $\rho_{XY.D}$, is defined by

$$\rho_{XY.D} \equiv \frac{\rho_{XY} - \rho_{XD}\rho_{YD}}{[(1 - \rho_{XD}^2)(1 - \rho_{YD}^2)]^{1/2}} . \tag{2}$$

An estimate of $\rho_{XY.D}$ is given by

$$r_{XY.D} \equiv \frac{r_{XY} - r_{XD}r_{YD}}{[(1 - r_{XD}^2)(1 - r_{YD}^2)]^{1/2}} , \tag{3}$$

where r_{XD} and r_{YD} are defined in a similar manner to r_{XY} in expression (1).

As an illustration, the 20 sets of measurements presented in Table 1 can be shown to give rise to sample correlation coefficients,

$$r_{XY} = 0.2994, \quad r_{XD} = 0.5043 \quad \text{and} \quad r_{YD} = -0.5272.$$

Table 1. Depth, resistivity, and biolog at 20 levels in stratigraphical column. Also given are statistics for investigating concordance of directions of change of resistivity and biolog

Level i	Depth d_i	Resistivity x_i	Biolog y_i	Concordance in direction of change z_i	Cumulative Statistic μ_i
1	108.8	2.40	6.85	1	3
2	105.7	1.80	6.45	1	6
3	105.1	1.85	6.54	1	9
4	101.4	1.50	6.37	1	12
5	99.6	1.80	6.54	1	15
6	96.9	2.00	7.51	1	18
7	96.0	2.20	8.97	1	$21 = \mu^*$
8	95.0	1.75	8.14	0	5
9	93.2	2.15	7.65	1	8
10	92.4	2.75	8.74	1	11
11	91.5	1.60	7.66	1	14
12	90.0	1.70	7.79	1	17
13	88.1	1.55	7.65	1	20
14	87.2	1.68	8.11	0	4
15	86.6	1.78	8.02	1	7
16	85.7	1.70	7.50	1	10
17	84.8	1.80	8.05	1	13
18	83.9	1.50	7.59	0	-3
19	83.0	1.20	7.83	1	0
20	82.1	0.75	7.09		

Substituting these values into expression (3) yields

$$r_{XY \cdot D} = 0.7704.$$

In this situation, the size of correlation between X and Y would appear to have been masked by the effect of D.

It is important to stress that $r_{XY \cdot D}$ provides a measure of correlation between X and Y under the assumption that changes in X and Y can be represented accurately by a linear expression in D. One would usually expect the relationships to be much less simple, for example, a polynomial expression in D. The effects of $D, D^2, D^3, D^4, \ldots$ on the correlation between X and Y could be eliminated successively by a series of operations of the type defined in expression (3), although more efficient procedures for achieving this aim are available. A fuller discussion of partial correlation, including a description of tests that an observed value differs significantly from zero, is presented by Fisher (1941, Chapter VI) and Kendall and Stuart (1973, Chapter 27).

In many situations, there will be uncertainty associated with the precise form of a relevant expression in D; on occasion, information on the behavior of X and Y as functions of time may be obtained, but the relationship between time and depth might not be known precisely.

An alternative to the previous procedure is the use of rank correlation coefficients, in which measurements recorded for each variable are ranked in order of magnitude, and a measure of agreement between two variables is obtained by comparing the two sets of ranks. An example of a rank correlation coefficient is Kendall's τ (Kendall, 1970), and the partial rank correlation coefficient of X and Y, conditional on eliminating the effect of D, can be defined in an analogous manner to expression (3):

$$\tau_{XY \cdot D} \equiv \frac{\tau_{XY} - \tau_{XD}\tau_{YD}}{[(1 - \tau_{XD}^2)(1 - \tau_{YD}^2)]^{1/2}}. \tag{4}$$

Evaluation of rank correlation coefficients for the data presented in Table 1 leads to a result similar to the use of correlation coefficients: a small positive rank correlation between X and Y is increased when the depth variable effect is removed.

A fuller discussion of partial rank correlation is presented by Kendall (1970, Chapter 8). It is difficult to derive statistical significance tests on observed values of $\tau_{XY \cdot D}$, but some results are presented by Johnson (1979).

Concordance of Directions of Change

The statistics described in the previous section provide a global measure of the agreement between two stratigraphically-recorded variables. In some situations, it is more pertinent to consider variations over smaller periods of time; thus one is led to examine the change between a pair of stratigraphically-neighbouring measurements,

$$\begin{aligned}\Delta x_i &\equiv x_{i+1} - x_i \\ \Delta y_i &\equiv y_{i+1} - y_i\end{aligned} \quad (i = 1, \ldots, m, \quad \text{where} \quad m \equiv n - 1) \tag{5}$$

It can be relevant to ask if, over the same interval of time, changes in the variables were concordant (i.e. the variables both increased, or both decreased), or discordant (i.e. one variable increased, while the other decreased).

In this formulation, one makes no use of the size of the changes between successive readings, regarding this information as less important or possibly less reliable than the direction of change. In addition, the assumption has been made that the sizes of time/depth intervals, over which changes are studied, are appropriate to the investigation.

It is convenient to code the concordances by defining a set of variables $\{Z_i (i = 1, \ldots, m)\}$:

$$Z_i \equiv \begin{cases} 1 \text{ if } (\Delta x_i)(\Delta y_i) > 0; \text{ changes concordant} \\ 0 \text{ if } (\Delta x_i)(\Delta y_i) < 0; \text{ changes discordant.} \end{cases} \tag{6}$$

For simplicity in the exposition, it is assumed that $(\Delta x_i)(\Delta y_i) \neq 0$, although appropriate generalizations are available.

The following model will be postulated for the set $\{Z_i (i = 1, \ldots, m)\}$: each Z_i, independent of any other Z_j $(j \neq i)$, takes the value 1 with probability p, and the value 0 with probability $(1 - p)$. The value of p is an indication of the degree of concordance of the two sequences of changes.

The assumptions of this model merit some discussion. $(\Delta x_{i-1})(\Delta y_{i-1})$ and $(\Delta x_i)(\Delta y_i)$ are clearly related, in that they involve the values of x_i and y_i. However, only the signs of these quantities are used in the definition of Z_{i-1} and Z_i (expression (6)), and if the time/depth interval between successive measurements is sufficiently large, it may be reasonable to regard Z_{i-1} and Z_i as independent of one another. Therefore, an investigator should ensure that this is a reasonable assumption for any data set which is analyzed by this method.

The parameter p could be estimated by

$$\hat{p} \equiv R_m/m, \quad \text{where } R_m \equiv \sum_{i=1}^{m} Z_i. \tag{7}$$

The statistic $\hat{p}$ gives a 'point', or best, estimate of p. An indication of the accuracy with which p can be estimated is provided by stating a (classical) confidence interval for p. Under the specified model, R_m is a binomial random variable, and an approximate $100(1-2\alpha)\%$ confidence interval for the parameter p is given by

$$\left(\frac{\hat{p} + d_\alpha^2/(2m) - d_\alpha\{\hat{p}(1 - \hat{p})/m + d_\alpha^2/(4m^2)\}^{1/2}}{(1 + d_\alpha^2/m)}, \right.$$

$$\left. \frac{\hat{p} + d_\alpha^2/(2m) + d_\alpha\{\hat{p}(1 - \hat{p})/m + d_\alpha^2/(4m^2)\}^{1/2}}{(1 + d_\alpha^2/m)} \right), \tag{8}$$

where d_α is the value above which $100\alpha\%$ of the area of the standard Normal distribution lies. For example, if $\alpha = 0.025$, a 95 percent confidence interval is obtained by using $d_{0.025} = 1.96$. As an illustration, the 20 sets of observations summarized in Table 1 have $m = 19$ and $R_{19} = 16$. By substituting these quantities into expressions (7) and (8), the point and interval estimates of p are calculated to be 0.84 and (0.62, 0.94).

More elaborate statistics are possible if one wishes to distinguish concordances in which both variables increase from those in which both variables decrease, or if one wishes to incorporate the idea of thresholds, in which no change is deemed to have occurred unless the amount of change exceeds a specified threshold value.

This formulation uses stratigraphical information only in the identification of neighbouring pairs of measurements in order to allow the evaluation of Z_i's. It is, however, possible that the relationship between the pair of variables alters over time. This can be investigated by a test of the null hypothesis that the probability of concordance, p, remains unchanged over time, against the alternative hypothesis that there is a change in p. If the null hypothesis is rejected, it would be necessary to identify the position at which the change is believed to have occurred.

The following test procedure, which is illustrated using the 20 pairs of measurements presented in Table 1, was proposed by Pettitt (1979). The values of $\{Z_i \, (i = 1, \ldots 19)\}$ are shown in the fifth column of the table: it can be seen that out of the 19 changes, $n_1 = 16$ were concordant and $n_0 = 3$ were discordant. A sequence of statistics $\{\mu_j \, (j = 0, 1, \ldots, 19)\}$ is defined by

$$\mu_0 = 0; \quad \text{and} \quad \mu_j = \begin{cases} \mu_{j-1} + n_0 & \text{if} \quad Z_j = 1 \\ \mu_{j-1} - n_1 & \text{if} \quad Z_j = 0 \end{cases} \quad (j = 1, \ldots, 19) \qquad (9)$$

Clearly, $\mu_{19} = 0$, since by this stage the amount n_0 will have been added a total of n_1 times, and the amount n_1 subtracted a total of n_0 times. If most of the concordances occur early in the sequence, the μ_j's will increase to a maximum before decreasing; if most of the discordances occur early in the sequence, the μ_j's will decrease to a minimum before increasing. The largest absolute value of the μ_j's (without regard to sign) is the statistic employed to test the null hypothesis, and to indicate the most likely place at which the sequence could be divided. A large value of

$$\mu^* \equiv \max_j (|\mu_j|) \qquad (10)$$

will cause the null hypothesis that p has not changed to be rejected. Pettitt (1979) shows that $\mu^*/(n_0 n_1)$ is precisely $D_{n_0 n_1}$, the Kolmogorov–Smirnov two-sample statistic (Campbell, 1974, Chapter 3), for which the distribution under the null hypothesis has been tabulated extensively. Tables of critical values for $n_0, n_1 \leq 25$ are provided by Pearson and Hartley (1972, Table 55), and approximations are available for higher values (see: Pettitt, 1979). For $n_0 = 3, n_1 = 16$, the 5 percent critical value of μ^* is 39. The observed maximum value of 21 is likely to occur under the null hypothesis of no change in the binomial parameter, which accordingly cannot be rejected at this level of significance. It may be concluded therefore that, for these data, there is insufficient evidence to indicate any change over depth in this measure of the agreement between the two variables. It is worth remarking, however, that the Kolmogorov–Smirnov test is not powerful in detecting small departures from the null hypothesis.

If the conclusion reached from the test described previously had been that the stratigraphical column could be divided into sections, within each of which the binomial parameter could be regarded as constant, expressions (7) and (8) could be used to obtain point and interval estimates of the probability of concordance within each section. However, such estimates should be treated with caution, as the same data would have been used to identify boundaries of the sections and also to evaluate the estimates.

Piecewise Regression Models

In the previous two sections, methods were presented for comparing two stratigraphically-recorded variables. It is stressed that a large correlation or concordance should not automatically be interpreted in causative terms, for example, as implying that X causes Y. Such inferences cannot be made from the measurements themselves, but require external information about the data. The method which will be presented in this section assumes a more formal model relating the variables:

X is assumed to be recorded exactly, without measurement error. For a given value x of

$X, Y(x)$ is assumed to be a variable whose expected value, $E(Y(x))$, can be expressed as some function f of x:

$$E(Y(x)) = f(x). \tag{11}$$

A simple example, used for illustrative purposes, is the straight line

$$f(x) = a + bx. \tag{12}$$

In practical applications, X will rarely be measurable exactly. If errors in recording X are small compared to the variability of Y at a given value of X, however, the results will usually not be affected greatly by this departure from the model assumptions.

For a given set of data $\{x_i, y_i(i = 1, \dots n)\}$, estimates $\hat{a}$ and $\hat{b}$ of the parameters a and b, defining the intercept and slope of the line in equation (12), can be obtained by minimizing

$$S(a, b) \equiv \sum_{i=1}^{n} (y_i - a - bx_i)^2. \tag{13}$$

It can be demonstrated that the least-squares estimates are

$$\hat{b} \equiv S_{xy}/S_{xx} \quad \text{and} \quad \hat{a} \equiv \bar{y} - \hat{b}\bar{x}, \tag{14}$$

where S_{xy} and S_{xx} are defined in expression (1).

The predicted value of Y for a given value x is determined by

$$\hat{Y}(x) \equiv \hat{a} + \hat{b}x, \tag{15}$$

the least-squares regression line of Y on X. Equation (15) provides a best estimate of $Y(x)$. In order to obtain an indication of the reliability of this estimate, however, it is necessary to make assumptions about errors associated with $Y(x)$. It is usual to amplify the information provided in equations (11) and (12) by postulating the model

$Y(x_i) = a + bx_i + e_i$, where $\{e_i \ (i = 1, \dots, n)\}$ are independent Normal random variables with mean 0 and common variance σ^2. (16)

Under this assumption, confidence intervals for $Y(x)$ can be obtained (see Campbell (1974, Chapter 8) for a careful and detailed discussion).

The least-squares regression line of Y on X for the set of 81 pairs of readings is labelled $L(1,81)$ in Fig. 1. However, this procedure has completely ignored the information provided by depths of each pair of readings. This information can be incorporated into the analysis by modelling the data using a piecewise linear regression model. The data set is divided into subsets on the basis of the value taken by the depth variable and each subset comprises a collection of stratigraphically-neighbouring readings. Within each subset, a different linear regression of Y on X is obtained, for example, for depths d_1 to d_5, the regression line is

$$\hat{Y} \equiv \hat{a}_1 + \hat{b}_1 x$$

and for depths d_6 to d_{20}, the regression line is

$$\hat{Y} \equiv \hat{a}_2 + \hat{b}_2 x.$$

Regression lines based on the two subsets of data are also shown in Fig. 1 and it can be noted that predictions based on these lines could differ considerably from predictions based on the line derived from the complete data set.

Piecewise regression models have been the subject of investigation for some years

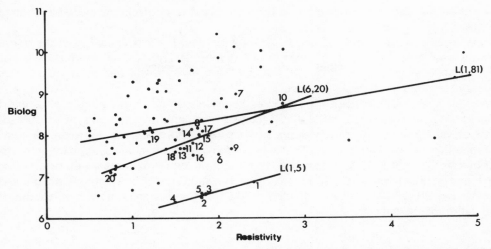

Fig. 1. A plot of biolog against resistivity for 81 levels in stratigraphical column. First 20 points are numbered in stratigraphical order. Also shown are least-squares regression lines of biolog on resistivity for parts of data: $L(i, j)$ denotes regression line based on points i to j inclusive

(Quandt, 1958). In some applications, variable X was also used to define the composition of subsets within which each regression has to be performed. For example, in econometrics, such models have been employed to describe the behaviour of a variable Y over time (X).

The composition of the subsets can be defined by markers which specify their boundaries; g subsets can be specified by $(g - 1)$ markers, denoted by $m_1, m_2, m_3, \ldots, m_{g-1}$, indicating that subsets comprise observations labelled

$$1 \text{ to } m_1; \quad m_1 + 1 \text{ to } m_2; \quad m_2 + 1 \text{ to } m_3; \ldots; \quad m_{g-1} + 1 \text{ to } n$$

respectively. It is convenient also to define $m_0 = 0$ and $m_g = n$.

If the error structure of Y is specified by expression (16), the data can be divided into subsets in the following manner; for a specified value of g, find the division which minimizes

$$S_g(a_1, \ldots, a_g; b_1, \ldots, b_g) \equiv \sum_{j=1}^{g} \sum_{i=m_{j-1}+1}^{m_j} (y_i - a_j - b_j x_i)^2. \tag{17}$$

A dynamic programming algorithm for achieving the division is described by Hawkins (1976). $\hat{a}_j$ and $\hat{b}_j$ are least-squares estimates of intercept and slope parameters in the jth subset, and are obtained from expressions similar to (14). More sophisticated models for the form of regression function $f_j(x)$ in the jth subset are possible within this framework if the data appear to require them.

The procedure finds an optimal (in terms of expression (17)) piecewise regression over g subsets for any specified value of g, but provides no advice on an appropriate value for g. The data can be fitted more closely, as measured by expression (17), by choosing larger values of g. However, adequate models are required which are economical in their use of parameters. Exact statistical tests for g are not available because the data have been employed in the choice of subsets, and so standard hypothesis tests cannot be used validly on the same subsets. Nevertheless, approximate criteria for investigating values of g are presented by Quandt (1958) and McGee and Carleton (1970); an examination of a plot of the data can assist in the decision together with relevant external information available from the stratigraphy.

Regression lines for points 1–5 and 6–20 shown in Fig. 1 are evaluated from the optimal division of the first 20 sets of readings into two subsets. The optimal three subsets comprised points (1–5), (6, 7), and (8–20) but the latter division was not regarded as corresponding to a genuine change of model.

Once the piecewise regression model has been constructed, an estimate of Y can be obtained for given values of x and d, the value of d being employed only in the specification of the relevant subset. The data within a single subset also enable an approximate idea of the reliability of this estimate for Y to be obtained, in an analogous manner to that described by Campbell (1974, Chapter 8) (although the entire data set would be employed to estimate σ^2 in expression (16)). However, such results should be treated with a measure of reserve, as they will have been affected by the fact that the same data were used in the specification of the subsets.

Caveat Lector

The three methods of analysis outlined in this paper indicate possible ways of measuring and modelling the relationship between two stratigraphically-recorded variables. The presentation has stressed assumptions made about the data in each method. Before employing a particular method of analysis, however, an investigator should be satisfied that the relevant assumptions are valid for the data.

Acknowledgment. I am grateful to Professor R. A. Reyment for discussions and correspondence on this work, and for providing data with which to illustrate the methods.

References

Campbell, R. C., 1974, *Statistics for Biologists*, Second edition. Cambridge University Press, London, 385 pp.

Fisher, R. A., 1941, *Statistical Methods for Research Workers,* Eighth edition. Oliver and Boyd, Edinburgh, 344 pp.

Hawkins, D. M., 1976, Point estimation of the parameters of piecewise regression models, *Appl. Statistics,* **25**(1), 51–57.

Johnson, N. S., 1979, Nonnull properties of Kendall's partial rank correlation coefficient, *Biometrika,* **66**(2), 333–337.

Kendall, M. G., 1970, *Rank Correlation Methods,* Fourth edition. Griffin, London, 202 pp.

Kendall, M. G., and Stuart, A., 1973, *The Advanced Theory of Statistics,* vol. 2, Third edition. Griffin, London, 723 pp.

McGee, V. E., and Carleton, W. T., 1970, Piecewise regression, *Jour. Amer. Stat. Assoc.,* **65**(331), 1109–1124.

Pearson, E. S., and Hartley, H. O., 1976, *Biometrika Tables for Statisticians*, vol II. Biometrika Trust, London, 385 pp.

Pettitt, A. N., 1979, A non-parametric approach to the change-point problem, *Appl. Statistics*, **28**(2), 126–135.

Quandt, R. E., 1958, The estimation of the parameters of a linear regression system obeying two separate régimes, *Jour. Amer. Stat. Assoc.,* **53**(284), 873–880.

Quantitative Stratigraphic Correlation
Edited by J. M. Cubitt and R. A. Reyment
© 1982 John Wiley & Sons Ltd.

Correspondence Analysis Used to Define the Paleoecological

Control of Depositional Environments of French Coal Basins

(Westphalian and Stephanian)

ISABELLE COJAN AND HAZEL BREMNER-TEIL

Ecole des Mines de Paris, Fontainebleau, France

Abstract

The objective of the study was to assess the efficiency of correspondence analysis in the identification of floral and fauna associations linked to lithological sequences of some French coal basins (Nord-Pas de Calais, Cevennes, Decazeville, Jura), and as a result, provide proposals concerning their depositional environments. Paleontological data from borehole cores (divided into units 25 cm thick) were coded, computer verified and analyzed by correspondence analysis. Several runs were necessary for each coal basin, placing eccentric variables in the factorial space as supplementary elements. The samples were grouped into different thicknesses to investigate variations in floral associations over the vertical dimension.

Floral associations were identified as a specific level based on the 25 cm-thick unit. Perenniality, however, was shown to exist covering thicknesses from a meter to a decameter, indicating that floral associations are organized into sequences whose amplitudes are comparable to those of sediments. Both flora and sediments are affected by vertical and lateral evolutions which may be interpreted as obliquities or specific localizations of certain species. The relationships identified allow the definition of depositional areas linked to coal-bearing material within a coal basin.

Introduction

The objective of this paper is to present the results of a pilot study on the application of correspondence analysis to quantitative stratigraphy. To our knowledge, this statistical technique (Benzecri, 1980) has not been employed in this field before although it has been applied to other types of geological data (David, Campiglio and Darling, 1974; Teil and Cheminee, 1975; Teil, 1976). Moreover, it is widely used in France for the analysis of large data sets resulting from questionnaires, polling votes, archeological digs and psychological experiments. We wished to assess the usefulness and efficiency of the technique in interpreting paleontological data for correlation purposes.

The data were obtained from borehole cores of French coal basins (Nord-Pas de Calais, Cevennes, Decazeville and Jura, Fig. 1) and studied using correspondence analysis. The time available only allowed study of paleontological data; subsequently, if the results proved to be satisfactory, lithological data could be introduced into the files for further

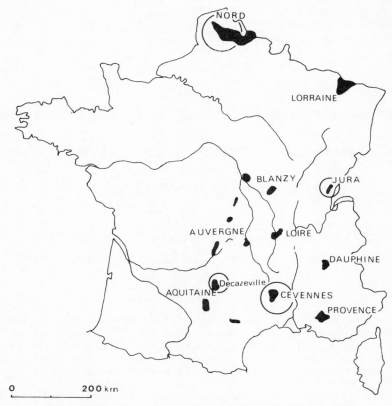

Fig. 1. Main coal basins in France (⊙ coal basins studied)

correlations. At this stage, a parallel between lithology and paleontology was applied in a qualitative manner.

The Nord-Pas de Calais Coal Basin of Westphalian (late Carboniferous) age has been exploited since the nineteenth century. 12 borehole cores were made available, and each log contained information on the lithological and paleontological contents related to the depth of the borehole. The total thickness of the Westphalian beds under study extends from the surface to 700 m. To obtain guidelines for our study, two of the boreholes were studied in detail. From their cores, approximately 800 samples containing flora identified at specific levels were recorded. Visual analysis of the data indicated that the analysis may be undertaken at the species level. This point is important because paleontological studies often use information at the generic or family level, thus masking, at times, environmental behaviors of different species within the same genus. However, the floral associations identified for the two boreholes could not be considered to be representative of the entire basin. Study of further boreholes was therefore necessary and the resulting large data set implied that an automatic statistical method was needed.

Correspondence analysis was chosen because it helps to identify associations between flora and fauna, and because it has already been shown to be an efficient and fast method (Teil, 1976). Moreover, interpretation of results is made more straightforward by the use of absolute and relative contributions computed for each variable. One of the advantages of correspondence analysis is that samples and variables can be represented simultaneously on the same set of factor axes. In this study, the samples were not plotted on graphs because of their large numbers. Subsequent studies could group specific samples together, and by employing their positions on factor axes (factor loadings), an ellipse can be plotted which represents the distribution of these samples in the factor space of variables (Jambu, 1979). This technique could be useful, for example, to group samples belonging to individual boreholes, or to group samples extending over a certain stratigraphic interval.

Systematic analysis of paleontological data from the Nord-Pas de Calais Basin had not been previously performed with the aim of identifying associations between flora and fauna. However, the associations found in this Basin, as a result of our study, were compared to those identified in three other coal basins (Cevennes, Decazeville and Jura) of Stephanian age. Their geology will be briefly outlined, followed by a description of the data employed in statistical analysis and a presentation of the method and results.

Description of the Coal Basins

Varied types of coal basins were chosen to determine if floral associations could be defined for all types of basins and to investigate interferences between sedimentation and distribution of the associations. Four Carboniferous coal basins were studied: Nord-Pas de Calais (limnic of Westphalian age), and Cevennes, Decazeville and Jura (paralic of Stephanian age).

The borehole data needed to provide a coherent description of the paleontology, and to come from the same stratigraphic zone to make comparisons possible.

In the Nord-Pas de Calais Basin, the sediments are rich in flora (one sample every 2 m), and, therefore, the vertical distribution of the flora distribution can be studied. This basin is too wide to enable the 12 boreholes to be related to distinct environments, because it corresponds to the central part of a basin stretching from U.K. to Belgium. On the contrary, the three Stephanian basins, each covering an area of about 400 km^2, show significant, identifiable sedimentary variation. Boreholes were drilled between 1950 to 1970 in the Cevennes and Decazeville Basins by mining companies, the French Geological Survey (B.R.G.M.) and Les Charbonnages de France. In the Jura Basin, only exploratory boreholes were drilled.

Paleontological data were recorded from the boreholes to define stratigraphical units but no studies on the distribution and interpretation of flora and fauna were performed. Lithologic correlations have been derived from the direct analysis of drill-logs or from observations on the sediments during exploitation.

Data

Preliminary studies on two borehole logs from the Nord-Pas de Calais Basin indicated that the paleontological data had to be treated at a specific level. Fauna is less abundant and varied (15 species) than the flora which includes about 200 different species. Only the occurrence of a species was noted because the borehole diameters did not allow definition of a quantitative distribution of the species (for example, relative size of the leaf and stage of perservation), The analysis did not require any *a priori* paleontological knowledge: statistical techniques are employed to identify associations.

Each borehole was divided into units 25 cm thick, and for each unit, data were recorded directly on paleontology, tectonics and stratigraphy. The choice of thickness was based on several criteria; it had to satisfy the definition of facies such that thin lithological beds (for example, coal seams and tonstein horizons) could be recorded, enable the distribution of flora and fauna to be noted in sufficient detail, and produce a file which would not be cumbersome. A thicker unit might have produced flora and fauna artificial groups resulting in unsatisfactory associations. Subsequent analyses performed on the flora did indicate, however, that the elementary unit could have been thicker with no loss of information (described later), but this result may not be valid on introducing lithological data into the file.

Coding and processing of the paleontological and tectonic data are described in Appendix 1. However, it should be mentioned here that names of flora and fauna were recorded as a four-letter mnemonic code in which the first two characters signify the genus, and the last two characters the species. A full list of codes for each genus and species are presented in Appendix 2, and an example of an input data sheet is seen in Table 1.

The number of occurrences of each species recorded in the initial file was computed; extremely-rare species were grouped under generic name to avoid losing their information completely (a cut-off frequency value of 1 per cent was chosen). This synthesized file built from the initial file was employed in the statistical study, with the data arranged into a suitable processing form. Presence of a flora or fauna was denoted by '1' in the column corresponding to the rank of its code, or '0' if absent. A logical description table k_{IJ} containing these Boolean values was prepared: $k(i, j) = 1$ signifies that the flora or fauna j is present in the sample i; $k(i, j) = 0$ signifies the contrary. Such a table is termed 'simple' (Jambu, 1979) and normally implies that presence of a characteristic carries more information than absence (was a flora or fauna absent, or was it not recorded in the log?). Furthermore, the notation of rare species and abundant species is equivalent ('1'). The probability, however, of recording a rare species is less than that of a frequent species, and Buzas (1970) has shown that presence-absence data give results similar to those from contingency data. Processing contingency data could assign too much importance to abundant data, masking possible associations with less abundant species, and could produce distortions of the factors of inertia in correspondence analysis.

Application of Correspondence Analysis

A mathematical description of correspondence analysis is presented in Benzecri (1980), and several applications of the method are described in Bastin and others (1980). The program employed in this study was written by Bordet (1972).

After building the data table k_{IJ} and performing an initial analysis with correspondence analysis, the flora or fauna that seem to distort the factor axes (being too eccentric from the center of gravity of the factorial space) were placed as supplementary elements i.e. they are no longer employed for calculating the factor axes, but they are projected on to the new factor axes. Several runs were necessary to determine the axes which showed a relatively homogeneous and interpretable distribution of data points.

6 factors were computed during each run because their cumulative percentage of inertia totalled 25 percent and differences between each factor's percentage of inertia was small (0.5 percent). The first factor had on average ~4–5 percent of inertia. These values may seem small, but considering the data set (70 species for the Cevennes Basin with an average of 5 species present per sample) and the fact that the table contains Boolean values, the factors are significant and interpretable.

The distribution of species on each factor was considered, with the help of the absolute and relative contributions computed for each species (Appendix 1B explains these terms). Floral associations which remained stable over the axes were considered to be significant. Plots were made of different combinations of axes, but only plots of axes 1 and 2, and 3 and 4 are presented here. These plots provide the majority of the visual information required to

Table 1. Partial listing of borehole. (C: basin code; AZ: borehole identification; 51270: height of top of borehole in cm; VIDE0019, ZOFA0018, PEPK ... : description of samples; —, F, T: definition of sample data; S5: stratigraphy; 05: rank of section of 125 m in borehole (from top); BE231: rank of the sample in borehole

C	AZ51270DMGA	FS5	BC469
C	AZ51270VIDE0003	S5	BC470
C	AZ51270CACISMOB	FS5	BC473
C	AZ51270VIDE0027	S5	BC474
C	AZ51270VIDE0016	S5	05BD001
C	AZ51270PEARANSH	FS5	BD017
C	AZ51270VIDE0005	S5	BD018
C	AZ51270COLIPEAR	FS5	BD023
C	AZ51270VIDE0001	S5	BD024
C	AZ51270COLIPEAR	FS5	BD025
C	AZ51270VIDE0031	S5	BD026
C	AZ51270COSPKSSP	FS5	BD057
C	AZ51270VIDE0083	S5	BD058
C	AZ51270ANST	FS5	BD141
C	AZ51270VIDE0007	S5	BD142
C	AZ51270CASPSISP	FS5	BD149
C	AZ51270VIDE0003	S5	BD150
C	AZ51270COLI	FS5	BD153
C	AZ51270VIDE0019	S5	BD154
C	AZ51270SISPPEARCOSP	FS5	BD173
C	AZ51270VIDE0019	S5	BD174
C	AZ51270CASP	FS5	BD193
C	AZ51270VIDE0003	S5	BD194
C	AZ51270KSSP	FS5	BD197
C	AZ51270VIDE0003	S5	BD198
C	AZ51270ODOBCYSP	FS5	BD201
C	AZ51270VIDE0299	S5	BD202
C	AZ51270VIDE0011	S5	06BE001
C	AZ51270ZOFA0046	TS5	BE012
C	AZ51270VIDE0036	S5	BE058
C	AZ51270ZOFA0025	TS5	BE094
C	AZ51270VIDE0055	S5	BE119
C	AZ51270PEHECOSP	FS5	BE174
C	AZ51270VIDE0009	S5	BE175
C	AZ51270CACI	FS5	BE184
C	AZ51270VIDE0003	S5	BE185
C	AZ51270CACIPELEPEHECOSPPEDEANSTANSH	FS5	BE188
C	AZ51270VIDE0003	S5	BE189
C	AZ51270PEHE	FS5	BE192
C	AZ51270VIDE0003	S5	BE193
C	AZ51270CACIPEPK	FS5	BE196
C	AZ51270VIDE0003	S5	BE197
C	AZ51270SMSP	FS5	BE200
C	AZ51270VIDE0019	S5	BE201
C	AZ51270PEPKSISP	FS5	BE220
C	AZ51270VIDE0010	S5	BE221
C	AZ51270VIDE0001	S0	BE231
C	AZ51270VIDE0001	S0	BE231

Table 2. Correspondence analysis—example of listing

Rank of axis 1		Eigenvalue 0.4389081		Percentage of inertia 3.168 : 3.168		
Name of variable	Number of occurrences of variable	Distance of variable from centre of gravity of factorial space	Factor loading	Contribution		
				Absolute	Relative	Cumulative
ALSP	114.000	15.0791	0.5100	1.032	1.725	1.725
ALCR	36.000	49.5343	−0.0010	−0.000	−0.000	0.000
ALDE	64.000	28.2164	0.8578	1.638	2.608	2.608
ALSE	148.000	10.4445	−0.3042	−0.476	−0.886	0.886
ANSH	84.000	20.5492	0.0279	0.002	0.004	0.004
ANST	212.000	8.6089	0.2683	0.531	0.836	0.836
ASEQ	44.000	45.1748	−0.8870	−1.204	−1.741	1.741
ASLY	110.000	17.8782	0.9358	3.351	4.898	4.898
BOSP	71.000	33.9885	−0.9053	−2.024	−2.411	2.411
CASP	523.000	3.5782	−0.4797	−4.187	−6.432	6.432
CAUN	84.000	23.5853	0.4734	0.655	0.950	0.950
COSP	418.000	5.0811	−0.7828	−8.909	−12.059	12.059
CRSP	36.000	48.7523	0.7296	0.667	1.092	1.092
CSWA	36.000	66.1121	0.8750	0.959	1.158	1.158
CYSP	92.000	17.6416	0.4844	0.751	1.330	1.330
DIFU	32.000	73.5546	1.4562	2.360	2.883	2.883
DINE	40.000	40.6606	−0.4559	−0.289	−0.511	0.511
DIST	65.000	27.1526	−0.3301	−0.246	−0.401	0.401
LIMU	61.000	36.2006	1.0670	2.416	3.145	3.145
LISU	318.000	6.8074	−0.7921	−6.940	−9.217	9.217
LPLA	118.000	14.0032	0.1706	0.120	0.208	0.208
MASP	159.000	11.0695	0.7428	3.051	4.984	4.984
MALA	98.000	16.1915	0.4002	0.546	0.989	0.989
MASA	127.000	12.0779	0.3267	0.471	0.884	0.884
NESP	147.000	15.9834	0.8453	3.654	4.471	4.471
NEFL	390.000	4.3037	0.3932	2.097	3.592	3.592
NEOB	168.000	10.4163	0.4798	1.345	2.210	2.210
NEPG	66.000	37.0030	2.1869	10.980	12.925	12.925
NERA	87.000	18.0564	0.0698	0.015	0.027	0.027
NETE	181.000	10.0653	0.0555	0.019	0.031	0.031
PESP	79.000	23.8144	−0.6598	−1.196	−1.828	1.828
PEAB	37.000	48.1478	0.1635	0.034	0.056	0.056
PEMI	27.000	67.6561	−1.0123	−0.962	−1.515	1.515
PEPD	46.000	32.3623	−0.3407	−0.186	−0.359	0.359
PICO	82.000	19.6766	0.4285	0.524	0.933	0.933
RDSU	24.000	56.6974	0.5034	0.212	0.447	0.447
SGSP	383.000	5.2357	0.0583	0.045	0.065	0.065
SISP	65.000	31.2678	−0.4755	−0.511	−0.723	0.723
SMSP	113.000	18.0443	0.0434	0.007	0.010	0.010
SMCU	218.000	7.4531	0.0517	0.020	0.036	0.036
SMMA	134.000	10.3618	0.0381	0.007	0.014	0.014
SMMY	39.000	54.0191	0.8102	0.891	1.215	1.215
SMSA	222.000	6.8174	0.4259	1.401	2.660	2.660
SSSP	212.000	8.1661	0.0389	0.011	0.019	0.019
SSNE	147.000	11.6256	0.5497	1.545	2.600	2.600
SSNU	30.000	56.4254	0.8086	0.682	1.159	1.159
SSSR	216.000	8.5292	0.3255	0.796	1.242	1.242
STSP	284.000	8.5665	−1.7149	−29.052	−34.330	34.330
TSSP	63.000	30.1415	−0.6694	−0.982	−1.487	1.487

interpret results. Visual inspection shows which points are spatially similar and thus correlated (resulting from a similar depositional environment). However, it should not be forgotten that these points are in fact two-dimensional projections of multidimensional space, and that two close projections on one graph may represent two distant points on another.

The correspondence table obtained from the analysis (the table containing factor loadings for the flora) could be analyzed subsequently by automatic classification methods, such as ascending hierarchical classification, to give clusters whose significance and validity could be tested by available computer programs (Jambu, 1979).

Results

A detailed description of the results can be obtained in Cojan (1980). In this paper, only the principal basin-by-basin results are presented. A comparison of the results is made, followed by correlation with lithology.

(a) *Nord-Pas de Calais*

An initial run with correspondence analysis indicated that the fauna (see key in Fig. 2a) were distinct statistically from the flora. The plot of factor axes 1 and 2 visually shows that the first axis is described by the fauna, whereas flora are distributed along the second axis. As a result, the fauna were placed as supplementary elements. In the new plot, certain flora (Lycopodiales) are distinguished from the main floral group along the second axis (Fig. 2b). Comparison of the two figures (Figs. 2a and b) indicates that these species occupy an intermediate position between fauna and flora. Furthermore, the species' internal structure suggests that the plants have grown in a marshy environment. Therefore, the first axis may correspond to an increasing degree of humidity starting from the flora through the Lycopodiales to the fauna.

The Lycopodiales were placed as supplementary elements in the next analysis to enable the behavior of flora grown in a drier environment to be studied.

The cloud of data points is well distributed (Fig. 3), and four associations (A, B, C and D) were defined.

Previously data were analyzed as independant items without reference to position within the borehole. Samples now were grouped through different thicknesses to understand how floral associations (A, B, C and D) evolve according to the thickness of sediment. With a thickness of less than 2.5 m (10 consecutive samples grouped together), the influence of grouping has no effect on the associations; they remain in the same positions on the factor axes. With a thickness of greater than 12.5 m (50 samples grouped), grouping obliterates associations because species group around the centre of gravity of the factor space,. It must be remarked that the data table still contains Boolean values, occurrences of flora were not cumulated on grouping the samples.

Grouping the samples in thickness of 2.5 m and 12.5 m made little difference to the relative positions of associations A, B, and C in the plan of factor axes (Fig. 4). This can be interpreted as indicating a gradient from a calm environment (for example, a swamp, expressed by association A) to a perturbed one (for example, a channel mouth expressed by association C). In the same plan, association D is nearer to association B and C on grouping the samples. An explanation for this is that association D is composed of plants which grew further away from plants in associations B and C but with influence of external phenomena (for example, water level variations or transport) all these plants were fossilized close together.

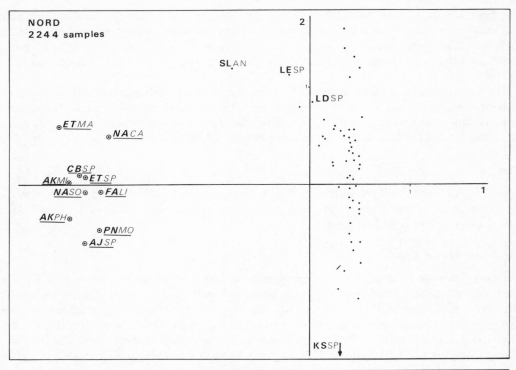

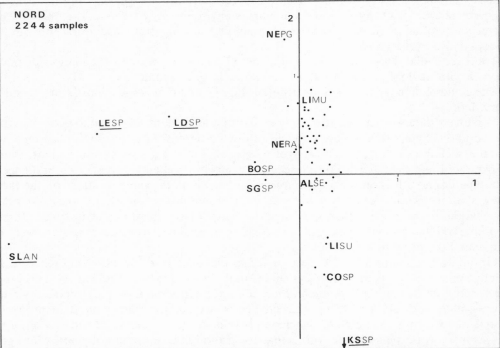

Fig. 2. Nord-Pas de Calais coal basin: Projection of variables on axes 1–2. (a) Fauna and flora, (b) Flora. (• flora, ⊙ fauna, *ETMA* variable underlined is supplementary element in following analysis)

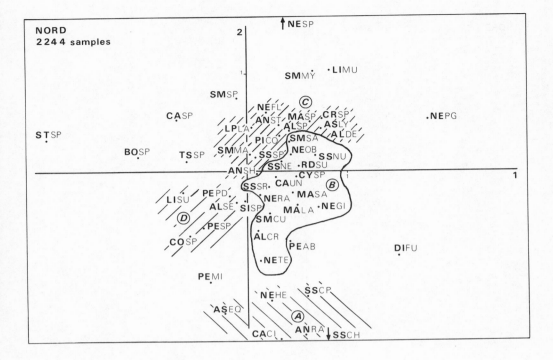

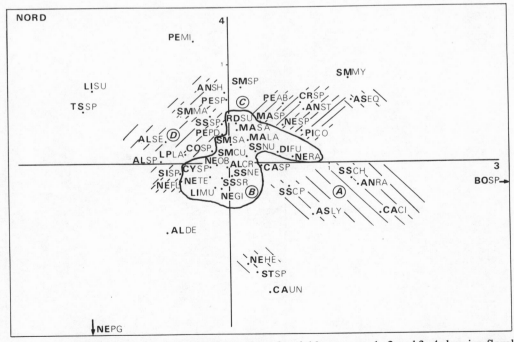

Fig. 3. Nord-Pas de Calais coal basin: Projection of variables on axes 1–2 and 3–4 showing floral associations: A, B, C, D

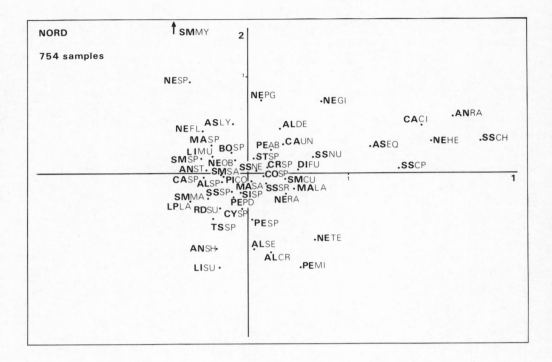

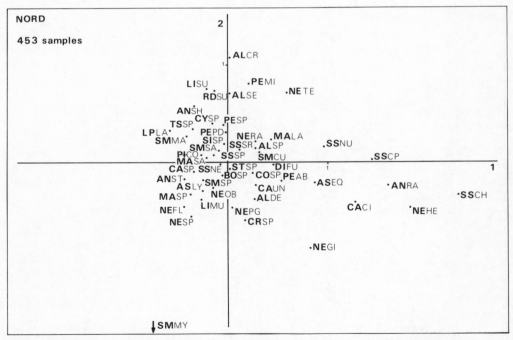

Fig. 4. Nord-Pas de Calais coal basin: Projection of variables on axes 1–2. Influence on associations of sample grouping. (a) Elementary sample: 6.25 m, (b) Elementary sample: 12.50 m

The floral associations, even though they have been defined at the thinnest level (25 cm), are shown to be perennial in time over the larger sampling intervals, indicating slow and calm evolutions of the environment. Naturally, it was important to know if this permanence occurred in other coal basins, and if the associations in the Nord-Pas de Calais Basin were related to a specific depositional environment. To answer these questions, comparison was made with three stephanian coal basins after statistical analysis.

(b) *Stephanian coal basins*

The number of samples studied from these basins is approximately half the number studied in the Nord-Pas de Calais Coal Basin for the same thickness of sediments. This variation is explained by the abundance in the Stephanian sequences of coarse sediments which were not favorable to the preservation of floral elements at the time of deposition. 1350 samples were obtained from 25 boreholes for the Cevennes, 1950 samples from 24 boreholes for Decazeville and 860 samples from 19 boreholes for the Jura.

Comparing floral occurrences in the three basins, it is recognized that their cumulated frequencies are higher in Decazeville, than either Cevennes or the Jura and the average number of species per sample decreases from 5, 4 to 3 respectively in the three coal basins. The high percentages for certain flora in Decazeville do not correspond necessarily to the most frequent species over the three basins.

Each coal basin was analyzed in the same manner as for the previous basin, and with the same objectives.

(1) *Cevennes basin.* The paleogeographic environment of the Cevennes basin can be compared to a flood plain. The boreholes studied came from a part of the basin only partially affected by later tectonics (Gras, 1970). Initial application of correspondence analysis indicated eccentric flora as supplementary elements in the subsequent analysis. This analysis distinguished two principal associations (denoted (1) and (2) in Fig. 5) which are influenced by sampling intervals greater than 25 m (100 samples). In the graphical representation (Fig. 6), the group of flora in association (1) is more compact than in Fig. 5, whilst some flora in association (2) remain separate.

These varied evolutions of associations can be explained probably by the fact that association (2) corresponds to a group of plants which were disturbed before their fossilization by water currents more than the plants in association (1).

(2) *Decazeville basin.* As in the Cevennes Basin, boreholes were obtained from part of the basin which can be described as a graben, partially active during sedimentation (Vetter, 1968). Analysis of the flora and display of the results reveals a multispecific association (3), with isolated terms (Fig. 7). The influence of sampling occurs above a 12.5 m interval of sediments (Fig. 8), whereby association (3) is constant and the peripheral terms come closer to the central group. However, the stability of associations over a large thickness of sediment exhibits a consistency of environment.

(3) *Jura Basin.* Due to the smaller number of samples, grouping based on different sampling intervals was not made. Correspondence analysis produced projections in the factor space of the flora and show a major multispecific association surrounded by small associations of two or three flora (Fig. 9).

(c) *Comparison of Stephanian basins*

In interpreting the relationships between floral associations in the three basins, it is suggested that differences observed were induced by an hydrodynamic component. This

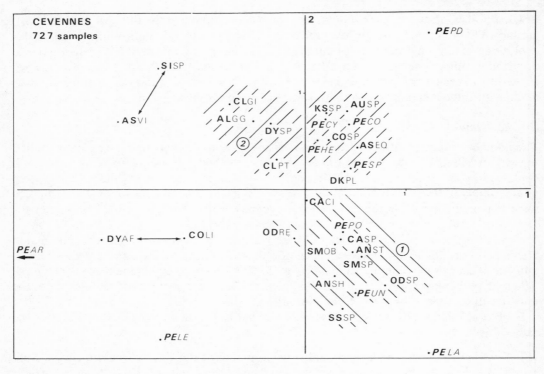

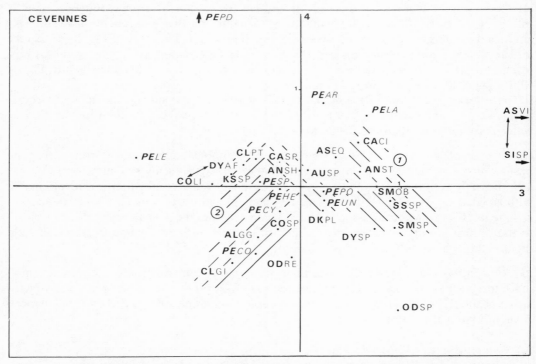

Fig. 5. Cevennes coal basin: Projection of variables on axes 1–2 and 3–4 showing floral associations: (1), (2)

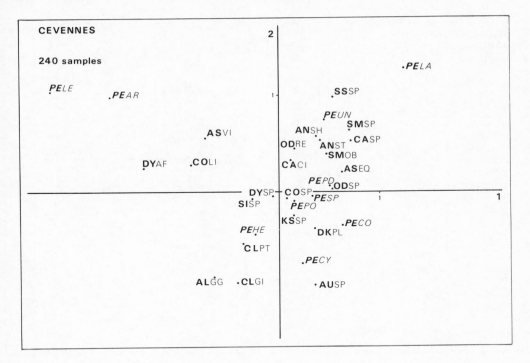

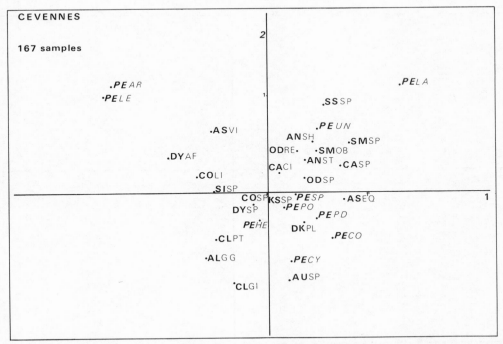

Fig. 6. Cevennes coal basin: Projection of variables on axes 1–2. Influence on associations of sample grouping. (a) Elementary sample: 25 m, (b) Elementary sample: 50 m

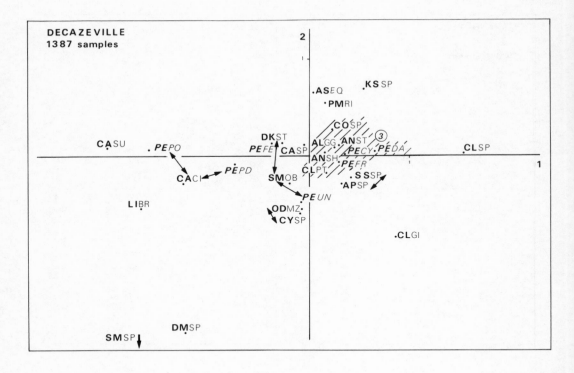

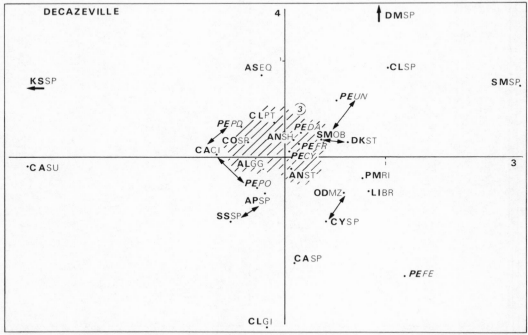

Fig. 7. Decazeville coal basin: Projection of variables on axes 1–2 and 3–4 showing floral associations

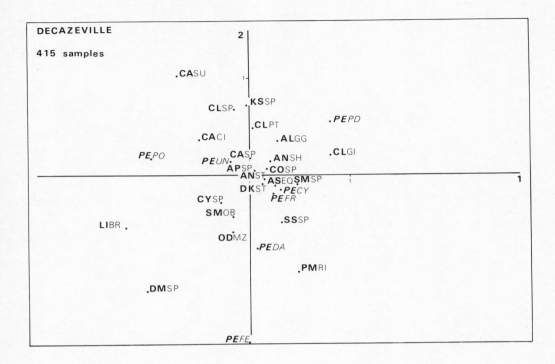

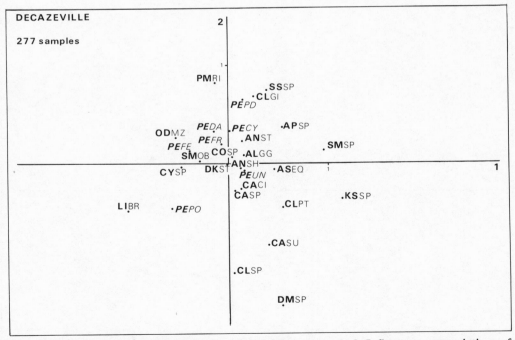

Fig. 8. Decazeville coal basin: Projection of variables on axes 1–2. Influence on associations of the sample grouping. (a) Elementary sample: 12.5 m, (b) Elementary sample: 25 m

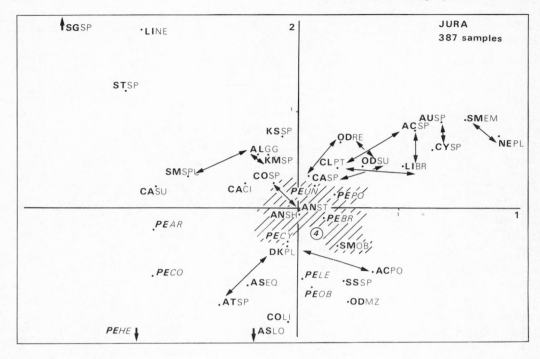

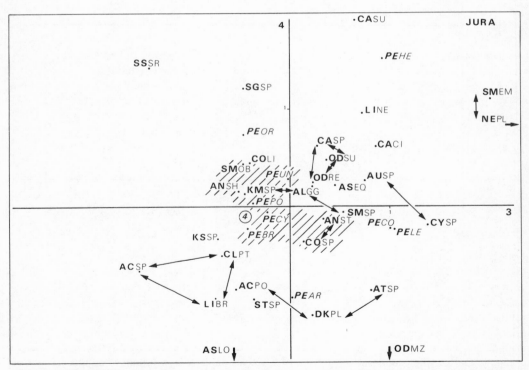

Fig. 9. Jura coal basin: Projection of variables on axes 1–2 and 3–4 showing floral associations

transport phenomenon helps to explain the concentration of certain variables in the factor space and the scattering of others. Most of the basic constituents of association (1) from the Cevennes occur in associations (3) and (4) from the Decazeville and Jura Basins respectively. As a result, two main associations can be proposed: association I = ANSH, ANST, SMOB, CASP, PEPO, PEUN; association II = ALGG, ASEQ, COSP, PECO, PECY. To interpret further such associations, a comparison with the paleomorphology is necessary.

The three sedimentation areas under study range from a proximal zone (Jura) (Lienhardt, 1962), through a source zone (Decazeville) towards a more distal zone in Cevennes. Such a sequence indicates a decreasing rate of perturbation towards the Cevennes, which is in agreement with conclusions drawn from the distribution of floral associations. At this scale, a lateral evolution is seen between the associations and perturbations they are submitted to. However, to understand more precisely the relationships between sediment and floral associations, correlations on the Nord-Pas de Calais Basin were studied qualitatively at a small scale.

(d) *Sediment-floral associations*

The boreholes from the Nord-Pas de Calais Basin were correlated in detail, based on a tonstein bed whose time of deposition is considered as synchronous (Fig. 10a). Lateral evolutions are rapid extending generally over only a few kilometres particularly in sandstone and coal seams. Comparing lithologic logs with the associations identified statistically, it is seen that well-developed coal seams are linked with associations B or C, and to two flora (SISP and ALSE) belonging to association D.

A map representing the thickness of the Tonstein Patrice coal seam (Fig. 10b) shows the relation between coal thicknesses in each borehole and the floral associations previously identified by correspondence analysis. The area of greatest coal thickness is linked with association C and flora ALSE (in association D). According to the decreasing thickness of coal seam, the associations can be placed in the following order: C (maximum thickness), B and D (least thickness or lack of seam). Thus the areas, in which phytogenic material accumulated and was preserved, can be identified by certain floral associations or flora.

A second comparison performed with the sediment data is based on the vertical dimension. Floral associations remain stable over sediment thicknesses of 10 m to 20 m (Fig. 11), for example association D is scattered along the boreholes without any precise relationship, whereas B and C are incorporated into sequences having the same thickness and polarity as the sedimentary sequence. The term 'sequence' can be extended therefore to define a more complete facies which includes sediment and floral associations.

Conclusions

This methodological study reveals that analysis of paleontological data needs to be performed at a species level and that species are organized into associations which can be employed as markers for coal seams. These points are important, because paleontologists frequently work at the generic or family level to define assemblages even though it is realised that species within the same genus can have different ecological settings. Statistical analysis by correspondence analysis provides a quick and clear representation of multispecies associations which would have been impossible by analyzing the data visually. Correlations with the sedimentary data were based on results obtained from exploitation of the basins. Further studies could be envisaged in which these correlations could be obtained quantitatively after lithological data are introduced into the computer file.

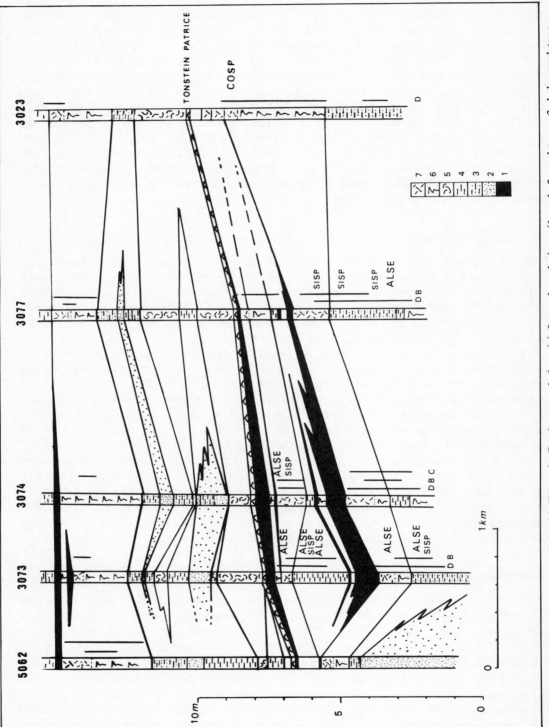

Fig. 10 Nord-Pas de Calais coal basin: Tonstein Patrice correlation. (a) Lateral evolutions (1-coal; 2-sandstone; 3-shaly sandstone; 4-shale; 5-shale with limnic fauna: 6-rooted shale: 7-coal)

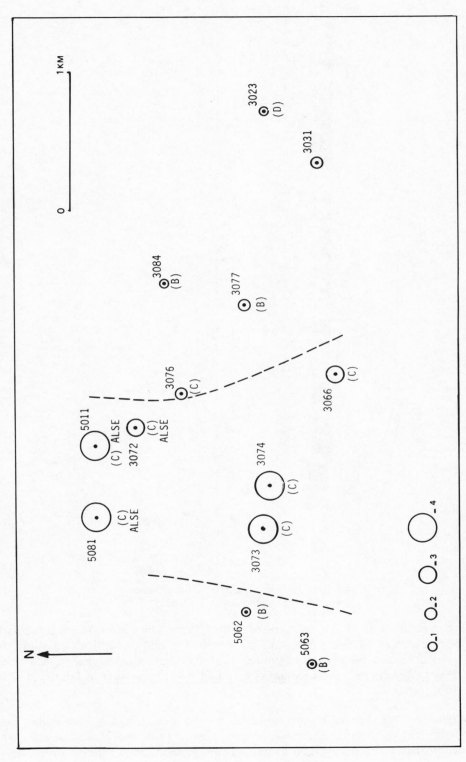

Fig. 10. Nord-Pas de Calais coal basin: Tonstein Patrice correlation. (b) Relative distributions of floral associations and thicknesses of coal seams ((c); Association *c*; Coal seams with thicknesses: under 0.5 m = 1, over 0.5 m = 2, 1.0 m = 3, 2.5 m = 4)

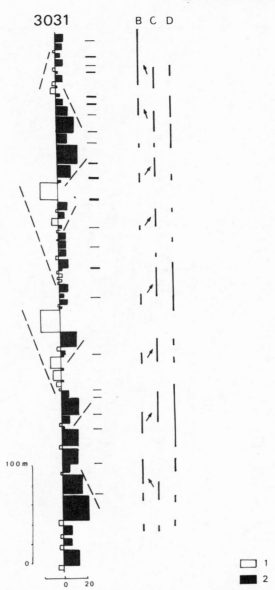

Fig. 11. Nord-Pas de Calais coal basin: floral
sequences (1—Sandstone; 2—shale)

This method could be applied to any type of sediments for which information is regularly distributed over a vertical dimension, and for which it would be difficult to study the associations due to the variety of paleontological data and thicknesses of sediments involved. Types of sediments such as carbonates and phosphates rich in fauna could be processed in this manner.

Acknowledgments. We are extremely grateful to the Charbonnages de France and B.R.G.M. who lent us their borehole log reports, also to Professor H. Pelissonnier who enabled this project to be undertaken as part of the second author's thesis. Professor B.

Beaudoin provided many helpful remarks and ideas during the project. Many thanks also go to Mme M. T. Lesage who spent long hours programming, testing and verifying the borehole data, putting them into a form suitable for statistical analysis.

Appendix 1

Data coding

Each borehole was divided into 25 cm-thick units for coding the following log information:

(1) For paleontologic data, a four-letter code (for example, COLI: CO signifies the genus, and LI the species; COSP signifies that the species within the genus CO are unspecified) was employed to record the presence of a species (Appendix 2). A maximum of 15 codes could be employed for each sample (but this number was never reached). One input card was used for each sample containing paleontological data.

(2) Tectonic data were coded (ZOFA = faulted zone, DISC = unconformity). The tectonic zone could range over several samples. Therefore, samples were grouped together on one card with an indication of the total sample number.

(3) Stratigraphic data were recorded on each card (for example, S = Stephanian, SØ = basement, and W = Westphalian).

(4) Samples having no tectonic or paleontological data were indicated by the term VIDE (empty). Consecutive 'empty' samples were grouped together with a label on the card indicating their number.

The samples were identified according to their position in a borehole. Details of the coding and processing can be obtained in Cojan (Appendix 1, 1980). An example of an input data sheet is provided (Table 1).

Definition of contributions

To help interpret factor k, two coefficients are calculated for sample i and variable j.

The 'absolute contribution', $ca_k(i)$ or $ca_k(j)$ expresses the part given by sample i or variable j to the dispersion along factor k.

As the variance of the loadings $f_k(i)$ or $g_k(j)$ along k is s_k:

$$ca_k(i) = \frac{pi \cdot f_k^2(i)}{s_k}; \qquad ca_k(j) = \frac{p \cdot jg_k^2(j)}{s_k}$$

The absolute contribution allows the samples or variables responsible for the appearance of factor k to be determined.

The 'relative contribution', $cr_k(i)$ or $cr_k(j)$, expresses the part given by factor k to the dispersion of sample i or variable j.

The dispersion of sample i is evaluated by the distance from point i to the center of gravity G of the cloud:

$$d^2(i, G) = \sum_{k=1}^{p} f_k^2(i)$$

thus:

$$cr_k(i) = \frac{f_k^2(i)}{d^2(i, G)}; \qquad cr_k(j) = \frac{g_k^2(j)}{d^2(j, G)}.$$

Appendix 2

Paleontological codes

AC–*Acitheca*
SP–*A.* sp
PO–*A. polymorpha*
AJ–*Anthraconata*
SP–*A.* sp
AK–*Anthraconauta*
MI–*A. minima*
PH–*A. philipsi*
AL–*Alethopteris*
SP–*A.* sp
CR–*A. corsini*
DE–*A. decurrens*
GG–*A. grandini*
SE–*A. serli*
AN–*Annularia*
RA–*A. radiata*
SH–*A. sphenophylloides*
ST–*A. stellata*
AP–*Aphlebia*
SP–*A.* sp
AS–*Asterophyllites*
EQ–*A. equisetiformis*
LO–*A. longifolius*
LY–*A. lycopodioides*
VI–*A. viticulosus*
AT–*Asteroteca*
SP–*A.* sp
AU–*Aulacopteris*
SP–*A.* sp
BO–*Bothodendron*
SP–*B.* sp
CA–*Calamites*
SP–*C.* sp
CI–*C. cisti*
SU–*C. suckowi*
UN–*C. undulatus*
CB–*Carbonita*
SP–*C.* sp
CL–*Callipteridium*
SP–*C.* sp
GI–*C. gigas*
PT–*C. pteridium*
CO–*Cordaites*
SP–*C.* sp
LI–*C. lingulatus*

FU–*D. furcatum*
DK–*Dicksonites*
PL–*D. pluckeneti*
ST–*D. sterzeli*
DM–*Dicranophyllum*
SP–*D.* sp
DY–*Dorycordaites*
SP–*D.* sp
AF–*D. affinis*
ET–*Estheria*
SP–*E.* sp
MA–*E. mathieui*
FA–*Faune*
LI–*F. limnique*
KM–*Calamostachys*
SP–*C.* sp
KS–*Cardiocarpus*
SP–*C.* sp
LD–*Lepidodendron*
SP–*L.* sp
LE–*Lepidophyte*
SP–*L.* sp
LI–*Linopteris*
BR–*L. brongniarti*
MU–*M. münsteri*
NE–*L. neuropteroides*
SU–*L. subbrongniarti*
LP–*Lepidophyllum*
LA–*L. lanceatum*
SP–*M.* sp
LA–*M. latifolia*
SA–*M. sauveri*
NA–*Naiadites*
CA–*N. carinata*
SO–*N. sowerbi*
NE–*Neuropteris*
SP–*N.* sp
FL–*N. flexuosa*
GI–*N. gigantea*
HE–*N. heterophylla*
OB–*N. obliqua*
PG–*N. pseudo gigantea*
PL–*N. planchardi*
RA–*N. rarinervis*
TE–*N. tenuifolia*

PE–*Pecopteris*
SP–*P.* sp
AB–*P. abbreviata*
AR–*P. arborescens*
ER–*P. bredovi*
CO–*P. cyatheoide*
CY–*P. cyathea*
DA–*P. daubreei*
FE–*P. feminaeformis*
FR–*P. fructifié*
HE–*P. hemitelioides*
LA–*P. lamurensis*
LE–*P. lepidorachis*
MI–*P.* cf. *miltoni*
OR–*P. oreopteridius*
PD–*P. plumosa-dentata*
PO–*P. polymorpha*
UN–*P. unita*
PI–*Pinnularia*
CO–*P. columnaris*
PM–*Pseudomariopteris*
RI–*P. ribeyroni*
PN–*Planolites*
MO–*P. montanus*
SG–*Sigillaires*
SP–*S.* sp
SI–*Sigillaria*
SP–*S.* sp
SL–*Sigillariophyllum*
AN–*S. anthemis*
SM–*Sphenophyllum*
SP–*S.* sp
CU–*S. cuneifolium*
EM–*S. emarginatum*
MA–*S. majus*
MY–*S. myriophyllum*
OB–*S. oblongifolium*
SA–*S. saxi fragae folium*
SS–*Sphenopteris*
SP–*S.* sp
CH–*S. chaerophylloides*
CP–*S. crepini*
NE–*S. neuropteroïdes*
NU–*S. nummularia*
SR–*S. striata*

CR–*Corynepteris*	OD–*Odontopteris*	ST–*Stigmaria*
SP–*C.* sp	SP–*O.* sp	SP–*S.* sp
CY–*Cyclopteris*	MZ–*O. minor-zeilleri*	TS–*Trigonocarpus*
SP–*C.* sp	RE–*O. reichi*	SP–*T.* sp
DI–*Diplotmema*	SU–*O. subcrenulata*	

References

Bastin, Ch., Benzecri, J. P., Bourgarit, Ch., and Cazes, P., 1980, Pratique de l'analyse des Données, Vol. II. *Abrégé Théorique. Etude de cas modèle.* Dunod, Paris, 464 pp.

Benzecri, J. P., 1980, Pratique de l'analyse des Données, Vol. I. *Analyse des correspondances: Exposé élémentaire.* Dunod, Paris, 424 pp.

Bordet, J. P., 1972, *Correspondence analysis program.* Ecole des Mines de Paris. Unpublished document.

Buzas, M., 1970, On the quantification of biofacies. *Proc. N. Am. Paleont. Conf. Chicago 1969 B*, pp. 101–116.

Cojan, I., 1980, *Approche paléoécologique de bassins carbonifères Français*, Thèse Docteur-Ingénieur, Ecole des mines de Paris, 178 pp.

David, M., Campiglio, C., and Darling, R., 1974, Progresses in R- and Q-mode analysis: correspondence analysis and its application to the study of geological processes, *Can. Jour. Earth Sci.,* **II** 131–146.

Gras, H., 1970, *Etude géologique détaillée du Bassin Houiller des Cevennes,* Thèse, Université de Clermont–Ferrand, 305 pp.

Jambu, M., 1979, Classification automatique pour l'analyse des données, Vol. 1, *Methodes et Algorithmes,* 320 pp.; Vol. 2, *Logiciels.* Dunod, Paris, 400 pp.

Lienhardt, G., 1962, Géologie du Bassin Houiller Stéphanien du Jura et de ses morts-terrains, *Mémoires du B.R.G.M.*, no. 9, 460 pp.

Teil, H., and Cheminee, J. L., 1975, Application of correspondence factor analysis to the study of major and trace elements in the Erta Ale Chain (Affar, Ethiopa), *Jour. Math. Geology,* **7**(1), 13–25.

Teil, H., 1976, Use of correspondence analysis in the metallogenic study of ultrabasic and basic complexes, *Jour. Math. Geology,* **8** (6), 669–681.

Vetter, P., 1968, Géologie et paléontologie des Bassins Houillers de Decazeville, de Figeac et du Détroit de Rodez, *Edition Houillères du Bassin d'Aquitaine—Albi.* Tome I, 403 pp.

III LITHOSTRATIGRAPHY

Quantitative Stratigraphic Correlation
Edited by J. M. Cubitt and R. A. Reyment
© 1982 John Wiley & Sons Ltd.

Quantitative Correlation of a Cyclic Limestone—Shale Formation

W. Schwarzacher

Department of Geology, The Queen's University of Belfast, Belfast BT7 1NN, Northern Ireland

Abstract

A strictly time periodic model with superimposed stochastic component is employed to describe the sedimentation of a Carboniferous limestone-shale series in the northwest of Ireland. The model can be used to examine the significance of detailed stratigraphic correlation in the area.

Introduction

Many quantitative methods of stratigraphic correlation are based on a procedure of matching. A chosen parameter referring to fossil content or lithology is measured in two sections which are separated geographically and matching is achieved by calculating corresponding positions of maximum agreement between the two sets. However, it is impossible to evaluate the correlations statistically unless basic assumptions are made about the processes of deposition. For instance, when conventional matching is performed, it is tacitly assumed that two separated sections represent records of the same sedimentary history. Clearly, the reliability of a stratigraphic correlation must depend upon this hypothesis which, in turn, can only be tested for situations where the stratigraphic correlation is already known. This impasse being unavoidable, it has been suggested (Schwarzacher, 1980) that it might be advantageous to examine possible stratigraphic correlations using theoretical models to predict the sequence to be expected at a given locality. Then, matching can be performed on this expected sequence. On the other hand, when stratigraphic correlation has already been established, a check on the sedimentation model is available by this method.

Basic assumptions have to be made in stratigraphic correlation and these will be discussed for the two-dimensional case. Consider a section through a sedimentary sequence, the horizontal coordinate, s, fixes the geographical location and the vertical coordinate, t, determines the stratigraphic position where t is assumed to be related to time. Each point in the section is thought to be linked to a variable, $X_{s,t}$, which determines the lithology or fossil content of the sediment. Matching procedures attempt to obtain the position of best agreement between a series of measurements taken at two localities for example at s and $s + h$, which refer to the horizons $t = 0$ to $t = n$. A suitable measure of matching is provided by the variogram (Schwarzacher, 1980):

$$\gamma_h = \frac{1}{2n} \sum_{t=0}^{n} (X_{s,t} - X_{s+h,t})^2.$$

275

A necessary but not sufficient condition of obtaining a stratigraphic correlation by matching is:

$$\gamma_h < \text{Var } X_{s,t},$$

which implies that similarity between adjacent sections must be greater than expected from random variation. Similarity between localities can be developed by a number of processes and a general model for the generation of $X_{s,t}$ can be written as:

$$X_{s,t} = L(s,t) + \varepsilon_{s,t} + G(t) + e_t + T(s)$$

where L, G, and T are physical processes responsible for lithology at points s, t.

The model assumes that L is a spatial temporal process which accounts for the local correlation structure of the variable, which may be caused by local current conditions, biological activity or similar factors. Random processes of this type can be represented by the general autoregressive process:

$$L(X_{s,t}) = \Sigma \, a_{ij} X_{s-i,t-j} + \varepsilon_{s,t},$$

where $\varepsilon_{s,t}$ is a random variable depending on space and time and the a_{ij}'s are weights which determine the effect of points which surround s, t.

$G(t)$ is associated with a random term e_t and, for the purpose of matching, is the most important variation. It should be noted that G and e are independent of the geographical position and G is, therefore, a basin-wide (global) process.

In the absence of basin-wide variation, stratigraphic correlation by matching becomes impossible and the first step must be to establish whether such a component exists. $G(t)$ can either be a random process (such as a Markov sequence) or a combination of deterministic function together with random process. A separation of a possible trend in time is not essential if $G(t)$ is independent of s.

The final component $T_{(s)}$ represents a regional trend which is assumed to be deterministic. Whether this is justified, depends upon how much is known about the sedimentary history which also determines whether the three components L, G, and T can be separated in practice. In order to be useful, a sedimentation model must be designed for a specific situation and an attempt will be made to construct a model for the Lower Carboniferous limestone-shale sequence in the north west of Ireland.

Sedimentological Background

The investigated area is located as shown in Fig. 1 and the 14 sites at which sections have been measured are illustrated in Figure 2. The profiles comprise the middle part of the Glencar Limestones which is of Asbian age and contains well-developed cyclicity (Schwarzacher, 1964). Limestone beds occur in bundles of 1–3 m in thickness, separated by layers of marly shale. In the study, the 8 most easily-recognizable cycles (cycles two to 9) were used and a complete record of the data is shown in Fig. 3. Measurements were stored in a mini computer and Fig. 3 is based on a photograph of the monitor screen displaying the data. Limestone is shown in black and shale in white for this diagram. The topmost limestones in the more massive sequences were chosen as cycle boundaries, indicated for each preceding cycle by a dotted line. Columns for each cycle from left to right, represent localities 12, 7, 13, 14, 4, 5, 10, 2, 11, 0, 15, 16, 17, and 1. This is approximately the sequence obtained by projecting localities on to a line which runs from southwest to northeast.

The diagram shows clearly that there is a thinning of individual cycles in this direction,

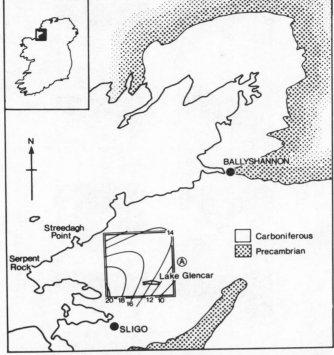

Fig. 1. Regional setting of middle Glencar Limestone
Isopachytes are given in metres

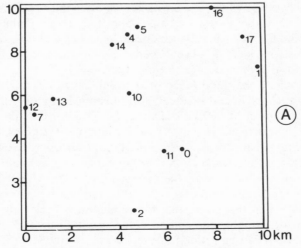

Fig. 2. Distribution of localities in studied area (see
Fig. 1)

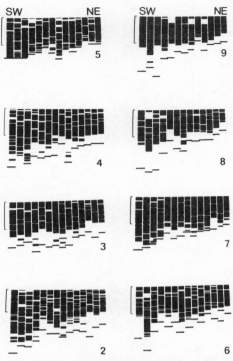

Fig. 3. Measured cycles of middle Glencar Limestone. Cycles are numbered in strati-graphically-ascending order and localities are arranged to provide a SW–NE profile (see text). Black = limestone, white = shale, scale on left is 100 cm

complimented by a similar decrease of thickness in the complete middle Glencar Limestone interval, the isopachytes of which are illustrated as a second order trend surface in Fig. 1. The data suggest that the limestone was deposited in an embayment of the larger Donegal–Sligo basin. Other facies changes occur together with thickness variation includ-ing, in particular, a strong increase in shale percentage towards the northeast and changes in fauna such as the disappearance of *in situ* coral colonies.

In order to investigate the importance of the regional trend, a first-order trend surface was calculated for each cycle and residuals were examined by two-dimensional variogram analysis using the program MAREC 2 (David, 1977). Figure 4 illustrates results for the direction southwest to northeast and with a field width of 90 degrees in each direction. The experiment did not reveal any local correlation of trend surface residuals and variation from locality to locality can be explained by the overall variance. It is likely that this result is due to insufficient data but, with the available material, the spatial component of correlation in $L_{s,t}$ is undetectable for distances in excess of 1000 and can be ignored.

It should be stressed, at this point, that the sedimentary cycles are mappable units and, therefore, the lithostratigraphic correlation is firmly established. The problem is whether stratigraphic correlation can be refined to the tracing of individual beds. Naturally, this is linked to the question of origin and definition of sedimentary bedding.

It has been claimed that bedding in the Carboniferous and other limestones is of diagene-

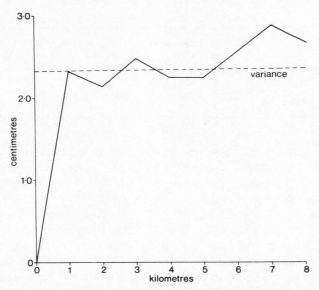

Fig. 4. Variogram of first-order trend-surface residues

tic origin and does not record primary events (Duff, Hallam, and Walton, 1967). Although it is likely that diagenesis accentuates sedimentary bedding, there is petrographic evidence which can be explained only by primary changes. This includes variation in fossil content of the marl and limestone layers which cannot be attributed to different preservation. The limestone-shale boundary shows features which are absent in the limestones, such as laminations, occasional ripple marks, and a distinct dependance of burrows on bedding. In addition, the bedding planes are strictly parallel to the cycle boundaries which can be traced over the whole area.

It has been claimed (M. Walther pers. communication, 1980) that the average thickness of limestone beds remains constant, even when overall thickness of a time-equivalent sequence changes. Similar observations were reported from Jurassic sediments by Hallam (1964) and in both situations, this was taken as evidence for the secondary origin of bedding planes. However, the 'average thickness' without a precise definition of beds, can be meaningless when nonequivalent phenomena are compared. This is illustrated diagramatically in Fig. 5 where three sections (A, B, and C) have the same average bed thickness but differ in

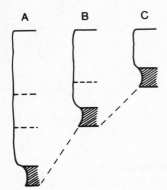

Fig. 5. Thickness change of cycle with formation of additional bedding planes

Table 1. Average thickness of limestone beds

Localities	Mean thickness of beds	No. of beds	Average thickness of cycles
12, 7, 13,	14.68 cm	190	205.7 cm
2, 11, 0,	13.40 cm	164	151.6 cm
16, 17, 1,	13.32 cm	124	125.8 cm

the number of beds which have been developed. Whether a bedding plane is observed, can be an accident of weathering and detailed petrographic analysis reveals that discontinuities exist, even when no bedding planes are visible in the field. The condensed sequences usually show a less-detailed pattern of bedding. This is supported by analyzing the extreme southwestern (localities: 7, 12, and 13 in Fig. 1), central (localities: 0,2, and 11) and extreme northeastern areas (localities: 16, 17, and 1). Table 1 shows that the average thickness of limestone beds does decrease parallel to the regional trend but not as much as expected from isopachytes of the middle Glencar Limestone. On the other hand, increase in the number of bedding planes is noticeable and clearly indicates that nonequivalent beds have been compared.

Further information about the beds' origin can be derived from the shape of the thickness distribution of the beds (Fig. 6). This has been discussed in detail in earlier papers (Schwarzacher, 1975, 1976), where it was suggested that a single bed originated by randomly-

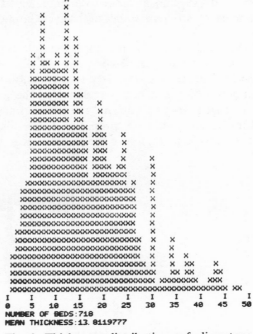

Fig. 6. Thickness distribution of limestone beds. Scale is in centimetres

distributed bursts of sedimentation occurring with constant frequency. The time interval representing a bed, contains on average three sedimentation steps, which are responsible for the three modes in the distribution curve. Consequently, times represented by the beds are of equal length on average. The model of step-like sedimentation is supported by evidence of fossil preservation in life position and also by the distribution of burrows.

The earlier analysis did not take into account how bed formation is related to the larger phenomenon of sedimentary cycles and this will be examined further.

If one accepts that the sedimentary cycles are a result of time-periodic phenomena, then all the cycles should be identical, irrespective of geographic or stratigraphic position. Any regional or time trend can be eliminated by reducing the cycles to a constant thickness, which for convenience is taken as unity. This transformation was performed for the measurements and is shown in Fig. 7. In these 'reduced cycles', the position of any bedding plane can be fixed by a number in the range of zero to one. The frequency of all the bedding plane positions is shown in Fig. 8a. In this diagram, a bedding plane is defined either as a discontinuity in the limestone or as a limestone–shale or shale–limestone boundary.

In interpreting shale–limestone sequences, the question always arises as to whether the deposition of either rock type is independent, or follows a common dictate. The classical problem of whether one rock type dilutes the second type during formation, or they vary independently, is difficult to answer. For comparison, relative positions have been calculated by setting shale thicknesses to zero as shown in Fig. 8b. In this situation, the limestone-shale boundaries coincide with shale-limestone boundaries. The two diagrams are essentially similar with four well-developed maxima at 0.28, 0.40, 0.56, and 0.78. Two

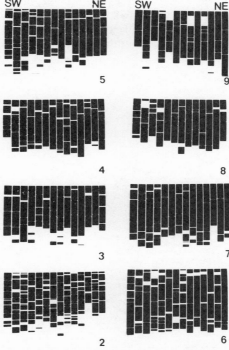

Fig. 7. Cycles shown in Fig. 3 reduced to
unit thickness

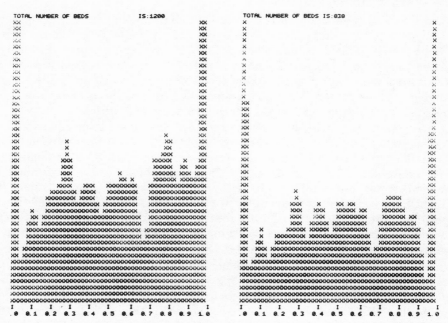

Fig. 8. Distribution of bedding plan positions in reduced cycles. (a) All bedding planes inclusive of shale, (b) Bedding planes with shale reduced to zero thickness

minor maxima occur at 0.10 and 0.88, the former being more prominent on the plot without shale while the latter is prominent on the plot including shale.

In order to obtain data on the persistency of relative positions, a 150 m-long continuous exposure at locality 2 was studied and the relative position of the first bed in each cycle determined at 1.5 m intervals. The variogram for cycle 2 is illustrated in Fig. 9. The mean

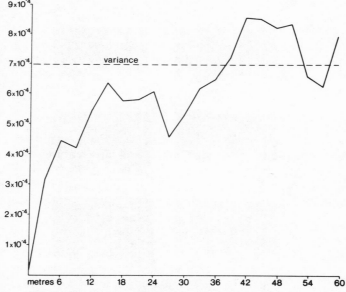

Fig. 9. Variogram of local relative thickness variation

position of the first bed in the example is 0.143 with a variance of 6.967×10^{-4}. The variogram has a range of approximately 40 m and the variance is approached rapidly. There is an indication of periodicity in the correlation structure which may be caused by a regular thinning and swelling of beds. Nevertheless, reduced profiles which are separated by more than 50 m are not affected by any regional trend.

Interpretation of the Sedimentary Record

Before attempting to construct a sedimentation model, a tentative interpretation of the reported observations will be given.

A time-series analysis of a longer Carboniferous sequence, suggested that the cycles under discussion were caused by regular phenomenon and the 21 000 year period of the equinoxes was put forward as a possible cause. An analysis of the cycle internal structure in terms of relative positions of the bedding planes, suggests that there are four preferred positions which divide each cycle into 5 beds. The result is similar to that obtained from an analysis of Triassic dolomite limestone cycles in the calcareous Alps (Schwarzacher 1954, 1975) where the '5 division of cycles' is so distinct that it was noticed during field work. Similar cycles which can be divided into 5, have been observed in other formations (Schwarzacher, 1975; Fischer, 1980) and the pattern is particularly obvious in the Cretaceous cycles figured by Gilbert (1895). It must be noted that in the Glencar Limestone, however, a clearly-visible division into 5 beds is the exception rather than the rule.

One of the reasons is that limestone beds are subdivided into a number of sedimentation steps which can lead to the formation of bedding planes and produce a sequence in which the number of beds has apparently increased. In addition, the transition from shale to lime sedimentation seems to be partly governed by random disturbances, implying that the transition does not always have to occur at the same time.

Maxima in the position distribution curve (Fig. 8) indicate, therefore, the most likely occurrence of a bedding plane. It is possible that more than one bed may be formed near this position. Unfortunately, it is difficult to estimate standard deviation in a polymodal distribution. If normal distributions are fitted approximately to the data, estimated variances of 6.10×10^{-3} to 1.10×10^{-2} are obtained. The central maximum at 0.56 seems to have the widest spread. The variances around the preferred position of the bedding planes are higher by a factor of at least 10 than the variance which was earlier estimated in the study of a single bed, because planes which are not identical have been compared in the position diagram. If this explanation is accepted, then a definite answer to the previous question is provided: can stratigraphic correlation be refined to the tracing of individual beds? Individual bedding planes cannot be correlated from locality to locality. However, it emerges that within a cycle, there are positions in which bedding planes are more likely to occur and such levels can be correlated. Nevertheless, this result must be taken in conjuction with the accuracy of the observations. Here, bedding planes are features which were observed macroscopically and recorded during field work. It is known that refined observation, particularly by petrographic examination of peels, produces more discontinuity surfaces and it is possible that more accurate stratigraphic correlation can be achieved with additional data.

The Sedimentation Model

If the model is based on the preceding interpretation, it should contain a deterministic element which reflects the cyclic nature of the sequence. It should also contain a stochastic

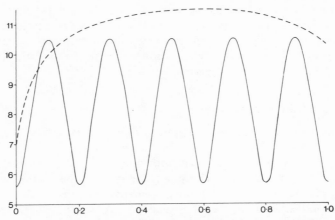

Fig. 10. Low and high frequency component of periodic time
dictate

component which affects the distribution of limestone and shale layers. The periodic element of the model operates in true time and various shapes for the function were fitted. Best results were obtained by combining two exponentials so that the function increases rapidly at the beginning of the cycle and decreases more slowly at the end of the cycle (Fig. 10). It was assumed that the time dictate determines the probability of lime or shale deposition at any moment. Clearly, when the probability of limestone formation is high, limestone is likely to be deposited. If the high probabilities are discrete (as in a step function, 5 massive limestones separated by shale layers would be created. Because the function is assumed to be sinusoidal, alternations between limestone and shale are generated during the transitions from low to high and therefore bedding planes occur in groups; they are more frequent near the four shale horizons which cause the division of the cycle into 5.

In practice, normally-distributed random numbers were generated and when these fell below the function of the time dictate, a unit thickness of shale was deposited. When the random number was larger than the time dictate, a gamma-distributed random variable was used to determine how many units of limestone thickness should be deposited. The latter insured that the thickness distribution of limestone beds could be created similar to the observed distributions.

However, the model contains many unknown parameters such as the shape of the low-frequency function, the amplitude of the sine waves, the variance of the normally-distributed random numbers and the parameters of the gamma distribution. A simulation run was regarded as successful when it produced a distribution of bedding planes which appeared to be similar to the observed material. In addition, bed thickness distributions of limestone and shale beds of the simulated series could be compared with the field data.

Although the model is simple in terms of the deterministic time component, the resulting stratigraphic series is a stochastic process which would be difficult to treat analytically. For this reason, Monte Carlo methods were employed to explore the model behaviour. It was discovered that shape of the low-frequency function determines the height and width of the distribution maxima. Simulations based on the time-dictate illustrated in Fig. 10, yielded the distribution illustrated in Fig. 11 and this agrees fairly well with the observed distribution (Fig. 8). Examples of simulated cycles, reduced to unit thickness, are illustrated in Fig. 12 and should be compared with Fig. 7.

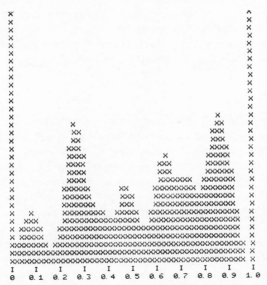

Fig. 11. Relative positions of bedding planes in
simulated cycles

During simulation runs, the standard deviations corresponding to time intervals in which bedding plane formation is likely can be calculated. For the example illustrated, values of between 0.10 and 0.13 were obtained. This is in agreement with values estimated from field data.

If the model is accepted, standard deviations can be employed to construct confidence limits for the position maxima. Under this assumption, stratigraphic correlation can become quantitative.

A standard deviation of 0.1 and an alpha error of 10 percent implies that only beds belonging to the stratigraphically-lowest maximum can be regarded as significantly different

Fig. 12. Simulated cycles reduced to unit thickness

from beds of the highest maximum. Adjacent groups cannot be separated. On a more practical level, wrong correlations (meaning beta error) can be controlled by narrowing the confidence limits. An approximation suggests that with a beta error of 10 percent, 30 percent of all bedding planes can be classified as belonging to one of the maxima.

Conclusions

Analysis of the middle Glencar Limestone yielded two interesting results. It was discovered that a deterministic model based on a 5-fold cycle coupled with a stochastic sedimentation process, produced a reasonable explanation of the sequence. Using the model, it must be concluded that individual bedding planes cannot be identified in different localities, implying that positive tracing of individual beds is impossible within the present data. However, one can attribute bedding planes to one of four discrete groups within a cycle in most situations. Stratigraphic correlation on a lower level than sedimentary cycles is possible therefore.

The model discussed, is of considerable general interest because it illustrates the great ease with which a periodic time-series can be transformed into a disorderly stratigraphic record by introducing small variations during sedimentation. Furthermore, it demonstrates how splitting of minor cycles can make it difficult to decide how many beds constitute a cycle. Comparative studies in different formations have shown that the 5-division cycle is a frequent pattern of sedimentation, but the number 5 is not constant necessarily and more beds per cycle may be present. It is not yet decided whether this results from variation in the time dictate or sedimentary processes; It seems likely that both factors are important.

References

David, M., 1977, *Geostatistical ore reserve estimation*. Elsevier Publ. Co., Amsterdam, 364 pp.

Duff, P.McL. D., Hallam, A., and Walton, E. K., 1967, *Cyclic Sedimentation*. Elsevier Publ. Co., Amsterdam, 280 pp.

Fischer, A. G., 1980, Rhythmic changes in the outer Earth, *Geol. Soc. London News Letter*, **8**(6), 2–3.

Gilbert, G. K., 1895, Sedimentary measurement of Cretaceous time, *Jour. Geology*, **3**(3), 121–127.

Hallam, A., 1964, Origin of the limestone-shale rhythm in the Blue Lias of England: a composite theory, *Jour. Geology*, **72**(2), 157–169.

Schwarzacher, W., 1954, Die grossrhytmik des Dachstein kalkes von Lofer, *Tschermaks Miner. Petrogr. Mitt.*, **4**(1–4), 44–54.

Schwarzacher, W., 1964, An application of statistical time-series analysis of a limestone-shale sequence, *Jour. Geology*, **72**(2), 195–213.

Schwarzacher, W., 1975, *Sedimentation Models and Quantitative Stratigraphy*. Elsevier Publ. Co., Amsterdam, 382 pp.

Schwarzacher, W., 1976, Stratigraphic implication of random sedimentation, in Merriam, D. F. (ed.), *Random Processes in Geology*. Springer Verlag, New York, 168 pp.

Schwarzacher, W., 1980, Models for the study of stratigraphic correlation, *Jour. Math. Geology*, **12**(3), 213–234.

Quantitative Stratigraphic Correlation
Edited by J. M. Cubitt and R. A. Reyment
© 1982 John Wiley & Sons Ltd.

A Proposed Geologically Consistent Segmentation and Reassignment Algorithm For Petrophysical Borehole Logs[1]

C. M. GRIFFITHS,

Ocean Engineering, School of Marine Technology, Armstrong Building, Queen Victoria Rd, Newcastle upon Tyne, England

Abstract

A nonparametric approach to the segmentation and reassignment of petrophysical log data has been developed. The algorithm employed uses a modified Kulinkovich technique and is applicable to multivariate data in an on-line environment. The approach is illustrated with examples from two boreholes.

Introduction

The ever increasing cost of drilling and coring boreholes, in all fields of geological investigation, has led to a corresponding increase in the use of petrophysical logs, both as a substitute for core retrieval, and as a valuable source of additional information. Petrophysical logs are essentially a record of variations in lithology as defined by the transmission of an electric current, the transmission of sound waves and the absorption and emission of nuclear radiation. Originally presented as an analogue trace, their use by computational systems depends on either the digitization of such traces at discrete intervals, or the use of the more recent digital tapes. The further use of these data sets for either stratigraphic correlation or bed evaluation depends on:

(a) the ability to recognize bed boundaries;
(b) the ability to identify beds of interest within the boundaries.

A major challenge within the field of log analysis is the automation of the bed recognition/selection process, and it is to this task that this paper is directed.

Many of the published segmentation techniques are reviewed in Hawkins and Merriam (1975), Shaw and Cubitt (1978) and Hawkins and ten Krooden (1978). Most of the available techniques rely on parametric statistics and (as Hawkins and ten Krooden point out) the implicit assumption that the data within segments are 'in some sense homogeneous' and that if data following a non-normal distribution is used, then 'one should generally attempt to transform them to normality to make use of existing methods'. This raises the question of whether such assumptions are valid in the geological context of a borehole, or in other words, are such assumptions geologically consistent?

---

[1] Ocean Engineering Report No. 51.

One segmentation technique which does not use such an implicit model is that of Kulinkovich, Sokhranov and Churinova (1966), who proposed a 'numeric differentiation' method, segmenting the trace at points of numerically-high derivatives. This method has been little used due to sensitivity to noise (Hawkins and Merriam, 1975). However, the simplicity of the method and non-parametric approach, combined with low computational requirement suggested that a modification, which introduced a runs-test qualification for choice of slope, could reduce the noise sensitivity and increase the utility of the technique.

Having chosen the segmentation points, the data within the segment reflects lithological boundaries rather than the lithology, due to the averaging effect of the combination of logging speed, source-receiver distance and time constant (for nuclear tools), and thus to enhance the possibility of relating further analysis to lithology rather than boundaries of that lithology, data within the segments are reassigned.

The algorithm described here is currently employed in the form of a FORTRAN IV computer program running on an IBM 370.

Criteria for Geological Consistency

Consideration of the nature of geological borehole sections suggests that the following factors should be considered when processing petrophysical data:

(1) The technique should not assume the existence of a lithological population. It is arguable that each sample of a lithology is divisible continually and, thus, the concept of a lithological population blurs into that of an individual.

In sociometrics and branches of geostatistics, the individual can be identified readily as, for instance, a person, fossil or pebble. In petrology, the individual is not only divisible but is also definable nonuniquely, as witness the multiplicity of igneous and sedimentary classification systems. An 'homogeneous' group of lithologies defined by grain size measurements, need have no homogeneity with respect to sonic velocities or gamma ray count, and it is as valid to define a '10 micro-second transit time interval' individual, as it is to employ a '10 cm core length interval' individual. There is, thus, no inherent reason why a classification based on fossil content or texture should correlate consistently with petrophysical log response.

(2) The geologist should consider whether any geological environment can be duplicated, as, even if the depositional environment remained similar, the material to be deposited must change with time as it derives from a finite source. The very act of deposition also must alter the depositional environment in some manner. It is, therefore, arguable that each act of deposition and diagenesis is unique and need follow no predictable distribution pattern when considered in vertical section.

If the instrument measuring the section was halted in progress up the hole and made a series of replicate measurements on a fixed section of sediment, then each measurement could justifiably be considered to be selected randomly from the infinite population of all measurements on that length, and thus follow a predictable distribution. However, the series of measurements on a sequence of different environments recorded as the instument moves up the hole cannot be considered as constituting a random sample without adding the question 'A random sample of what?'. In addition, the following should be considered:

(i) A noneroded, intruded, or tectonically-modified environmental sequence possesses the property of memory over short intervals; that is, the random element is absent. To invoke an environmental analogy, the presence of a random system would infer that next year's depositional environment at the Thames Estuary is equally as likely to be deltaic, glacial, desert or intermontagne.

(ii) The statistical concept of 'sample' requires that a sample is a randomly-selected subset of a population and each item in the population has an equal opportunity of inclusion in the sample.

A broadly-defined population in the log context could be 'all possible measurement interval lengths of rock in all possible boreholes'. Given such a population, does the borehole constitute a 'sample' because elements within it are distributed non-randomly? A sample should be selected also on the basis of an independent parameter. The only manner in which this could arise in borehole logging is if a regular grid were placed over the surface of the globe and boreholes drilled solely in relation to such a grid; that is, an oil well 'sample' of lithologies is preselected on the basis of lithology, structure or stratigraphy. Consider also the implications for multivariate analysis. Each has a differing measurement interval and depth of penetration and, therefore, according to this definition of a population, each log is sampling a different population and studies of correlation between logs is multipopulation analysis rather than multivariate.

With these considerations, calculation of statistics such as means, should be seen in a nonclassical statistical light; that is, they are valid calculations given a defined measurement window, but should be treated as variables rather than as population estimates. To extend Eisenhart's (1935) analogy, a borehole sequence can be compared to a line of objects such as a duck, an iron bar, a horse, a tree, a cat and a feather on which measurements of weight have been made. It is possible to calculate a mean weight for the entire sequence and for subsets within the sequence, but can this statistic be used validly as an estimator of a population parameter, indeed what population? It may be an interesting additional point that the mean weight of the sequence is as calculated, but further implications should be considered with great care, because this is an over-simplification of the geological case. In petrology, the duck would blend imperceptibly into the iron bar in one direction and into a horse in another.

(3) Because each lithology is sampled once, it becomes difficult to distinguish between instrument noise (which may have a predictable distribution), and lithological variation. Replicate log runs become the only mechanism of distinguishing between precision and accuracy, and the decision as to whether extreme values represent lithological variation rather than instrument noise, must be a geological rather than a statistical one.

(4) The existence of log trends is an embarrassment to many log-handling techniques. Indeed the removal of trends is frequently practised based on the assumptions stated by Mann and Dowell (1978). 'Removal of survey trends does not degrade geological results obtained from mathematical analysis.'

Provided instrument drift, if present, has been adequately corrected, the geologist ought to question this assumption on the basis that except for the change in drilling fluid resistivity with temperature (and thus depth) and reduction in invasion times, each log responds to a unique combination of lithology, pore fluids, pressures, and compaction states, all of which are significant geologically.

To summarize then, it is suggested that any technique for segmenting petrophysical logs:

(1) should be based on a distribution free approach;
(2) should allow for trends;

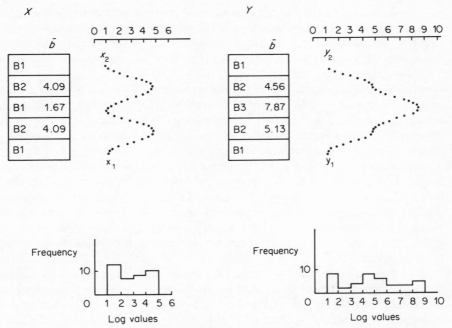

Fig. 1. Means and frequency distribution for situations X and Y. Individual block thickness is 1 m. Instrument is assumed to average over 1 m and to sample every 10 cm. Values are any arbitrary units. ($\bar{b}$ is mean log value for block averaged over 1 m)

(3) should employ descriptors other than mean and variance for defining 'homogeneity', if indeed the concept of lithological homogeneity is of value in many circumstances.

Log Response

Petrophysical logs such as sonic velocity, gamma ray and neutron tools, are a record of average response of a particular instrument to a lithological variable monitored over a set distance or time period defined by probe separation, source/receiver distance and logging speed.

Consider the two cases in Fig. 1. Cases X and Y represent log runs through two different sequences of man-made materials such as concrete and ash blocks. The blocks are assumed to have consistent properties within each block and between different blocks of the same material. As the same logging tool passes up through the sections at the same constant velocity, the instrument has measured values as plotted.

In the situation Y, the blocks $B1$ and $B2$ have the same properties as in situation X, but there is an additional block $B3$. The resultant curves are shown. Consider the result of applying standard statistical concepts to these two sets of data:

(1) standard scores for $B1$ and $B2$ in the two situations depend on the presence or absence of block $B3$ and not solely on intrinsic properties of the blocks;

(2) only if a window for mean calculation were equal to one data value would the block values recorded by a mean-derivation procedure equal the 'true' values;

(3) consider the use of the digitized curve $x_1 - x_2$ in a classification procedure designed

to derive the number and characteristics of the blocks involved in the vertical section. The fact that there are only 5 discrete units is disguised because the instrument records transition from one block to the next, as a function of the logging speed and averaging distance. If the purpose of log analysis is to correlate the 'picture' of one section of a log with a 'picture' (mathematically or visually defined) of a section of another log, there is no reason not to use all values in the analysis, because the boundary sequence is as useful as the lithological sequence. If, however, a classification/analysis or correlation of the characteristics of a given unit or sequence of units is involved, then, the mean value is a function of the transition between two lithologies rather than the lithologies themselves;

(4) in situation Y, the lower block $B2$ has a different calculated mean value due to the difference in number of digitization points that fall within the length of blocks $B2$ and $B3$. In other words, if the position of the measurement point changes relative to a bed boundary a different mean value results.

Suggested Procedure

The suggested approach is to segment the curves using a criteria that does not depend on assumptions of distribution, population or random sampling. The underlying assumptions of the suggested approach are:

(a) that there are basically three types of instrument response:
  (i) impersistent variations in log value due to instrument 'noise'.
  (ii) impersistent variations in log value due to small scale lithological variation, and
  (iii) relatively persistent changes in instrument reading due to lithological variation of either greater magnitude or consistency than the averaging distance of the tool.
(b) that the maximum slope of most log response occurs at the transition from one lithology to the next (Schlumberger, 1974).
(c) that maxima and minima of log curves are the closest approximation to the 'true' characteristic of the lithologies logged. This is equivalent to the oil field practice of drawing a 'shale line' that touches as many log peaks as possible (Wyllie, 1963).
(d) that trends are a geologically-significant feature of a log curve especially in a geotechnical context.
(e) that the data are spaced equally with respect to depth.

With respect to assumption (a), a further factor arises in that whether an event is persistent in a given log depends on the measurement/digitization interval. Therefore, the choice of 'persistence' cut off must be related to the digitization interval, maximum bed thickness of interest, and the purpose of the investigation.

The technique can be equated to the testing of runs up and down (but without a pre-selected sample window to test for significance), Kulinkovich's (Kulinkovich, Sokharanov, and Churinova, 1966) monotonicity interval, with subsequent reassignment of the values between identified boundaries.

Outline of the Algorithm

The algorithm is designed for use by a small system in an on-line environment and, therefore, has minimal memory and processing requirements.

The procedure, briefly, is to examine every data value, as it is logged, and assign a +1 or

−1 score depending on whether it is greater or less than the previous value. Data collection and scoring continues until, after a preset run-length has been exceeded, there is a change of sign. At this point, the run start is identified by searching the memory (usually less than 20 data points) and the run-length defined as a persistent event.

The aim of the process is to ensure that 'noisy', impersistent log variation is not treated as significant lithological variation (a difficulty with the Kulinkovich method). For any given run-length (say 5) a window could be chosen for which a run of five was highly significant. For instance, if the data interval was 0.2 m and a window of 1.2 m was chosen, then a run of 5 (5 out of 6 data points moving in the same direction) would indicate a low probability (less than 0.003 (Murdoch and Barnes, 1974, Table 21)) that the recorded event was due to random noise. However, the choice of window length is arbitrary, and implicit in the choice of 'persistence level' or run length cut-off which can be related to the bed thickness of interest.

Values involved in the event are reassigned according to the steepness of slope along the run. Every data value is subtracted from the previous point until a large linear distance surrounded by two smaller distances is recognized. The values before and after the position of maximum slope are assigned values of data points at either end of the slope. In situations where there is more than one inflection point within a run (for example Fig. 1, situation *Y*), the existence of a slope minimum is assumed to be geologically significant, and values are reassigned accordingly.

Having reached the start of a persistent event and reassigned values involved in the run, the algorithm examines any zone of impersistent events which preceeded the run.

Reassignment of data in impersistent zones is more difficult. The problem stems from assumptions made on the cause of impersistence; that is, if the zones are 'noisy', assuming a predictable distribution may be valid and the mean value could be calculated as the 'block' value for that interval. If, however, impersistence is due to small-scale lithological variation, the mode, maxima or minima may provide a geologically-consistent measure of bulk lithological response across the interval. Large amplitude, impersistent events, such as fractures or quartz veins, may have geological significance also and, therefore, methods of handling extreme values within impersistent zones (such as the improved Tukey test (Neave, 1979)) may be of importance.

Examples of Application

Figure 2 is an illustration of an application of the algorithm to hypothetical situations *X* and *Y* in Fig. 1. Note the effect of reassignment on the frequency distribution and block values available for multivariate classification, and also the fact that blocks *B*1 and *B*2 retain their values regardless of the neighbouring blocks. Figure 3 shows the detailed actions of the segmentation and reassignment algorithm on a sonic log from Well 98 (see later section for details).

Figures 4a and 4b illustrate the effects of the segmentation algorithm on two hand-digitized gamma logs from an 85 m section of an oil well. One gamma log was recorded with the ISF/SONIC log and the other with the FDC/CNL log over the same section. Log values are expressed as percentage of data maximum over the digitized interval; the raw data curve and reassigned values are indicated in each situation. Employing a persistence cut-off/run-length/monotonicity interval of three, the number of boundaries selected in each example differs by 6 with an average bed thickness of 0.55 m. Increasing run-length results in allocation of fewer boundaries and, consequently, greater average bed thickness (for example a run of 5 gives an average bed thickness of 0.9 m). Similarity between the two logs in

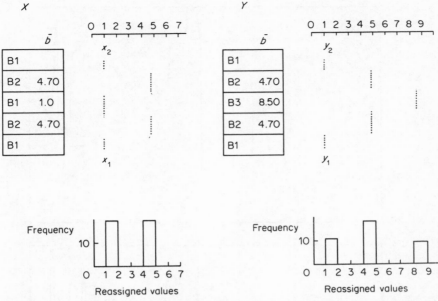

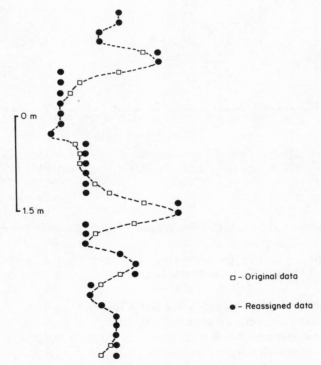

Fig. 2. Illustration of operation segmentation and reassignment algorithm. Data is
from Fig. 1

Fig. 3. Detail of operation of segmentation algorithm
as applied to sonic log data from Well 98 (Dorset) for
run-length of three

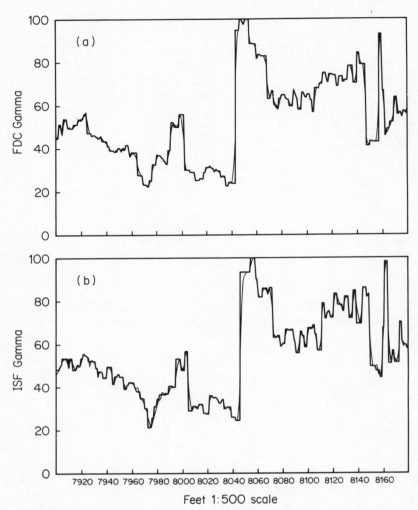

Fig. 4(a) and (4(b). Examples of application of algorithm to gamma log from
(a) ISF/SONIC log, and (b) FDC/CNL log, over same depth interval

positioning of boundaries by the algorithm suggests that instrument noise is a minor feature
of the logs.

Figures 5, 6 and 7 illustrate the use of the algorithm over a 76 m section of Middle
Jurassic strata from Dorset. The 5 logs involved are:

(P3FT)   sonic velocity averaged over 3 ft. (0.8 m),
(S3FT)   shear wave velocity over 3 ft. (0.9 m),
(P6FT)   sonic velocity over 6 ft. (1.8 m),
(FDC)   formation density compensated
(GAMMA) gamma log.

Figure 5 shows raw data as digitally recorded at 0.15 m intervals. Note the variation in

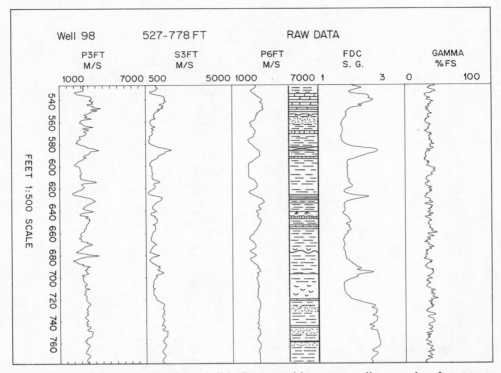

Fig. 5. Raw digitized log data from well in Dorset, with corresponding core log. Log types and units are as described in text

curve smoothness for the various logs, the effect of increasing the receiver spacing for sonic velocity, and the nature of data points across boundaries identified on the core log.

Figure 6 illustrates the effect of the algorithm with a run-length of three. Note the position of the boundaries on each log with relation both to one another and the core log.

Figure 7 demonstrates the effect of choosing a longer run-length (5) as the cut-off value. In these situations, the algorithm has employed a mean value in the zones of impersistent events.

The effect of increasing the run-length can be seen clearly on the gamma log where small scale variation is greatly reduced—to be replaced by mean values between persistent log variations. The positions of dominant formation changes and marker horizons are retained while reassigning the impersistent areas. For logs with less impersistent variation, the run-length can be increased to 9 or 10 without flattening due to reassignment. The FDC log is a good example of this.

The use of this technique allows the characteristic 'shape' of the log to be retained for correlation purposes while ensuring that the actual data values reflect lithological rather than erosional events.

Selection of run-length is dependent on the geological prerequisites of the analysis. If a clean, porous 1 m thick bed of sandstone is of economic significance, the tool averages over a 1 m section and records values at 20 cm intervals, then a run of 5 would only just identify a bed of value. If, however, a run of three were chosen, a bed of interest would be identified together with less significant beds.

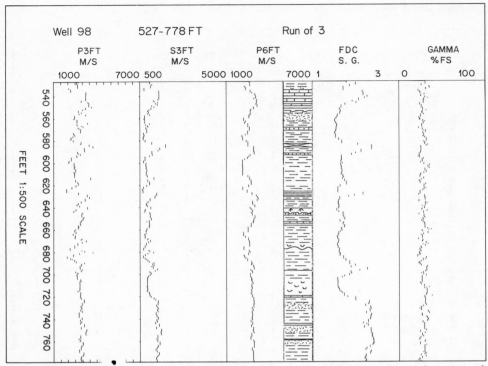

Fig. 6. Appearance of data set from Well 98 after segmentation and reassignment of data using persistence cut-off/run-length of three

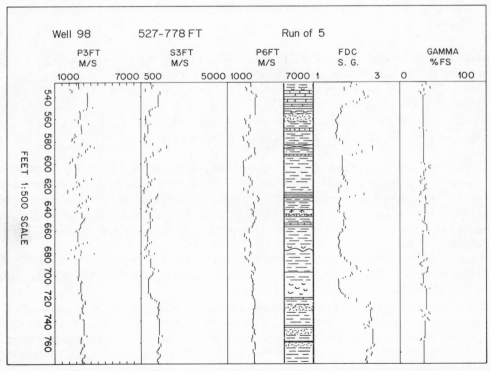

Fig. 7. Appearance of data set from Well 98 after segmentation and reassignment of data using persistence cut-off/run-length of 5

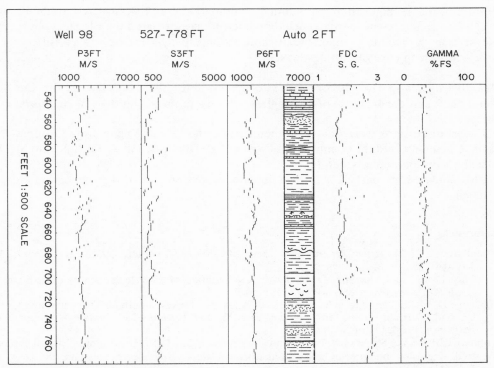

Fig. 8. Appearance of data set from Well 98 after segmentation and reassignment of data using automatic selection of persistence cut-off/run-length to give average bed thickness of 2 ft (0.6 m)

In practice, a run-length of three results in minimum loss of log information while ensuring that impersistent variations do not obscure persistent lithological change.

Because selection of run-length is related to sensitivity of the tool as well as lithological variation, it may be desirable to adjust the persistence cut-off for each logging tool. The computer program offers the option of preselecting a minimum average bed thickness of interest and adjusting the cut-off for each log to give a number of boundaries below a preset maximum (required minimum thickness of beds implies a maximum number of boundaries).

The effect of automatic run-length selection is indicated in Fig. 8 where run-lengths have been selected by the program for each log to derive less than 125 boundaries over the 76 m section, after a request to select an average bed thickness of 0.6 m.

Conclusions

The use of a geologically-consistent segmentation and reassignment algorithm gives the following advantages when the ultimate aim is a multivariate classification of lithological parameters measured at equal intervals.

(1) The possibility of deriving a classification based on lithologies rather than transition between lithologies is increased.
(2) Assumptions on the distribution of lithological measurements are avoided.
(3) Geologically-significant trends are not removed from the data.
(4) The assumption that the borehole log represents a random sample is avoided.

Furthermore, the algorithm has minimal computational and storage requirements, is applicable to small on-site systems including microprocessor-based data-loggers, and renders the section immediately amenable to multivariate analyis for on-site classification and comparison with core or cuttings logs.

Acknowledgments. The author wishes to acknowledge the assistance of the Occidental consortium and the Institute of Geological Sciences in providing the raw data used in this study.

Useful discussions were held with members of the Ocean Engineering Group at Newcastle University, and the comments of Professor B. Denness, Mr. A. G. Judd and Mr. P. C. Machen were particularly helpful.

This study forms part of a project funded by the Science and Engineering Research Council.

References

Eisenhart, C., 1935, A test for the significance of lithological variation, *Jour. Sed. Petrol.*, **5** (3), 137–145.

Hawkins, D. M., and Merriam, D. F., 1975, Segmentation of discrete sequences of geologic data, *Geol. Soc. America*, Mem. no. 142, 311–315.

Hawkins, D. M., and ten Krooden, J. A., 1978, A review of several methods of segmentation, in Gill, D., and Merriam, D. F. (ed), *Geomathematical and Petrophysical Studies in Sedimentology*. Pergamon Press, Oxford.

Kulinkovich, A., Ye, Sokhranov, N. N., and Churinova, I. M., 1966, Utilization of digital computers to distinguish boundaries of beds and identify sandstones from electric log data, *Int. Geol. Rev.*, **8**(4), 416–420.

Mann, C. J., and Dowell, T. P. L., 1978, Quantitative litho-stratigraphic correlation of subsurface sequences, *Computers & Geosciences* **4**, 295–306.

Murdock, J., and Barnes, J. A., 1974, *Statistical Tables*. Macmillan, London.

Neave, H. R., 1979, Quick and simple tests based on extreme observations, *Jour. Quality. Technol.*, **11**(2) 66–79.

Schlumberger, 1974, *Log Interpretation*, Vol. 11, *Applications*: Schlumberger Ltd., New York, pp. 20–22.

Shaw, B. R., and Cubitt, J. M., 1978, Stratigraphic correlation of well logs: an automated approach in Gill, D., and Merriam, D. F. (ed.), *Geomathematical and Petrophysical Studies in Sedimentology*. Pergamon Press, Oxford.

Wyllie, M. R. J., 1963, *The Fundamentals of Well Log analysis*. Academic Press Inc., New York, p. 39.

Index

299